Biomarkers: A Pragmatic Basis for Remediation of Severe Pollution in Eastern Europe

NATO Science Series

A Series presenting the results of activities sponsored by the NATO Science Committee. The Series is published by IOS Press and Kluwer Academic Publishers, in conjunction with the NATO Scientific Affairs Division.

General Sub-Series

A. Life Sciences	IOS Press
B. Physics	Kluwer Academic Publishers
C. Mathematical and Physical Sciences	Kluwer Academic Publishers
D. Behavioural and Social Sciences	Kluwer Academic Publishers
E. Applied Sciences	Kluwer Academic Publishers
F. Computer and Systems Sciences	IOS Press

Partnership Sub-Series

1. Disarmament Technologies	Kluwer Academic Publishers
2. Environmental Security	Kluwer Academic Publishers
3. High Technology	Kluwer Academic Publishers
4. Science and Technology Policy	IOS Press
5. Computer Networking	IOS Press

The Partnership Sub-Series incorporates activities undertaken in collaboration with NATO's Partners in the Euro-Atlantic Partnership Council – countries of the CIS and Central and Eastern Europe – in Priority Areas of concern to those countries.

NATO-PCO-DATA BASE

The NATO Science Series continues the series of books published formerly in the NATO ASI Series. An electronic index to the NATO ASI Series provides full bibliographical references (with keywords and/or abstracts) to more than 50000 contributions from international scientists published in all sections of the NATO ASI Series.
Access to the NATO-PCO-DATA BASE is possible via CD-ROM "NATO-PCO-DATA BASE" with user-friendly retrieval software in English, French and German (© WTV GmbH and DATAWARE Technologies Inc. 1989).

The CD-ROM of the NATO ASI Series can be ordered from: PCO, Overijse, Belgium.

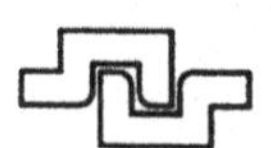

Series 2: Environmental Security – Vol. 54

Biomarkers: A Pragmatic Basis for Remediation of Severe Pollution in Eastern Europe

edited by

David B. Peakall

St. Mary's Road,
Wimbledon, London, United Kingdom

Colin H. Walker

Victoria Road,
Mortimer, Berkshire, United Kingdom

and

Pawel Migula

Department of Human and Animal Physiology,
Katowice, Poland

Kluwer Academic Publishers

Dordrecht / Boston / London

Published in cooperation with NATO Scientific Affairs Division

Proceedings of the NATO Advanced Research Workshop on
Biomarkers: A Pragmatic Basis for Remediation of Severe Pollution in Eastern Europe
Cleszyn, Poland
21–25 September 1997

A C.I.P. Catalogue record for this book is available from the Library of Congress.

ISBN 0-7923-5643-8 (HB)
ISBN 0-7923-5644-6 (PB)

Published by Kluwer Academic Publishers,
P.O. Box 17, 3300 AA Dordrecht, The Netherlands.

Sold and distributed in North, Central and South America
by Kluwer Academic Publishers,
101 Philip Drive, Norwell, MA 02061, U.S.A.

In all other countries, sold and distributed
by Kluwer Academic Publishers,
P.O. Box 322, 3300 AH Dordrecht, The Netherlands.

Printed on acid-free paper

Printed in the Netherlands

TABLE OF CONTENTS

Part 1. KEY-NOTE PRESENTATIONS

Investigations of Pollution Problems in Eastern Europe and Russia

Part 2. Working Groups Reports

Part 3. ABSTRACTS OF POSTERS

Introduction

With the political changes at the end of the 1980s there has been much greater contact between scientists from eastern European countries and their counterparts in western Europe and North America. In this new atmosphere of open discussion, exchange of ideas and co-operation there has been growing interest and awareness about the serious pollution problems in certain areas of the former Soviet Union and other countries of eastern Europe. Industries and mining operations which had given insufficient attention to the control of effluents and emissions have been especially implicated. Many of these pollution problems are now receiving more attention than heretofore, but the scale of the problem should not be underestimated. Human health hazards have, understandably, been given priority and relatively little has been done to assess damage caused to natural populations and ecosystems.

It seemed, therefore, opportune to bring together scientists from the former Soviet Union and eastern Europe with those from western countries to discuss pollution problems and to study in depth a few selected areas which are heavily polluted and have already been the subject of investigations. An important question was whether new techniques being developed in the west could provide the basis for biomarker strategies to identify harmful effects on natural populations and point the way to effective remedial action.

The immediate objective was to set in place collaboration so that joint projects could be formulated and funding sought to develop biomarker strategies in selected study areas. The implementation of effective remedial action depends upon political and economic considerations which lie outside the competence of scientists concerned with defining and quantifying the effects of pollutants. In many cases work is still required using biomarker strategies before there can be a clear picture of the damage that is being caused by particular pollutants or combinations thereof. Thus, the present workshop was seen as an essential preliminary step towards the definition of priorities for bioremediation.

Five areas were selected for detailed considerations. These were the Katowice Administrative District and Upper Vistula Catchment Area in Poland, Northern Bohemia in the Czech Republic and the Kola Peninsula and Kuzbass in the Russian Federation.

Katowice Administrative District in Poland.
This District is the largest and oldest industrialised region in Poland. All the major power stations are located here and since it contains 10% of the Polish population in only 2% of the land there are many small sources of pollution. The pollutants that have caused the greatest concern are air pollutants, especially sulphur dioxide and ozone. Widespread damage of the forest of the region have been caused by these pollutants. Large areas have been heavily contaminated by a wide variety of industries, including cokeries, chemical plants and refineries.

The Working Group recommended that biomarkers and bioassays be used to augment existing monitoring programs to make them more comprehensive. The use of biomarkers would complement existing programs which currently focus on the measurement of levels of pollutants. Specifically they recommend the following:

(a) Enhanced site-specific risk assessment, along the lines used by the USEPA, at major industrial sites.
(b) The use of biomarkers to identify hotspots and assess the effects of complex mixtures.
(c) The group expressed concern about the limited amount of data available on the levels of PCBs, dioxins and pesticides.

The Upper Vistula Water Catchment area in Poland. The Vistula River is the largest river in Poland. The upper portion of this drainage area, the Upper Vistula Basin comprises some 50,000 sq. kms which represents 15% of the area of Poland. The western part of the basin has deposits of hard coal, zinc and lead ores, natural gas and sulphur and has been heavily industrialised. Pollution has been caused by salt water inputs from coal mines, from pesticides and other agricultural pollutants and for a variety of industrial pollutants. The environmental quality of this region has deteriorated seriously over past decades and effects on population and community structure have been shown although some recent improvements have been made following the political change of 1989. The main areas of concern discussed by the Working Group and their recommendations are:

(a) Current monitoring efforts of forests, especially the Ratanica forest catchment, should continue. This should include measurements of biomarkers in soil invertebrates. Details of the work already carried out in this field in Poland are given in the paper in this volume by Migula and co-workers. The group considered that the monitoring of plants was largely appropriate but that measurement of the effects of nitrogen oxide and ozone pollution should be added.
(b) The group considered that the current monitoring of environmental quality of urban areas, focused on the city of Krakow, provided a good dataset on air quality. The group suggested the addition of biomarkers on soil invertebrates and plants.
(c) It was suggested that the monitoring of aquatic ecosystems could be focused on retention reservoirs. However not enough data were available to put forward detailed plans. It was suggested that preliminary screening be carried out to identify hot-spots for more detailed study.

Northern Bohemia in the Czech Republic.

Northern Bohemia stretches along the northwest border with Germany and Poland. Over one and half million persons live in an area of 16,000 sq. kms. The major environmental problems of the region are associated with the emission of sulphur dioxide, smoke and other particulates by power plants operating on brown coal, devastation of the landscape by strip mining for coal, and pollution from a variety of chemical industries and mining.

There has been considerable monitoring activity in this area in recent years. Soil pollution has been studies; vertebrate populations have been monitored with a view to using them as bioindicators of pollution; effects of pollutants on the forest canopy have been assessed and there have been integrated studies on soil ecology. In keeping with the requirements of recent Czech laws, coal-mining companies have been restoring habitats devastated by open cast mining. Notwithstanding these activities, much remains to be done to establish the environmental impact of particular pollutants (e.g. those arising from chemical industry), as an essential preliminary to taking any remedial action that may be required.

The Working Group considered that biomarkers should be used which measure early responses, that are closely related to the toxicity of pollutants and have been well validated. Specifically they recommend the following:

(a) The P450 1A1 enzyme system for organic pollutants especially at sites contaminated with PAHs, PCBs and dioxins.
(b) The metal-binding protein metallothionein (phytochelatin in plants) to assess the impact of heavy metal pollution.
(c) Stress proteins, which are induced by a wide variety of pollutants, and thus can be used as an integrated signal for environmental stress.
(d) Structural modifications of DNA (adducts, strand breakage) which can be caused by a variety of genotoxic pollutants (e.g. PAHs) and can lead to serious pathological conditions.

Kola Peninsula in the Russian Federation.

The major sources of pollution in this area is from mining and smelting. The main pollutants are SO_2 and heavy metals. Decreased biodiversity and biomass of plants and animals have been observed and community structure and nitrogen-fixation are known to be affected.

Kuzbass in the Russian Federation.

Contamination by radionuclides caused by accidents and atomic weapons testing has been a serious problem. The most serious contaminant in recent years was plutonium-239, following an accident at the Siberian Chemical Plant in 1993. Investigation of tooth enamel by electronic spin resonance showed a good correlation with cytogenetic aberrations of the lymphocytes. Pollution of the environment was also caused by metallurgical plants and the coal industry.

Due to logistic problems scientists from the Kuzbass area were unable to attend and thus there was no working group to consider the problems of the area. However two papers detailing the problems and the studies undertaken were submitted and are included in this volume (Chapters 6 & 7).

The workshop was structured so that plenary lectures by scientists from eastern Europe which described the polluted areas and the work done in them preceded those from western scientists which reviewed biomarker assays and strategies. Subsequently the participants divided into workings group to consider the problems of the specific areas and to produce detailed reports making recommendations for biomarker strategies to obtain clarification of pollution problems. The sequence of this volume follows the pattern described above giving the text of the plenary lectures first followed by the reports of the working groups and finally the abstracts of the posters that were presented at the meeting.

Kola Peninsula in the Russian Federation.

The major sources of pollution in this area are from mining and smelting. The main pollutants are SO_2 and heavy metals. Decreased biodiversity and biomass of plants and animals have been observed and community structure and nitrogen-fixation are known to be affected.

The Working Group recommended the use of biomarkers to enable the effects of the massive contamination of the area on environmental and human health to be determined. Biomarkers can be used to monitoring and map the worst affected areas and to document improvements that result from control measures. In particular it is recommended that plant and invertebrate biomarkers should be used as part of a comprehensive biomonitoring programme.

Recommendations of NATO ADVANCED RESEARCH WORKSHOP

1. For all of the four study areas the working groups concluded that while there is a good deal of data on the levels of heavy metals and gaseous pollutants (SOx, NOx, O_3) there is relatively little data on organic pollutants such as PCBs, dioxins and PAHs despite the likelihood that certain of these compounds are present in substantial concentrations. This data gap needs to be rectified before an overall hazard assessment can be made. Biomarkers are considered to be a cost-effective way of determining if these compounds are present in concentrations that are exerting a significant effect. For example, use could be made of biomarkers (e.g. CALUX assays, induction of cytochrome $P_{450}1A1$) that give estimates of the Ah-receptor mediated toxicity of planar polyhalogenated compounds expressed as toxic equivalents relative to dioxin.

2. Biomarker assays can be utilised in two distinct ways. First, in conjunction with bioassays (e.g. the Ames test and MICROTOX assay) they are useful for inexpensive initial screening of relatively large areas to identify "hot spots", where organisms are being adversely affected by pollutants. Second, they have a critical role in the next stage of environmental impact assessment, the establishment of the harmful effects of particular types of pollutants. By following this approach the most cost effective way of carrying out bioremediation can be defined.

3. Quality assurance is as important in the use of biomarkers as it is in analytical chemistry and it is vital that any programme that is set-up takes this into account. International co-operation is essential both in the exchange of personnel and gift of reference material. Full advantage of programmes such as the NATO Collaborative Research Grants, Linkage Grants and Expert Visits should be made. The possibility of a NATO Advanced Study Institute Grant to assist in technology transfer should be considered.

4. It should be brought to the attention of regulatory bodies that the concept of biomarkers can give information on the seriousness of the effects of toxic chemicals. In contrast residue levels merely tell us what is there. Only rarely is it possible, through the use of toxicological data, to assess the significance of the residue data. The complexity of organic

pollution, and the fact that many organic pollutants have only limited biological persistence, greatly limit the extent to which chemicals residues can be related to their biological effects. Validated biomarkers are available in a wide variety of species and recommendations of the Working Groups include the use of biomarkers in soil invertebrates and plants.

5. Specimen banks have proved invaluable in aiding the identification of geographical and temporal trends of new and previously unmeasured pollutants. A survey should be undertaken (by national bodies such as Academies of Science) in eastern Europe to determine the current availability of material in museums, and medical and veterinary institutions. Based on this information plans should be developed for improving and co-ordinating the collection of specimens in eastern Europe bearing in mind the specific requirements of preservation of specimens for future biomarkers assays.

6. Where physical reclamation is undertaken it is important to ascertain that wildlife moving into the area remains healthy. Some studies have shown that, while species are attracted into the area, they are unable to reproduce successfully. Thus the area can actually serve as a "black hole" while appearing to have a substantial wildlife population.

7. It should be recognized that in some old mining areas there are well documented cases of the existence of resistant strains of plants. The identification and correct analysis of plant resistance can give valuable information about the environmental impact of heavy metals and other pollutants in highly polluted areas.

8. For the Kola Peninsula, where the sources of pollution, although massive, are concentrated in discrete areas, it is considered valuable to study biomarkers along long (up to 100km) gradients so that improvements occurring due to remedial action can be plotted. In other areas an approach based on comparison with control sites where the level of pollution is much lower is considered best. In many areas the level of pollution is so severe that, although improvements are being made, a pristine environment is far away.

In addition to the general recommendations above, which were made by the entire workshop, specific recommendations were made by the Working Groups and are included in their reports.

Acknowledgements

We are grateful to NATO and the Polish Committee for Scientific Research for their support of this Advanced Research Workshop.

The success of the Workshop was due, in large measure, to the excellent logistical support given by the University of Silesia and the tireless work of Professor Migula and his staff.

We would also like to thank Uniglobe Top Flight Travel for their work in making the travel arrangements for the western scientists and to Mrs Gill Bogue for her preparation of the final manuscript. The efficiency of these persons was an important factor in the success of the Workshop.

LIST OF PARTICIPANTS

Jan Fredick BORSETH,
Elf Akvamiljo; Mekjarvik 12,
N4070 Randaberg, NORWAY
e-mail JanFredik.Borseth@algemeen.tox.wau.nl

Lubos BORUVKA
Department of Soil Science and Geology
Czech University of Agriculture in Prague
6 Suhdol, 16521 Prague, CZECH REPUBLIC
e-mail boruvka@af.czu.cz

Jiri CIBULKA
Department of Soil Science and Geology
Czech University of Agriculture in Prague,
6 Suohdol, 16521 Prague, CZECH REPUBLIC
e-mail cibulka@bohem-net.cz

Tracy COLLIER
NOAA-NMES 2725 Montlake Blvd, East,
Seattle, WA 98112, USA
e-mail Tracy.k.collier@noaa.gov

Krzysztof DMOWSKI
Department of Ecology, University of Warsaw,
Krakowskie Przedmiescie 26/28,
00-927 Warszawa, POLAND
e-mail KADM@plearn.edu.pl

Wilfred H.O. ERNST
Department of Ecology and Ecotoxicology
Vrije Universiteit, De Boeislaan 1087
1081 HV Amsterdam, THE NETHERLANDS
e-mail wernst@bio.vu.nl

Galina EVDOKIMOVA
Kola Science Centre of the Russian Academy of Science,
14, Fersmana St., Apatity 184200, Murmansk Region
RUSSIAN FEDERATION
e-mail galina@inep.ksc.ru

Stefan GODZIK
Institute for Ecology of Industrial Areas,
Kossutha 6, 40-007 Katowice, POLAND
e-mail godzik@ietu.katowice.pl

Jesus GONZALEZ-LOPEZ
Department of Microbiology, University of Granada,
E-18071 Granada, SPAIN
e-mail Jcarril@goliat.ugr.es

Krystyna GRODZINSKA
Institute of Botany, Polish Academy of Sciences,
Lubicz 46, 31-512 Krakow, POLAND
e-mail Grodzin@ib-pan.krakow.pl

Richard HANDY
Department of Biological Sciences,
University of Plymouth
Drake's Circus, Plymouth UK PL4 8AA
UNITED KINGDOM
e-mail RHandy@plymouth.ac.uk

Ludo HOLSBEEK
Laboratory of Ecotoxicoogy and Polar Ecology,
Free University of Brussels (VUB),
Pleinlaan, 7 F405; 2B-1050, Brussels, BELGIUM
e-mail lholsbee@vub.ac.be

Anne KAHRU
Institute of Chemical Physics and Biophysics,
Akadeemia tee 23, Tallinn EE0026, ESTONIA
e-mail Anne@kbfi.ee

Andrzej KEDZIORSKI
Department of Human & Animal Physiology,
University of Silesia,
Bankowa 9, 40-007 Katowice, POLAND
e-mail nakon@us.edu.pl

Sean KENNEDY
National Wildlife Research Centre,
Canadian Wildlife Service,
Hull, Quebec KIA OH3 CANADA
e-mail sean.kennedy@ec.gc.ca

Sergei KOTELEVTSEV
Department of Biology, Moscow State University,
RU - 119899 Moscow; RUSSIAN FEDERATION
e-mail serg@kot.bio.msu.su

Paulina KRAMARZ,
Department of Ecosystem Ecology, Jagiellonian University
Ingardena 6, 30-060 Krakow, Poland
e-mail Kramarz@eko.uj.edu.pl

Jana KUBIZNAKOVA
Institute of Landscape Ecology
Czech Academy of Sciences
Na sadkach 7, 37005 Ceske Budejovice
CZECH REPUBLIC
e-mail janaku@dale.uek.cas.cz

Werner KRATZ,
Fachbereich Biologie FU Berlin FB Biologie
WE5, Grunewaldstrasse 34,
D-12165 Berlin, GERMANY
e-mail Kratz@fub46.zedat.fu-berlin.de

Laurent LAGADIC.
Unite d'Ecotoxicologie Aquatique
INRA, 65, rue de Saint-Brieuc
F-35042 Rennex Cedex, FRANCE
e-mail lagadic@roazhon.inra.fr

Albert LEBEDEV
Department of Chemistry, Moscow State University
Leninskije Gory, 117234 Moscow
RUSSIAN FEDERATION
e-mail Lebedev@rsu.rdn.su

Mirdza LEINERTE
Ministry of Environmental Protection and Regional Development
Rupniecibas iela 25, LV-1045, Riga, LATVIA
e-mail kmc@main.vvi.gov.lv

Levonas MANUSADZIANAS
Department of Hydrobotany, Institute of Botany,
Z Ezeru 47, 2021 Vilnius, LITHUANIA
e-mail Leones@pub.asf.lt

Victoria MARTINEZ-TOLEDO
Department of Microbiology, Institute of Water Research
Rector Lopez Argueta s/n; University of Granada,
E-18071 Granada, SPAIN

Pawel MIGULA,
Dept. of Human and Animal Physiology,
University of Silesia,
Bankowa 9, 40-007 Katowice, POLAND
e-mail migula@us.edu.pl

Tinka MURK
Department of Toxicology, Agricultural University
P.O. Box 8000, 6700 HD Wageningen, THE NETHERLANDS
e-mail Tinka.murk@algemeen.tox.wau.nl

Miroslaw NAKONIECZNY,
Centre for Studies of Human and Natural Environment
University of Silesia
Bankowa 9, 40-007 Katowice, POLAND

David PEAKALL
17 St Mary's Road
Wimbledon, London, UK SW19 7BZ
e-mail dpeakall@compuserve.com

Janeck SCOTT-FORDSMAND
Department of Terrestrial Ecology,
National Environmental Research Institute
Vejlsovej 25, P.O. Box 314,
DK-8600 Silkeborg, DENMARK
e-mail vs@dmu.dk

Vladimir SERGEEV
Department of Zoology, Kemerovo State University,
Krasnaja 6, 650043 Kemerovo,
RUSSIAN FEDERATION
e-mail bios@kemgu.kemerovo.su

Lee SHUGART
L.R. Shugart & Associates, Inc.
Consultants in Genetic & Molecular Ecotoxicology
P.O. Box 5564, Oak Ridge, TN 37831-5564, USA.
e-mail Lrshugart@icx.net

Jose Paulo SOUSA
Inst. Ambiente e Vida
Universty of Coimbra
Largo Marques de Pombal,
P-3000, Coimbra, PORTUGAL

Colin WALKER
35, Victoria Road, Mortimer,
Berkshire, UK RG7 3SH

Jason WEEKS,
Institute of Terrestrial Ecology, Monk's Wood,
Huntingdon, Cambridgeshire, UK PE17 2LS
UNITED KINGDOM
e-mail j.weeks@ite.ac.uk

Adam WOZNY
Institute of Experimental Biology
University A. Mickiewicz
Fredry 10, 60-701 Poznań, POLAND
e-mail adaw@hum.amu.edu.pl

participants who contributed to but were absent in Cieszyn:

Serve V. CHERNYSHENKO
Dnepropetrovsk State University
72 Gagarin Avenue, 320625 Dniepropetrovsk-10,
UKRAINE
e-mail svc@compten.dnie/iro/retvivsk.ua

Elena KOROLEVA
Laboratory of Bioindication, Dept. of Biogeography
Moscow State University, 19899 Moscow,
RUSSIA FEDERATION
e-mail koroleva@cs.msu.su

Nicolai ILYINSKIKH
Siberian Medical University
Tomsk
RUSSIA
e-mail root@ecogen.tomsk.su

1. DYNAMICS OF THE INDUSTRIAL TRANSFORMATION OF TERRESTRIAL ECOSYSTEMS IN THE KOLA SUBARCTIC

G.A. EVDOKIMOVA
The Institute of the North Industrial Ecology Problems
Kola Science Centre, Russian Academy of Science
Fersman street 14, 184200 Apatity, Russia

Abstract

Heavy metals and sulphur dioxide are the main pollutants of the Kola Subarctic. Soils are contaminated so that they have different chemical and biological properties from those of the zone soils at distances up to 25-30 km from the largest sources of chemical pollution. Chronic soil pollution by the emissions from the non-ferrous industry Severonikel has caused not only a decrease in the total number of microorganisms but also the destruction of the structure of the microbial community and the constriction of its species diversity. Homeostasis of microbial communities in podzol soil has been changed by concentrations of 300-400 mg/ kg copper and 600-700 mg/kg nickel. In the polluted soil the neutrophilic microbes disappear and a dominant position is taken by acid-tolerant organisms, mainly by soil fungi. Anthropogenic degradation of the soil is connected with the destruction of the normal activity of its living components, decreasing its mass and diversity. Taking this fact into account all methods of soil restoration must aim to increasing the activity of living microbial component of the soil and to save the polyfunctionality of its communities.

Keywords: soil, microorganisms, homeostasis, pollution, acidification, heavy metals, remediation, Kola Subarctic.

1.1. Introduction

In the Kola Subarctic region there is an acute problem of pollution of the environment by heavy metals and compounds of sulphur. About 80% of the sulphur dioxide and more than 30% of the heavy metals emitted into the air in the North European part of Russia fall within the Kola Peninsula [1]. Regional soil-climatic conditions that have an important influence upon the processes of soil pollution by the toxicants of industrial origin are the following:

- there is a weak intensity of energy-mass exchange in the ecosystems of the North caused mainly by the low temperatures and low self-cleaning ability of the natural media;

D. B. Peakall et al. (eds.),
Biomarkers: A Pragmatic Basis for Remediation of Severe Pollution in Eastern Europe, 1–14.

- concomitant with this is the rather limited biodiversity of the soil biota; and as a result there are limited possibilities of the biological utilization of wastes;
- the acid reaction of the soil solution of podzol and peat soils which dominate in soils of the Kola Peninsula, lead to increasing the mobility of pollutants and their subsequent toxicity;
- the excess of rainfall over evaporation contributes to the leaching of elements down the soil profile such that it may be the cause of the pollution of the underground waters.

One of the main polluters in their region is the enterprise of non-ferrous industry ("Severonikel", Kola peninsula), in production since 1935. At the time of foundation "Severonikel" was surrounded by the Northern-taiga forests. For 40 years now, there has been a uniformly barren landscape. These alterations were the basis of complex investigations of the soils and soil micropopulation in the zones subjected to the influence of air emissions from the plant of this non-ferrous industry. This monitoring scheme has been in place for more than 20 years. The following questions were posed: how quickly do the chemical, biological and physical-chemical properties of the soil change under the influence of pollutants and what are the functions of microorganisms in these processes? These small essences, according to L. Pasteur, play " an extremely large role in the nature and life of a human being".

The conceptual basis for the ecological-microbiological approach to the soil protection from the chemical pollution was the consideration of the vital activity of microbial communities in the following context:

- what is the role of microorganisms as indicators of environmental quality?
- what is the medium-regulating the function of microorganisms and how is this reflected in the processes of migration and transformation of pollutants?
- what is the role of the microbial component of the soil in decreasing the toxicity of polluted soils by the exo-and endo-immobilization of toxic components ?
- what is the destructive activity of the soil biota facilitating decomposition of the contaminants to the simple mineral compounds?

1.2. Materials and Methods

Sample plots were established by introducing uncontaminated soil into the zones with different levels of industrial transformation to the ecosystem. These sample plots were established in the centre of pollution (0.2 km from the source of emission), in the impact zone (at 5 km), in the buffer zone (at 15 km) and in the background zone (at 50 km from the source of emission). The initial composition of introduced soil was Cu 29mg/kg, Ni 83 mg/kg and Co 30mg/kg; pH water suspension - 6.3, pH salt suspension - 5.3. The following findings briefly characterize the state of the vegetation and the extent of forest soil pollution in the zones under observation.

Impact zone. Woody layer completely destroyed, only depressed regrowth of birch can be found here. The soil is eroded. Ground vegetation covers about 20% of the surface, mainly it is crowberry (*Empetrum hermafroditum*) and wavy hair-grass (*Deschampsia flexuosa*). Mosses and lichens are absent. The concentration of copper ranged from 600 to 1000 mg / kg and that of nickel - from 1700 to 2600 mg /kg in the upper organogenic horizon during the season.

Buffer zone. Vegetation is of the shrub type pine-forest (*Pinus sylvestris*) with the addition of the birch (*Betula subarctica*) and predominance of mountain cranberry (*Vaccinium vitis-idaea*), bilbarry (*Vaccinium myrtillus*) and bearberry (*Arctostaphylos uva-ursi*) in the ground cover. The copper content in the organogenic horizon ranged between 200 to 400 mg /kg and for nickel - from 500 to 1000 mg / kg during the season. Lichens were represented by *Cladonia rangiferina, Cladina mitis.*

Background zone. Pine forest of shrub-lichen type with the addition of birch *Vaccinium vitis-idea* and *Arctostaphylos uva-ursi* dominate among the shrubs; representatives of genera *Cladonia, Cladina* and *Cetraria* dominate among the lichens and *Hylocomium* and *Pleurozium* prevail among the mosses. In the organogenic horizon the content of copper is on average 40, nickel - 80, cobalt - 30 mg /kg of soil.

1.3. Results and Discussion

How quickly did the metals accumulate in the soils? Observations of the dynamics of the heavy metals content in the 10-cm layer of introduced soil show a quick and important accumulation of copper, nickel and cobalt at the centre of the pollution (Figure 1). This process was the most intensive during the first months of the experiment. At this time the soil-adsorbing complex of originally non-contaminated soil was capable of active exchange sorption of the pollutants falling onto the soil. Within 5 years the amount of copper was increased 60-fold in the centre of pollution, nickel by 40-fold, and cobalt 6-fold in comparison with the initial level. (In 1980 this sample plot was destroyed by a bulldozer). In the soil of the impact zone the amount of copper was tripled and the quantity of nickel and cobalt was doubled in 10 years. Naturally, the concentration of heavy metals and the speed of their accumulation were sharply reduced according to the distance away from the source of emission.

In 10 years after the establishment of the experiment predictions for the next 10 years were made. These 10 years have passed already and we are now able to compare calculated values with the experimental ones (Figure 2). Compared digital rows were very close for copper and especially for nickel to 1990. After this year the empirical figures become lower than predicted ones. This is correlated with a general decrease in the volume production of the chemical plant during this period.

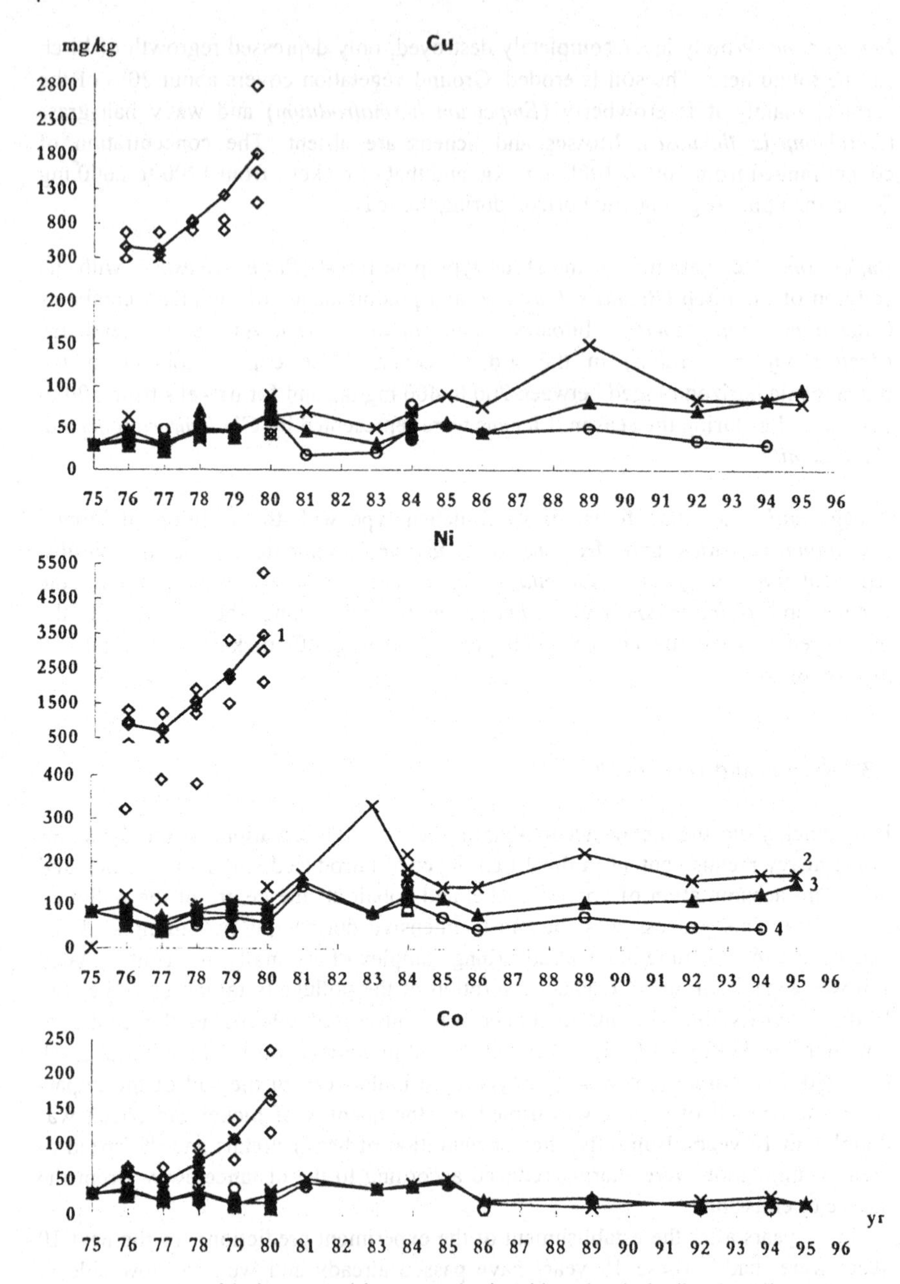

Figure 1. Dynamics of the heavy metals' concentration in cultivated podzol soil at the different distance from source of pollution. 1 - centre; 2 - impact zone; 3 - buffer zone; 4 - background zone.

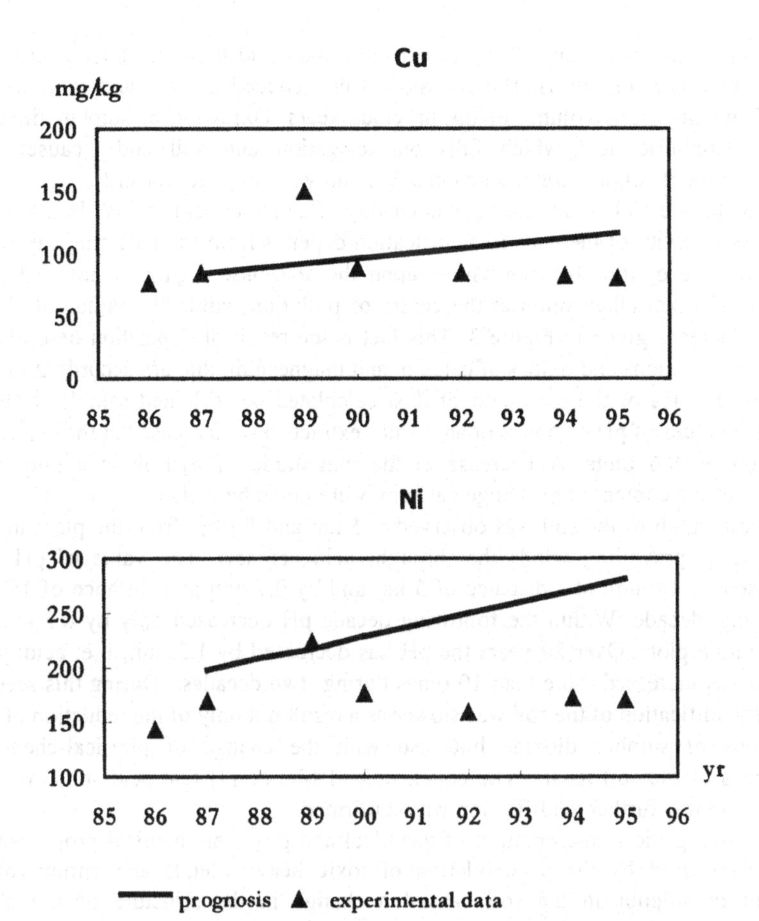

Figure 2. Dynamics of the heavy metals' concentration in cultivated podzol soil in the impact zone of pollution.

During many years the plant emitted more than 200 thousand tons of sulphur dioxide annually (Figure 3). The emissions were reduced to 1995 and came to the half of the emissions volume of the previous years. Oxidation of sulphur dioxide forms sulphuric acid which falls on vegetation and soil and causes the accumulation of sulphur compounds in soil, mainly of sulphate-ions [2].

How did the soil acidity change under the influence of emissions? It is known that the sensitivity of the soils to acidification depends from the buffering capacity. The influence of industrial emissions upon the acid-base regime of the soil was defined. The soil alkalization at the centre of pollution within the radius of about one kilometer is given in Figure 3. This fact is the result of deposition here of the heavy metal cations and oxides of calcium and magnesium that are according to our data up to 3-4% of the total sum of dust calculated per calcined sample. Over 5 years the values of pH of both salt and water extracts have increased at the centre of pollution by 0.6 units. A decrease in the magnitude of hydrolytic acidity and increase of the content of exchange calcium were noted here also.

Acidification of the soil was observed at 5 km and further from the plant in the direction of prevailing winds that brought acid aerosols. The value of pH has decreased by 1.0 unit at a distance of 5 km and by 0.7 unit at a distance of 15 km within one decade. Within the following decade pH decreased only by 0.2 unit at both sample plots. Over 20 years the pH has decreased by 1.2 unit, i. e. acidity of the soil has increased more than 10 times during two decades. During this second decade acidification of the soil was slower as a result not only of the reduction of the emissions of sulphur dioxide but also with the change of physical-chemical properties of the soil itself. The conclusion of McFee [3] that acid soils are less sensitive to the further acidification was confirmed.

Anthropogenic transformation of chemical and physical-chemical properties of the soils caused by the accumulation of toxic heavy metals and antimicrobial element of sulphur in the soil caused a change in the structure of microbial communities, with a resultant decrease in the biodiversity of soil biota. There is noted the effect of heavy metals upon biota of both direct and indirect characters. Indirect influence occurs as a result of the decline of the nutritional status of the acid soils.

We have found that the breakdown of homeostasis of microbial communities in podzol soil begins at a soil copper concentration of 300-400 mg /kg and a soil nickel concentration of 600-700 mg/ kg [4]. This occurs during the first year of the emission influence at the centre of pollution. Changes in microbial communities start at these metal concentrations with the inhibition and destruction of nitrogen-fixing bacteria genera *Azotobacter* and *Clostridium*, nitrifying bacteria, i.e. neutrophil microbes are observed. As the level of soil metal increases the number of ammonifying bacteria, cyanobacteria, actinomycetes, and yellow-green algae sharply decreases and the share of the fungi rises.

Acid-tolerant forms, mainly soil fungi, begin to dominate in the microbial communities. Metal-tolerant microorganisms generally are also acid-tolerant. Fungi of genus *Penicillium* were found in the forest soils even with a copper concentration of 3% and nickel of 5% with a pH lower than 4. These species were

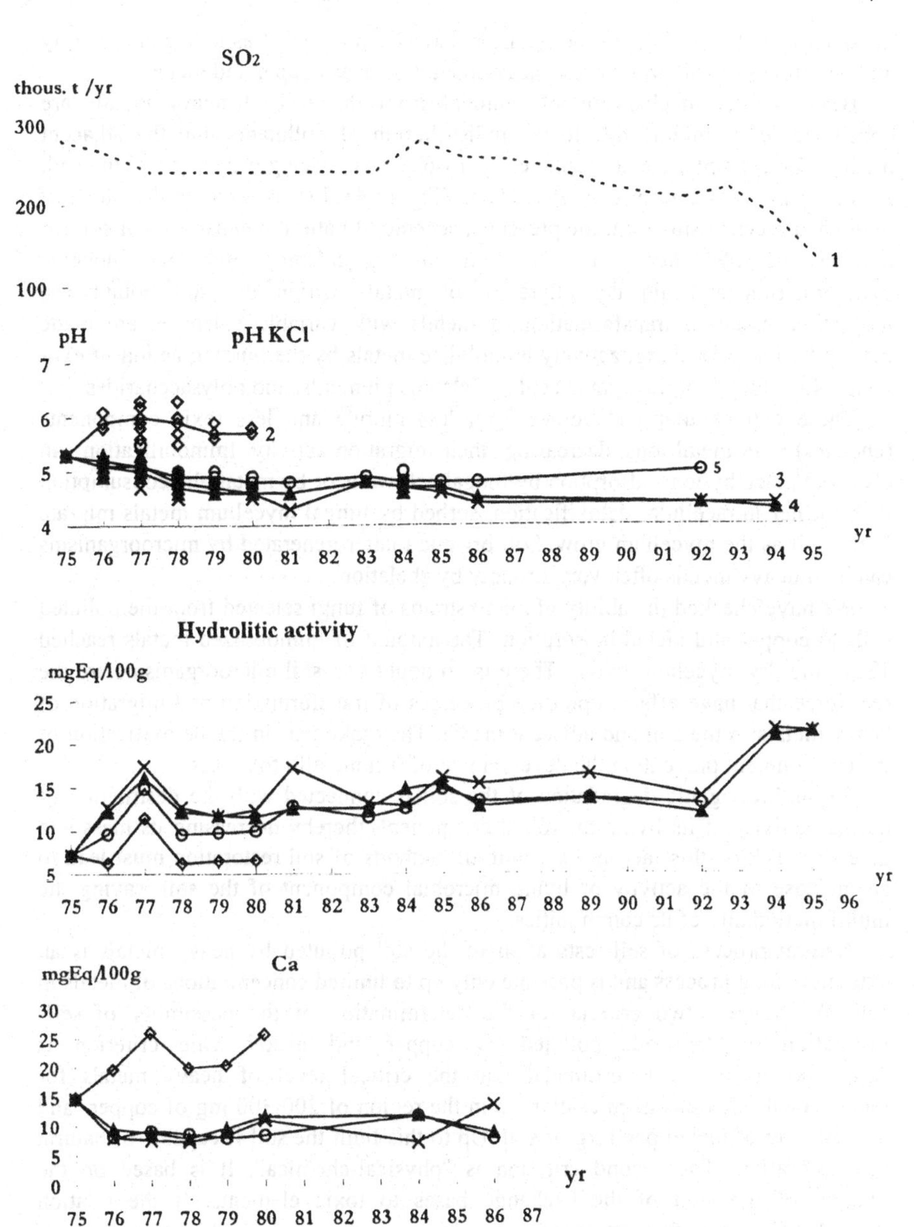

Figure 3. Acidity dynamic of cultivated podzol soil, SO_2 pollution, dynamics of exchangable calcium. 1 - SO_2; 2 - centre; 3 - impact zone; 4 - buffer zone; 5 - back ground zone.

the dominants there. So, microorganisms have been revealed as indicators of early and late stages of soil pollution by heavy metals such as copper and nickel.

Basic natural mechanisms of removal from the soil of heavy metals are concluded to be mainly due to the redistribution of pollutants into the adjacent media - for example, the atmosphere, hydrosphere and living organisms of the soil, including the root system of higher plants (Figure 4). Let us separate the block of biological mechanisms from the presented scheme of natural mechanisms of soil for removal of free heavy metals. Soil microorganisms carry out biogenic transformation and aid the migration of metals within the soil boundaries. Reduction-oxidation transformation of metals with variable valencies are made mainly by bacteria. Fungi actively immobilize metals by the binding action of exo-metabolites for example organic acids, melanins pigments, and polysaccharides.

These extracellular metabolites form less mobile and less toxic components (chelates) with metal ions, decreasing their migration activity. Immobilization can also take place by ions adsorption by fungal cell walls or by metabolic consumption with further intracellular detoxification, sorbed by fungal mycelium metals migrate in the soil as the mycelium grow. Soil organic matter generated by microorganisms can bind heavy metals often very strongly by chelation.

We have checked the ability of many strains of fungi selected from the polluted soils to copper and nickel biosorption. The amount of immobilized metals reached 1% of the dry mycelium mass. There is no doubt that soil microorganisms are the real force that have effects upon the processes of transformation and migration of heavy metals in the soil and adjacent media. They take part in the detoxification of the environment preventing the development of soil metallo-toxicosis.

So, anthropogenic degradation of the soil is connected with the destruction of normal activity of its living microbial component, thereby decreasing its mass and diversity. Taking this fact into account all methods of soil restoration must lead to an increase of the activity of living microbial component of the soil, saving the multifunctionality of its communities.

Natural process of self-restoration of the soil polluted by heavy metals is an extremely long process and is possible only up to limited concentrations of metals in soil. We suggest two criteria for the determination of the possibility of self-restoration of the soils polluted by copper and nickel. One criterion is "microbiological". It is estimated that the critical level of heavy metals for biochemical functions of microflora is in the region of 200-300 mg of copper and 400-500 mg of nickel per 1 kg of soil. Up to this limit the soil is capable of natural self-restoration. The second criterion is "physical-chemical'. It is based on the relation of the sum of the exchange bases to toxic elements. If the relation (Ca+Mg)/Cu is less than 10 and (Ca+Mg)/Ni is less than 5 the soil is not capable of self-remediation (Table 1).

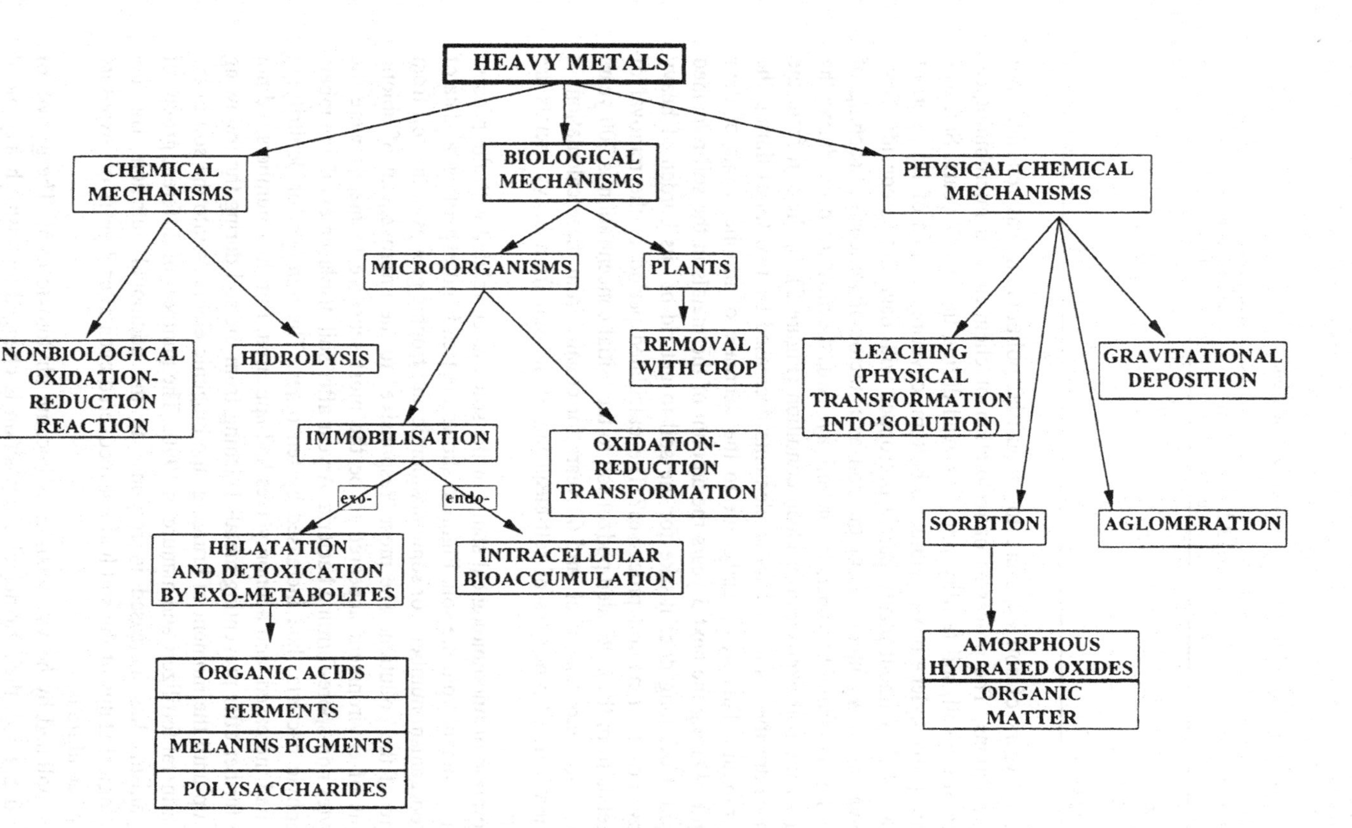

Figure 4. Natural mechanisms of soil self-cleaning from heavy metals.

TABLE 1. Scale for estimation of degree toxicity of cultivated podzol soil

Soil	$\frac{Ca + Mg}{Cu}$	$\frac{Ca + Mg}{Ni}$
Non-toxic	≥ 30	≥ 20
Poorly-toxic	10-29	5-19
Highly-toxic	≤ 9	≤ 4
Threshold value	10	5

After 10 years of soil exposure to industrial pollution it was returned into "clean" conditions. The processes of restoration of chemical and microbiological properties of the soil, soil fertility were observed during the next 7 years. Soil was placed into polyethylene boxes without a bottom, covering an area of $1m^2$. Part of the soil was held without plants, part was cropped with oats. Let us consider first, the removal of heavy metals from the soil as a factor of toxicosis. Leaching of copper and nickel from the upper, 10 cm soil layer was the most intensive during the first year of the soil exposition in clean condition (Figure 5). In the soil from the pollution centre the content of these elements decreased by two times during the first 3.5 months. This was mainly due to the decrease of mobile forms of their compounds. During the next 5 years the amount of the metals in the soil remained unchanged. How long does it take for the soil to get rid of its burden of heavy metals by means of natural processes? The calculated period of the removal of heavy metals from the cultivated podzol soil to the initial concentration is 200 years for copper, 90 years for nickel and 60 years for cobalt. From these calculations it follows that natural processes of self-cleansing of the soil from heavy metals are very long.

Restoration of microbiological status of the soil was correlated with the process of leaching metals from the soil. Figure 6 shows the trend and experimental data of the microorganism numbers of some taxonomic and trophic groups in the soil from the centre of the pollution. The most "responsive" to the improvement of abiotic conditions of environment appeared to be the most sensitive to metals were the saprophytes non-spore-forming bacteria. A year after soil displacement the numbers of saprophyic bacteria had increased by ten times. The numbers of denitrifying bacteria had increased, and actinomycetes had appeared. But their numbers did not increase further after preventing metals leaching from the soil during the following 5 years, up until the moment of bringing the biogenic elements into the soil in the form of mineral fertilizers and manure in 1990. The numbers of different groups of microorganisms has increased in response to the additional nutrition. But the microbiological status of the soil had not recovered even after 7 years of exposure to "clean" conditions.

When polluted by heavy metals soils become phytotoxic [5,6]. The growth of plants is depressed, their elemental composition is changed and ash, heavy metal and nitrate contents rise. The tendency for increasing contents of total and protein nitrogen and the sum of amino acids is found (Table 2). This change may be a defense mechanism, as these organic compounds form chelates with metals that are less toxic than free ions. Phytotoxicity of the soil introduced into "clean" conditions

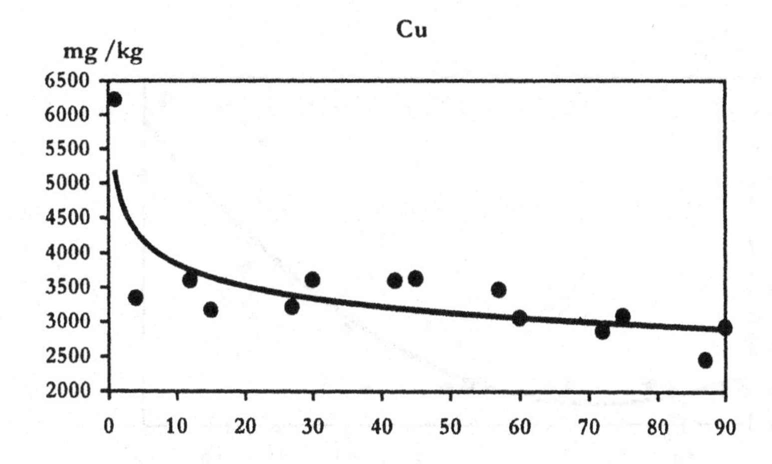

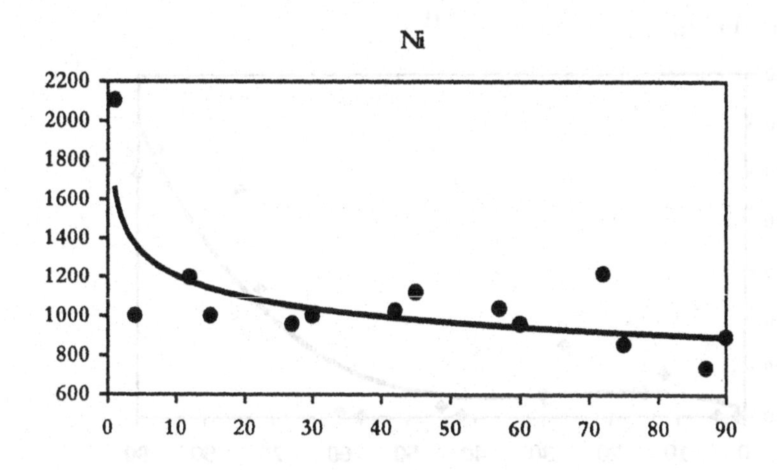

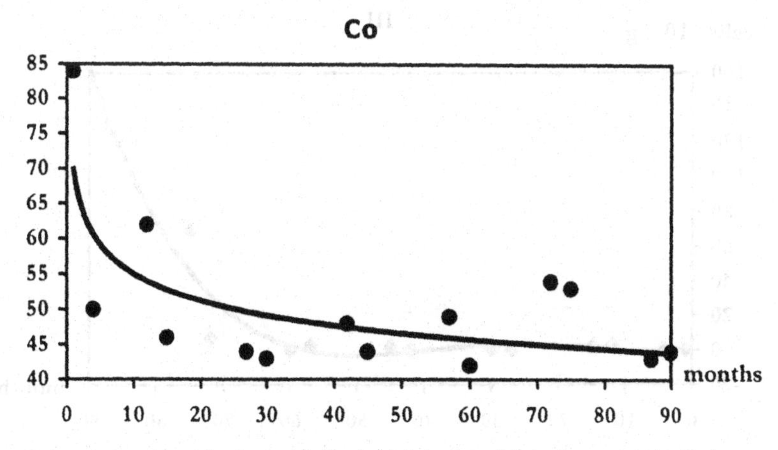

Figure 5. Leaching of copper, nickel and cobalt in the soil from the centre of pollution, after return of soil in "clean" condition

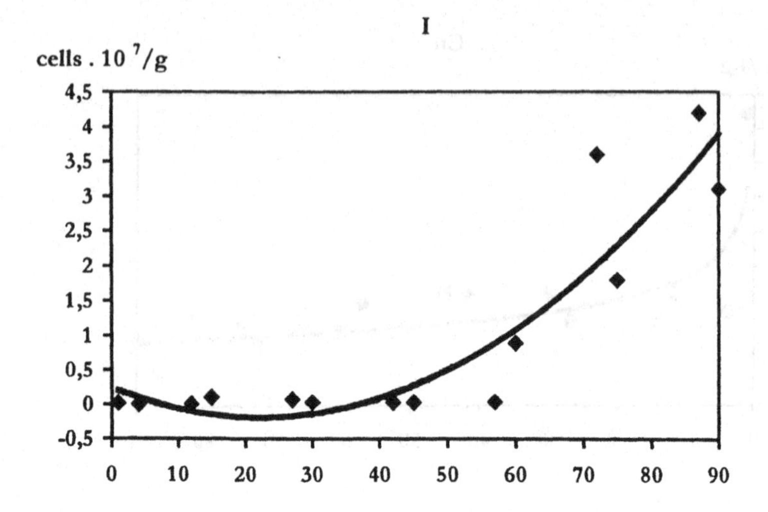

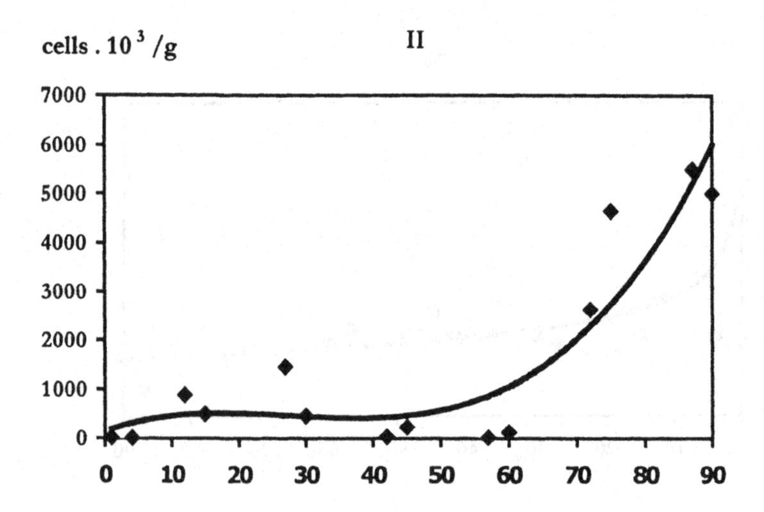

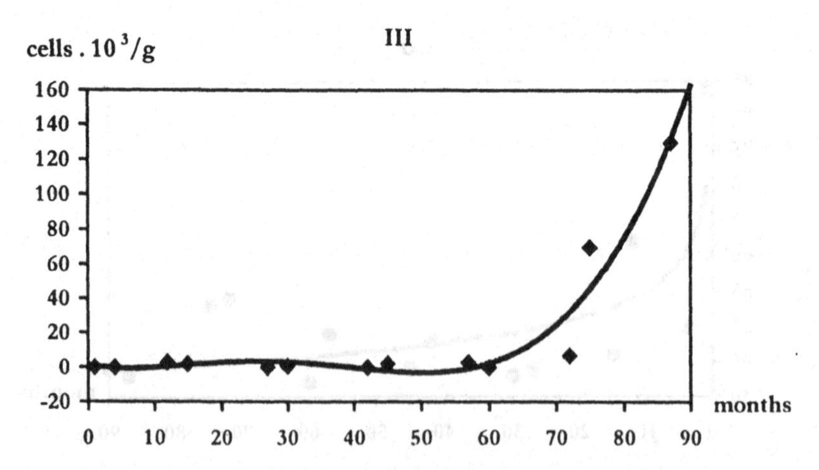

Figure 6. Restoration of microbiological status in the soil from the centre of pollution return of soil in "clean" condition. I - mineral nitrogentrophic bacteria; II - saprotrophic non-sporoforming bacteria; III - actinomycetes.

decreased rather slowly. The soil from the epicentre of pollution was not capable of producing a crop 7 years after transferring, despite the application of fertilizer. The crop of oats on the soil from the impact zone were also lower than controls and the quality of the plants was not in accordance with the accepted standards concerning soil nickel contents.

TABLE 2. Biological quality of oats grown on the fields with different degree of contamination by industrial emission

Index	Degree of contamination		
	High	Low	Control
Nitrogen total, %	2.7	2.4	1.9
Nitrogen albuminous, %	1.2	1.3	1.0
Nitrogen residual, %	1.5	1.2	0.8
Nitrate, ppm	7876.5	1609.5	1337.9
Sum aminoacid, $g \cdot kg^{-1}$	107.6	97.4	81.7
Ashes, %	12.1	7.3	7.7
Cellular, $g \cdot kg^{-1}$	29.4	32.8	31.5
Fat, $g \cdot kg^{-1}$	2.9	2.1	2.3
Carotene, $g \cdot kg^{-1}$	0.8	10.9	13.6
Copper, ppm Nickel, ppm	64.0	5.0	4.0
	77.0	6.0	4.0

Methods of regeneration of soils polluted by heavy metals have been developed to increase the capacity of cation exchange and regulation of the soil nutritional status by organic and mineral fertilizers that change the mobility of pollutants (Table 3). The most effective method of decreasing toxicity of the soils polluted by copper and nickel is liming to the value pH 6.4 - 6.8. Moreover this method decreases the migration from the soil to the plant not only of heavy metals but also of nitrates. On neutral media the toxicity of the soil decreases with the application of phosphates which form compounds of low solubility with metals. An effective method of decreasing the toxicity of soil polluted by heavy metals is application of organic fertilizers, in particular manure. We have found that the organic substances contained in manure effects the ability of the metals migrate in the soil in different ways [7]. Water soluble organic compounds in manure can bind metals in low molecular complexes, and these complexes can be leached through the soil profile. Later, with microbial transformation of manure poorly-soluble organo-mineral complexes with metals are created (e.g. as chelates), which immobilise the metals.

1.4. Conclusion

Biogenous elements (C, N, P, K) brought into the soil increase the vital activity of the soil microflora. Soil microflora decreases the metallo-toxicosis of the soil by means of the biosorption of heavy metals and the formation of chelates between metals and extracellular metabolites. All proposed measures increase the protective forces of the soil against pollution. But if we want to solve the problem of protection and restoration of soil cover it is necessary to organize the appropriate reconstruction of industrial enterprises in order to reduce or to remove completely the toxic emissions. It is highly desirable to form a high moral ecological philosophy for all levels of population according to the principle "what is not good for the environment - is amoral". To understand and to coordinate the laws of nature and development of mankind - is a real necessity at the moment.

TABLE 3. Measures contributing to the decreasing of toxicity of the soils contaminated by heavy metals

Measures	Dose t/ha
Manuring	100 - 150
Liming	up to pH 6.4 - 6.8
Introduction of superphosphate with average content of P_2O_5 - 45%	0.3 - 0.4
Magnesium fertilisers	0.03 - 0.08

Acknowledgments

The author thanks N. Mozgova, T. Ageeva for help in analytical work. This work was supported by the Russian Fund of Basic Sciences, Grant N° 04 - 48508.

References

1. Kruchkov, V.V. and Makarova, T.D. (1989) Aerotechnogenic pollution on ecosystems of the Kola North; Apatity, KSC RAS. 96p.
2. Lebedeva, E.V. (1993) Soil micromycetes of the non-ferrous metallurgy enterprise environs on the Kola Peninsula. *Mycology and Phytopathology*, **27**, N 1. 12-17.
3. McFee, W.W. (1983) Sensitivity ratings of soils to acid deposition a review. *Environ. and Exp. Bot.*, **23**, N 3. P. 203-210.
4. Evdokimova, G.A. (1995) Effects of pollution on Microorganisms Biodiversity in Arctic ecosystems. *Global Change and Arctic Terrestrial Ecosystems.*. Proceedings of Intern. Conf.; Oppdal, 21-26 Aug. 1993. Brussels Luxembourg Publication European Union . 315-320.
5. Il'in V.B.(1985) Elementary chemical composition of plants. Irkutsk. 129 pp.
6. Kabata-Pendias, A. and Wiacek, K. (1985) Excessive uptake of heavy metals by plants from contaminated soils. *Roczniki Glebosnaveze*, Warzawa, V. 36. P. 33-42.
7. Evdokimova, G.A. (1995) Ecological-microbiological foundations of soil protection in the Far North. Apatity. 272 pp.

2. BIOMONITORING IN NORTHERN BOHEMIA MINING AREA, CZECH REPUBLIC

JIRÍ CIBULKA
Czech University of Agriculture,
Prague 6 – Suchdol, 165 21,
Czech Republic

Abstract

The paper summarises the main environmental problems identified in the Northern Bohemia region. The region is characterised by a high concentration of chemical industry and brown coal strip mines. The environment of the region is badly effected by a combination of unfavourable geographical factors and very high level of atmospheric pollution. There are signs of negative effects even on human health.

It discusses current relevant environmental legislation and current results of long-term biological monitoring being performed in the area during co-operation between the Bilina Mines and Czech University of Agriculture in Prague.

The biological monitoring is based on current environmental legislation, gathers basic facts on natural conditions in the monitored area (pedology, flora and fauna populations) and describes real or possible environmental effects related to mining activity. A university team of experts has been monitoring the environmental situation within the area of the Bilina Mines for 5 years. The main goals of the university team of experts are: (a) a biological inventory of areas of future mining and (b) monitoring of natural biological processes going on at newly created dumpsites in relation to their revitalisation and rehabilitation. Selected significant research results were discussed and others were shown on accompanying posters.

Keywords: Bohemia, Biomonitoring, Mining, Heavy metals, Biomarkers.

2.1. Introduction

The Northern Bohemia region stretches along the Northwest borders with Germany and Poland between Kadan and Liberec cities. It has an area of 7,817 km^2 with approximately 1 million inhabitants living in about 2,500 towns and villages. The Northern Bohemia region can be characterised as a highly industrialised area with prevailing mining, chemical and refinery industries [1,2]. Another typical feature is

D. B. Peakall et al. (eds.),
***Biomarkers: A Pragmatic Basis for Remediation of Severe Pollution in Eastern Europe,* 15–28.**

the high concentration of coal burning power plants (7 of them) producing electricity not only for local industrial and household consumption but also supplying energy to a substantial part of the Czech Republic. Northern Bohemia has been heavily burdened for several decades by the high concentration of these industries [3]. The environmental and health consequences of this situation have been studied there by many institutions. NATO and PHARE projects have been also realised there. Based on several new environmental laws, all investors are now obliged to perform a long term and complex biological monitoring within areas that are affected by their activities [4].

2.2. Coal Mining in Northern Bohemia

Brown coal is the main mineral deposit being mined in Northern Bohemia. Strip mining (or open cast mining) contrary to pit mining is the main technology being employed there. The technology has two main operational phases. The first stage is stripping off topsoil and underlying layers of different minerals (mostly silt, sand and clays in Northern Bohemia) and depositing them on temporary or permanent dumpsites. The permanent dumpsites prevail and their masses are enormous since rock increase in volume during the course of mining. Just one mine, with its mining operational area of 2.5 km^2 and producing yearly about 8 million tonnes of coal, has to excavate, transport and re-deposit approximately 60 million m^3 of minerals above a coal layer. A coal layer can be "only" about 30 m thick while being covered by 120–150 m of upper rock horizons. In fact, an operating mine represents a huge hole slowly moving across the landscape at a velocity of 100 m in a year and following an underground coal seam while destroying everything in front of it. Permanent dumpsites of deposited sterile rock surround a mine as unavoidable features of this mining technology. The total area being adversely affected in this manner within Northern Bohemia region is about 26,000 ha with approximately 8,000 ha of permanent dumpsites (see Figures 1 and 2). In general, coal strip mining in Northern Bohemia has declined during the past years since it is considered as unacceptable for the future. Mining should totally stop there by 2030 and even now all mines focus their attention upon the correction of environmental damage caused by them in the past decades.

This current effort is mainly concentrated on landscape re-creation and revitalisation employing sophisticated, scientific recultivation methods. In reality it would take decades for natural revitalisation processes to transform sterile heaps of rock into green hills, fields or forests. A special treatment must be applied to form an "artificial" topsoil layer on their surface that will provide water and nutrition for plants. Due to a lack of natural topsoil for this dumpsite recultivation, special procedures must be used for restructuralizing and fertilising rock material into fertile soil profile. Such pedological measures make possible usage of dumpsites as agricultural plots or, more often, planting them with various plant species (grasses, deciduous and coniferous trees, etc.). The main types of recultivated areas are

Figure 1. Open cast coal mining in Northern Bohemia

Figure 2. Open cast coal mining in Northern Bohemia

classified as agricultural, forest, and recreational. Artificial water reservoirs are always an part of a new landscape.

2.3. Main environmental and health problems in the area

The main environmental problems in the region arise from atmospheric pollution [5,6]. Due to a combination of unfavourable natural geographic conditions and a high concentration of industries emitting air pollutants (gaseous and solid), there are regular seasons of smog (October through February). Temperature inversion causes the creation of a closure above the area followed by rising concentrations of air contaminants. It must be stressed that the mines do not cause the majority of the atmosphere pollution. It is a consequence of chemical industry and power plant activities. The main air pollutant originating from the strip mines is dust that is not so harmful to human health and can be controlled to some extent. Generally, the most environment devastating air pollutants are nitrogen and sulphur oxides, apart from many different organic pollutants.

It is said that the most important consequence of unfavourable environmental conditions in the Northern Bohemia is shortening of human life span and life expectancy [7]. It is a fact that Czech women and men have shorter lives by approximately 5 and 7 years, respectively, compared to the average values for the developed world. This phenomenon is even more obvious among the Northern Bohemia population. On the other hand, there is another controversial explanation of this feature. A reason for shorter lives could be (partly) seen in many unhealthy nutritional and living styles and habits being widely spread among our population (heavy smoking, fatty food, alcohol abuse, lack of exercise, a low standard of awareness concerning preventive health care and hygiene, etc.) in combination with relatively good average salary in labour professions. These habits are very typical for some social groups which are numerous in Northern Bohemia. Our sociologists and health experts have been studying this aspect. There is also foreign co-operation on the problem.

2.4. Corrective and preventative measures in the region

The basic reasons for performing corrective and preventative measures in the region are the following facts: a) regional industrial activity cannot be immediately stopped since it is currently vital for our state economy; b) strip mining can be limited since there will be future alternative sources of energy (e.g. nuclear power plant near Temelín, Ceské Budejovice, South Bohemia); c) mine operations must continue to a limited extent to accumulate sufficient financial resources for subsequent landscape rehabilitation [8,9,10,11].

The main corrective measure being realised in the region is the application of a scrubber technology on several local power plants. Currently, stripping off sulphur oxides, ash and other air pollutants using special stack filters has a very positive effect on the atmosphere quality. There is a distinct trend of decreasing amounts of atmospheric pollutants. However, this technology is rather costly and it will take several further years to equip all significant stacks and chimneys with these special scrubbers.

Concerning the mines, their activities are gradually being limited and slowed down. Doly Bílina will completely stop mining by 2030 and landscape re-creation and controlled rehabilitation should be finished by this date. Until then, all strip mines must respect strict environmental laws discussed below.

There is a significant environmental ("green") movement (Duha –Rainbow or Deti Zeme – Children of the Earth, etc.) working within the region and closely watching over all kinds of industrial activities there. As far as the Bílina Mines are concerned, it must be stressed that there is good co-operation between the mine management and the representatives of the movement. There are cases of successful joint rescue actions (flora and fauna rescue or emergency relocation, etc.) and of financial support to the movement. Also, all relevant planned activities of the mine are usually discussed with the movement.

2.5. Relevant new environmental legislation

As a consequence of political and economic changes in the Czech Republic after November 1989, our government is hoping to join the Czech Republic to the European Community (EC). One of the basic conditions for being accepted as a new EC member is to implement harmonised legislation with the rest of Europe. That is why an extensive collection of new laws has been implemented in the Czech Republic from 1990 [4]. A new feature concerning the environmental laws is the requirement for all subjects to perform environmental evaluation of their current or planned operations (environmental impact assessment, EIA). Such evaluation can be prepared in various ways defined in the laws (see below). Biological monitoring of affected landscape is always a compulsory part of such documentation. The EIA documentation must be prepared only by a licensed expert specialised in the environmental problem areas and accredited for this activity by the Ministry of the Environment.

The Act on the Environment (17/1992 Coll.) defines duties in its § 17 in respect of environmental protection for all who use landscape and natural resources. Its Enclosure N° 1 is a list of activities amenable to this Act. Pit and strip mining are listed there (par. 3.1.). Enclosure N° 2 defines a structure (table of contents) of the compulsory documentation and evaluation of planned activities with respect to the environment.

The Act on Environmental Impact Assessment (244/1992 Coll.) defines a procedure on environmental evaluation of new constructions and technologies. It has several enclosures describing its scope. Mining activity is again listed there but it is applicable only if a mine is opened as a brand new operation. The Ministry of the Environment does not apply this legislation to ongoing mining activities within currently defined mining areas. This is a very important official statement.

Apart from these major environmental laws, there is a number of valid measures regulating all aspects of strip mining. For example, there is a measure requiring that all mines must save and deposit a part of their financial profits in their special bank accounts to be used in future years for landscape revitalisation. Eventually coal mining will be stopped or substantially slowed down and the only mine activity will be closure and subsequent landscape reparation. A mining permit is usually issued for a period of five years and must be regularly updated. The issuing of a permit allowing prolongation is based on the evaluations of many technical and economic documents including the environmental report described above. Since mine management lacks the ability to perform scientific biological monitoring for preparing the report, they usually look for professional co-operation with specialised institutions or universities.

2.6. Co-operation between Czech University of Agriculture in Prague and Doly Bílina mines

The Czech University of Agriculture in Prague has been co-operating with the Bílina Mines shareholder company, for 5 years. The mining company belongs to one (Severoceské Doly) of two (Mostecká uhelná) existing coal strip mining companies in the Northern Bohemia. The mining company uses strip mining thus severely changing environmental conditions within the mining territory and its vicinity. On the other hand, the company is obliged to ensure landscape re-creation and recultivation after closure of the mine (see above). Currently, several tens of square kilometres of new, completely recultivated dumpsites are created and treated by the company.

The university team of experts (pedologists, botanist, entomologist, hydrobiologist and ornithologists) has been carrying out continuous and extensive monitoring of pedological conditions, flora and fauna in a landscape being prepared for mining [12,13,14,16]. Subsequently, mining will cause complete devastation of ecosystems. The main effort of the survey and monitoring during this phase of research work is to locate and possibly rescue rare or precious flora and fauna species from being lost for future. This work represents the first task for the university team.

The second task of the university team is to advise on the processes of permanent dumpsite recultivation and, consequently, to monitor the progress of biological succession going in these new biotopes. The posters showing and discussing results concerning botanical, insect and vertebrate inventories will deal in more detail with current research results of biological monitoring performed in the region. A presentation will be given on pedological findings and plant bioindicators.

2.7. Current results

2.7.1. LEGISLATIVE POSITION

The main goals of our co-operation are to meet all legislative requirements with respect to the environment. Our complex biomonitoring research has gathered an extensive set of scientific data for fulfilling legislative duties. The Bílina Mines has been given formal permission to continue in mining until 2003. This licence was granted and is based on preparing a complex proposal - A Plan of Opening, Preparing and Mining (POPD) - respecting the structure described in the Act on the Environment (N° 17/1992). This proposal has to include a chapter dealing with the environmental impact assessment of the planned mining activity. The chapter was prepared using our research results and it would not have been possible to write it without having continuous biological monitoring results available. The author of this workshop contribution is "a licensed professional" in the words of the Act of Environmental Assessment (N° 244/1992) and he is also entitled to prepare official statements dealing with the environmental issues defined in the Act of the Environment (17/1992) and the Act of the Environmental Impact Assessment (244/1992).

2.7.2. PEDOLOGY

Pedological research was performed by Prof. J. Kozák, Head of the Department of Pedology and Geology, and his team. Their results show that:

1. Soils (kambisols) surrounding the Bílina mine and its dumpsites can be classified as being of medium or low fertility. Their characteristics are dependent on the quality of soil-forming base rock and the depth of humus-forming horizon depth. Soils of higher quality are found on carbonaceous substrata (chernozem and rendzina).

2. The soil substratum quality is subsequently reflected even in original biological potential of the studied areas.

3. From the pedological point of view, dumpsite rock materials cannot be harmful to the area adjacent to a dumpsite (apart from occasional and temporary dusting). Spreading topsoil layer on dumpsite surface is obviously the best method of recultivating them. However, there is a lack of quality topsoil needed for this procedure, and mixing dumpsite rock with clay marl has been proved to be a good substitute recultivation procedure.

4. Based on pedological research, it is obvious that the studied landscape and its intact fragments surrounding the mine are quality areas. This is confirmed by monitoring of flora and fauna. Thus, rapid settlement of correctly recultivated dumpsites by a great variety of plant and animal species is to be expected.

2.7.3. HYDROBIOLOGY

Hydrological research was performed by Dr. J. Kurfürst, Department of Zoology, and it consisted of hydrochemical and hydrobiological monitoring of natural and artificial water streams and reservoirs within the studied area. The results obtained can be summarised as follows:

1. Generally, it must be stressed that the aquatic environment is a key component of any healthy landscape. It is important that recultivation plans include construction of both small and large water reservoirs within the monitored area.

2. However, great and extensive water basins may have insufficient flow and water turbulence. Still and deep water thus has low biological significance. This factor must be considered during planing and constructing such water basins in the monitored area.

3. A littoral zone of water reservoirs must be as wide as possible to provide ample opportunities for creating highly valuable transition ecosystems (e.g. nesting areas).

2.7.4. BOTANY

Botanical inventory and survey was done by Dr. V. Zelený, Department of Botany. His results were presented in detail on a poster. The main points are as follows:

1. Generally, there were no rare or unique plant species found within the Bílina Mine current territory. It has been proven that weed plants inhabiting the dumpsites during the first stages of their recultivation cannot spread from them to surrounding areas. These weed species are typical for exposed soils and they are not able to compete in continuous vegetation of forests, meadows and grassland.

2. Current recultivation provided good results. An agricultural type of recultivation can be used for cultivating non-traditional crops. A forestry type of recultivation supports renovation of original plant communities and substantially speeds up and stabilises the process of dumpsite revitalisation.

3. The successfully recultivated dumpsites should rapidly integrate into the surrounding landscape. Respecting and using the existing biocenters and bio-corridors will speed the process up.

2.7.5. ENTOMOLOGY

This part of monitoring has been performed by Prof. M. Barták, Head of the Department of Zoology, Czech University of Agriculture in Prague: His research is focused on evaluation of recultivation quality based on bioindicative features of *Diptera*. He monitored all dumpsites in the close vicinity of the mine. Detailed research results have been presented on a poster. The main points are as follows:

1. Generally, the area under study is of low quality considering the incidence and abundance of *Diptera* species.

2. Anthropogenic devastation of landscape has led to the existence of an area of low-pressure biodiversity of *Diptera* fauna. This low-pressure area is immediately saturated by immigration.

3. In particular, forest recultivations are populated by species that are considered as rare in the fauna of the Czech Republic.

4. Non-recultivated dumpsites are populated by a very special biocenose that is very species poor; however, it is very interesting from a faunistic point of view.

5. Dumpsite biological potential is lower than that of neighbouring ecosystems.

6. Correctly performed recultivation significantly speeds up the succession of *Diptera* communities, namely, by stabilisation of immigrants and by reducing their own emigration through offering food, shelter and habitat.

7. The quality of recultivated areas would be increased by higher diversity of surface formatting, for example, construction of artificial rock prairie, wetlands, etc.

8. A multicriteria method has been proposed for recultivation and other biotops quality evaluation based on bioindicative features of *Diptera* insect fauna.

2.7.6. VERTEBRATOLOGY

This survey was performed by Prof. K. Štastný, Head of Department of Ecology, and Dr. V. Bejcek, Department of Ecology. The inventory was focused on monitoring of populations of terrestrial and aquatic vertebrates. Detailed results have been presented and discussed in a special poster presented by Prof. K. Štastný. The principal results obtained can be summarised as follows:

1. The natural quality of landscape in the close vicinity of the Bílina mine is relatively good with high incidence of many plant and animal species. This

provides a sound basis for rapid re-population of correctly recultivated dumpsites.

2. In total, 32 species of protected animals were observed within the monitored area. Three species are classified as "critically endangered species" (*Rana ridibunda*, *Emberiza hortulana*, and *Milvus migrans*), eleven species are "strongly endangered" and eighteen species are classified as "endangered" species.

3. From a zoological point of view, the recultivated dumpsites are currently not very important due to their uniform and minimal diversity of plant coverage. However, it can be assumed that in future their biological value will increase during ageing of recultivation plant coverage.

4. Any kinds of small water reservoirs (both permanent and temporary) significantly accelerate biological rehabilitation of impaired landscape.

2.7.7. HEAVY METALS

This research was performed by the author of this paper and consisted of two parts: (a) monitoring heavy metal levels (lead, cadmium, mercury, chromium and arsenic) in forage (mixture of wheat and vetch) and weed plants growing on new recultivated areas and (b) monitoring of heavy metal content in wild edible and inedible mushrooms growing on recultivated dumpsites and in local forests.

Concerning (a), cultivation of forage crops is used as a process of encouraging formation and improving artificial topsoil quality. Currently, forage produced is not harvested and used for feeding animals as it may be done in future. A basic question was whether there is any significant accumulation of selected heavy metals in plant tissues. Accompanying weed plants were also sampled to monitor any possible selective heavy metal absorption.

Summarising, there is no danger of heavy metal accumulation in plant tissues originating from dumpsite rock materials. All heavy metal levels measured in the samples were below or equal to the valid permitted levels. Using such forage as a part of farm animal diet would not represent any health or hygienic problems [12].

Concerning (b), the main object of this part of monitoring was to follow heavy metal levels in wild mushrooms collected on new recultivated dumpsites and in local forests. Mushroom collecting is a very popular hobby in the Czech Republic and regular or high consumption of edible wild mushrooms contaminated with heavy metals could be a health hazard. Wild mushrooms are significant components of forest ecosystems (decomposers, symbiosis, etc.) and their incidence could reflex, in certain aspects, ecosystem status. Unfortunately, it has been established that the most popular edible mushrooms (*Boletus*, *Xerocomus* and *Agaricus* species) have a significant capacity for accumulating heavy metals. On the other hand, mushrooms are not considered as reliable organisms for monitoring the extent of the environment heavy metal pollution.

We compared our results obtained for mushrooms collected within the Northern Bohemian region (Dubí locality) with others based upon mushrooms sampled in relatively clean environments with healthy forests (Zbiroh and Chocen localities) in the South-western or Eastern regions of the Czech Republic, respectively. Our analyses document that the levels of cadmium, lead and mercury in the Northern Bohemia mushrooms (*Xerocomus*) were generally equal or even lower compared to those from the control region. Higher contamination was always detected in mushroom caps *versus* their stems. Detailed results of this study have been published [17] or prepared for publication.

2.8. Future goals

It is obvious that biological monitoring is very important for all strip mines operating in the region. According to the valid environmental legislation, biological monitoring within affected area must be considered as a responsibility of the mine owners.

In future years, we would like to with our successful co-operation which brings positive results to both of the participating parties.

In future, our main goal will be biological monitoring of the revitalising process going on at new dumpsites. The significance and extent of rescue monitoring and of biological inventories within mining territories will gradually decline during future years since there will be a slowing down and finally a total cessation of mining activity.

Based on the valid legislation, there will have to be significant post project activity – biological monitoring of new landscape, containing new rehabilitated dumpsites, will offer a unique opportunity for long term monitoring of their integration into the existing landscape.

Without exaggeration, this will be an interesting and important task for future generation(s) [18].

Acknowledgement

We appreciate the help provided by the Bílina Mines in preparation of this lecture and we also express our thanks to all quoted authors for their materials used in this lecture.

References

1. Štýs, S. and Helešicová L. (1992) Promeny meslcnl krajiny (Changes of Moon Landscape). Bílý slon Publ.

2. Anonymous (1995) Zelené promeny cerného severu (Green Changes of the Black North). Severoceské Doly, a.s., Chomutov, Czech Republic.
3. Anonymous (1995) Výsledky kontroly a monitoringu cizorodých látek - Situace v roce 1994. Yearbook of the Ministry of Agriculture of the Czech Republic. Ministry of Agriculture, Prague.
4. Anonymous (1993) Environmental laws of the Czech Republic. Vol. 1: General Environmental Laws, Ministry of the Environment of the Czech Republic, Prague.
5. Anonymous (1996) Zpráva o zivotním prostredí Ceské republiky v roce 1995 (A Report of the Environment of the Czech Republic in 1995). Ministry of the Environment of the Czech Republic, Prague.
6. Anonymous (1996) Hodnocení stavu zivotního prostredí - Monitoring cizorodých látek v potravních retezcích v roce 1995 (Evaluation of the Environment Status - Monitoring of Foreign Substances in Food chains). VŠCHT Publ., Prague.
7. Anonymous (1995) Monitoring zdravotního stavu obyvatelstva ve vztahu k venkovnímu a vnitrnímu ozvduší (Monitoring of populating health status in relation to outside and inside atmosphere). State Health Institute, Prague, 1996
8. Anonymous: State Environmental Policy. Ministry of the Environment of the Czech Republic, Prague, Czech Republic.
9. Soucek, V. (1995) Poznámky k posuzování vlivu na zivotní prostredí pri tezbe nerostu a jeho vyuzití pro úvahy pro sanaci dotcených pozemku (Notes on environmental impact assessment and its usage for considerations for rehabilitation of the affected plots). Hornická Príbram Conference, Príbram, Czech Republic, October 16-19, 1995.
10. Šedenka, K. (1995) Úhrady z dobývacích prostor, z vydobytých vyhrazených nerostu a financní rezervy na sanace, rekultivace a dulní škody podle horního zákona (Financial compensations for mining territories, for mined rocks and financial reserves for rehabilitation, recultivations and damages due to mining according to the Act on Mining). Hornická Príbram Conference, Príbram, Czech Republic, October 16-19, 1995.
11. Bendová H. (1995) Rekultivace a zákony o ochrane zemedelského a lesního pudního fondu (Recultivation and laws on protection of agricultural and forest soils). Hornická Príbram Conference, Príbram, Czech Republic, October 16-19, 1995.
12. Cibulka J. Kozák J., Valla M., Zelený V., Barták M., Smrz J. Kurfurt J., Pruzina I., Stastny K., Bejcek V. (1992) Orientacní zpráva o výsledcích pedologického, botanického a zoologického zhodnocení území v areálu Dolu Bílina (A preliminary report on the results of pedological, botanical and zoological evaluation of the Bílina Miňes area), UNICO AGRIC, Czech University of Agriculture in Prague, Prague, Czech Republic.
13. Cibulka J., Kozák J. Valla M., Zelený V., Barták M., Smrz J. Kurfurt J., Stastny K., Bejcek V. (1993) Záverecná zpráva o výsledcích pedologického, botanického a zoologického zhodnocení území v areálu dolu Bílina (A final report on the results of pedological, botanical and zoological evaluation of the

Bílina Mines area), UNICO AGRIC, Czech University of Agriculture in Prague, Prague, Czech Republic.

14. Cibulka J., Kozák J. Valla M., Zelený V., Barták M., Smrz J. Kurfurt J., Stastny K., Bejcek V. (1994) Pedologicko-biologické zhodnocení predpolí výsypky Radovesice (Pedological and biological evaluation of the Radovesice dumpsite vicinity), UNICO AGRIC, Czech University of Agriculture in Prague, Prague, Czech Republic1993.
15. Cibulka J., Kozák J. Valla M., Zelený V., Barták M., Smrz J. Kurfurt J., Stastny K., Bejcek V. (1995) Pedologicko-biologické zhodnocení predpolí výsypky Branany (Pedological and biological evaluation of the Branany dumpsite vicinity), UNICO AGRIC, Czech University of Agriculture in Prague, Prague, Czech Republic.
16. Cibulka J., Kozák J. Zelený V., Barták M., Smrz J. Kurfurt J., Stastny K., Bejcek V. (1996) Pedologicko-biologické zhodnocení území Dolu Bílina (pedological and biological evaluation of the Bílina Mines area), UNICO AGRIC, Czech University of Agriculture in Prague, Prague, Czech Republic1996.
17. Cibulka, J. Sisak L., Pulkrab K., Miholova D., Szakova J., Fucíkova´A., Slámová A., Stehulová I. (1996) Cadmium, lead, mercury and caesium levels in wild mushrooms and forest berries from different localities of the Czech Republic. *Sci. Agric. Bohem.*, 27, 1996, 2: 113-129.
18. Anonymous (1995) Landscape after terminating the Bílina Mines activity – computer aided landscape model. ArchModel, Teplice, Czech Republic.

3. THE ENVIRONMENTAL SITUATION IN THE UPPER VISTULA BASIN (SOUTHERN POLAND)

K. GRODZINSKA
W. Szafer Institute of Botany, Polish Academy of Sciences
Lubicz Str. 46
31-512 Kraków, Poland

Abstract

This paper provides general information on the upper Vistula basin and its climate, geology, soil, flora and fauna diversity as well as the diversity of land use patterns and industrialization. The environmental problems of this extensive area are also discussed (air pollution, surface and underground water contamination, soil contamination, changes in the land surface relief and the extent of anthropogenic alterations of the vegetation and fauna). It also presents a few examples of environmental problems in the strongly industrialized region of Kraków (forest catchment, retention reservoir, urban environments). The extent of environmental deterioration in the upper Vistula basin is demonstrated based on biological indication methods (sensitive indicators applied: changes in species composition and abundance of lichens; accumulative indicators: concentrations of heavy metals in mosses, sulphur content in the needles of the pine *Pinus sylvestris*). Activities aimed at improving the condition environmental of the upper Vistula basin are also discussed.

Keywords: environment deterioration, air pollution, water contamination, forest damage, plant bioindicators, environmental protection, southern Poland

3.1. Introduction

Poland is a typical representative of Central and Eastern Europe countries, where extreme environmental deterioration has occurred in the last half-century. Poland is among the largest and most densely populated countries in Europe. In terms of size (312 000 km^2) it is ranked ninth and its population (38 620 000) is eight in Europe [1] forty-nine administrative units called the voivodships. Most of these units are connected by the main river in Poland - the Vistula (Wisla). Its length is 1 047 km and the area of its basin 168 700 km^2, which corresponds to 54% of the total area of the country [1].

D. B. Peakall et al. (eds.),
***Biomarkers: A Pragmatic Basis for Remediation of Severe Pollution in Eastern Europe*, 29–48.**

3.2. General information on the upper Vistula basin

The upper section of the Vistula River (up to the mouth of the San River) is 394 km long. The upper Vistula basin is the richest water resource in the country and covers an area of 50 732 km^2, which is approximately 30% of the total drainage area and 15% of the area of Poland (Figure 1) [2].

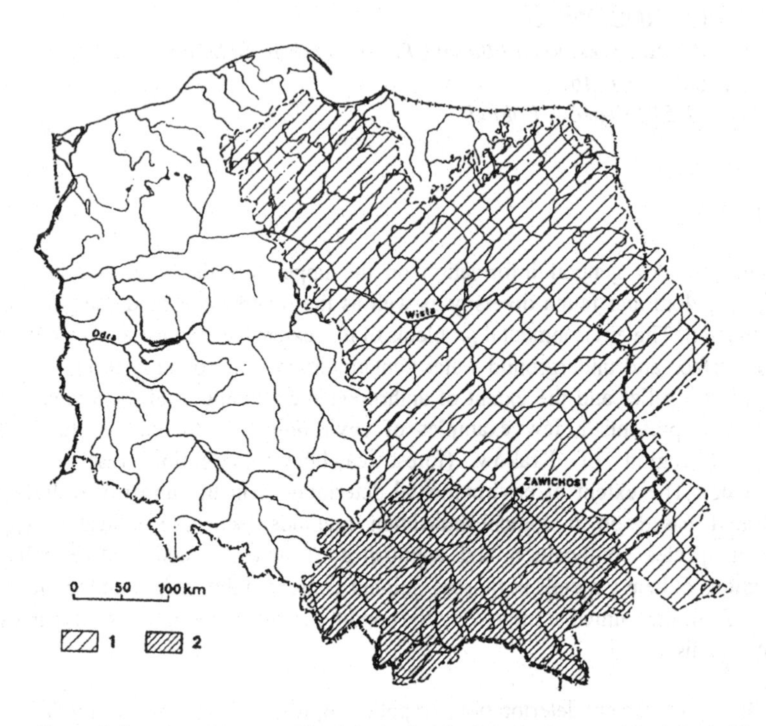

Figure 1. Location of the upper Vistula basin. 1 - Vistula basin; 2 - upper Vistula basin [2]

The right bank of the upper Vistula basin covers an area of 37 385 km^2, and the left bank is about three time smaller (12 934 km^2). The right bank tributaries are generally large mountain rivers with sources in Carpathian mountain ranges. The left bank tributaries are much smaller and originate in the Malopolska Upland.

The upper Vistula basin is located within three major geographical units: the Carpathians (45% of the basin's area), the Sub-Carpathian Basins (approx. 35%) and the Malopolska Upland (approx. 25%) (Figure 2). It stretches over the areas of ten voivodships (Bielsko-Biala, Katowice, Kielce, Kraków, Nowy Sacz, Tarnów, Tarnobrzeg, Krosno, Rzeszów and Przemysl) (Figure 3).

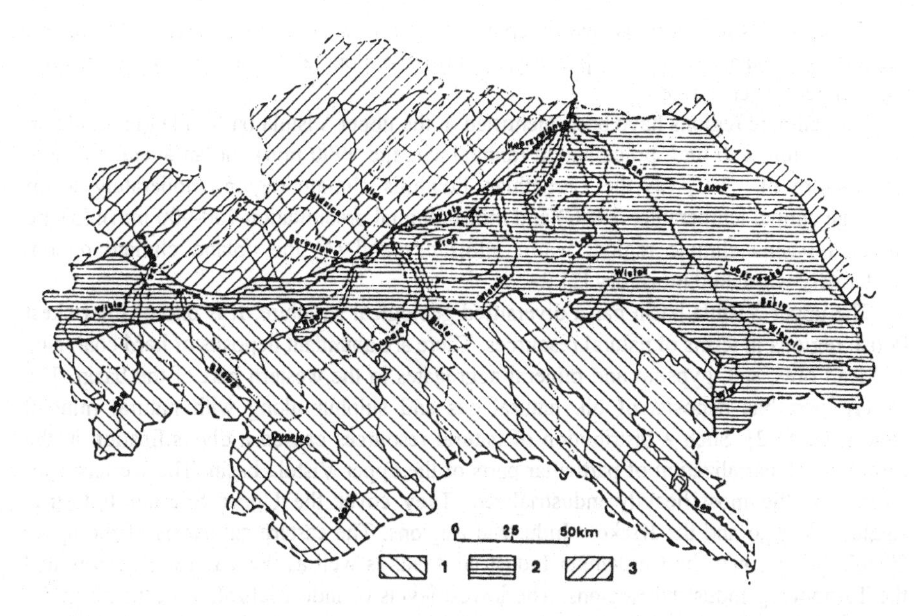

Figure 2. Geographical units within the upper Vistula basin. 1-Carpathian Mts; 2-Sub-Carpathian region; 3-Małopolska Upland [3]

Figure 3. Location of the upper Vistula basin within the administrative units

The upper Vistula basin is very diversified in terms of climate, geology, soil, flora and vegetation, fauna, land use patterns, industrialization and the extent of natural environment deterioration.

Four climate regions have been identified in the upper Vistula basin: (1) the mountain climate region with high precipitation (annual average >900 mm) and an average annual temperature of <7°C; (2) the Carpathian Foothills climate region, moderately warm (average annual temperature 7-8°C) and moist (annual precipitation 700-900 mm); (3) the foreland basins climate region, warm, moist/dry and (4) the upland climate region, moderately warm and moist/dry [2].

The upper Vistula basin has a diversified geological structure. It is one of the richest Polish regions in terms of natural resources. There are abundant deposits of hard coal, zinc and lead ores, salt, gypsum and mineral aggregates in the western part of the basin. The eastern part has deposits of oil, natural gas and sulphur along with natural mineral aggregates [1,2]. Such a distribution of natural resources caused a diversification in the extent of industrialization of particular parts of the upper Vistula basin. The western sub-basins are the most heavily industrialized. They cover the Upper Silesian Industrial Region, Bielsko and the Kraków Industrial Regions. The eastern subbasins of the upper Vistula basin cover the Carpathian Industrial Area, as well as the Tarnów-Rzeszów and the Tarnobrzeg Industrial Regions. The lowest levels of industrialization occur in the far eastern part of the upper Vistula basin, which has no significant resources, and also in its central part with valuable natural and tourist assets. The most common soils in the upper Vistula basin are brown soils and alluvial soils followed by rendzina soils, chernozems and podzolic soils [2].

As compared to the entire country area, the upper Vistula basin stands out with a lower percentage of agricultural land and a higher percentage of forest cover. Arable land accounts for 44% of the total area, meadows, grasslands and orchards make up 15% and forests cover over 32% [2].

The condition of the vegetation cover of the upper Vistula basin is diversified (Figure 4) [2, 3]. The areas with the best preserved natural vegetation are the mountain ranges (Bieszczady Mts., Beskid Niski Mts., Beskid Sadecki Mts.). In the Carpathian Foothills and the Sandomierz Basin vegetation is moderately deteriorated. The vegetation has been most affected in the Upper Silesian and Kraków Industrial Regions, where there are even large sites with no vegetation cover at all. Thus, the upper Vistula basin is an area of strong environmental contrasts. This area covers sites with the most deteriorated environments in Europe but also with the most valuable natural assets, protected by numerous reserves, national and landscape parks and protected landscape areas (Figure 4) [2].

3.3. Environmental problems in the upper Vistula basin

3.3.1. AIR POLLUTION

The upper Vistula basin is affected by various levels of gaseous (SO_2, NO_2) and dust emissions (heavy metals and other pollutants) from local and long-distance sources [4].

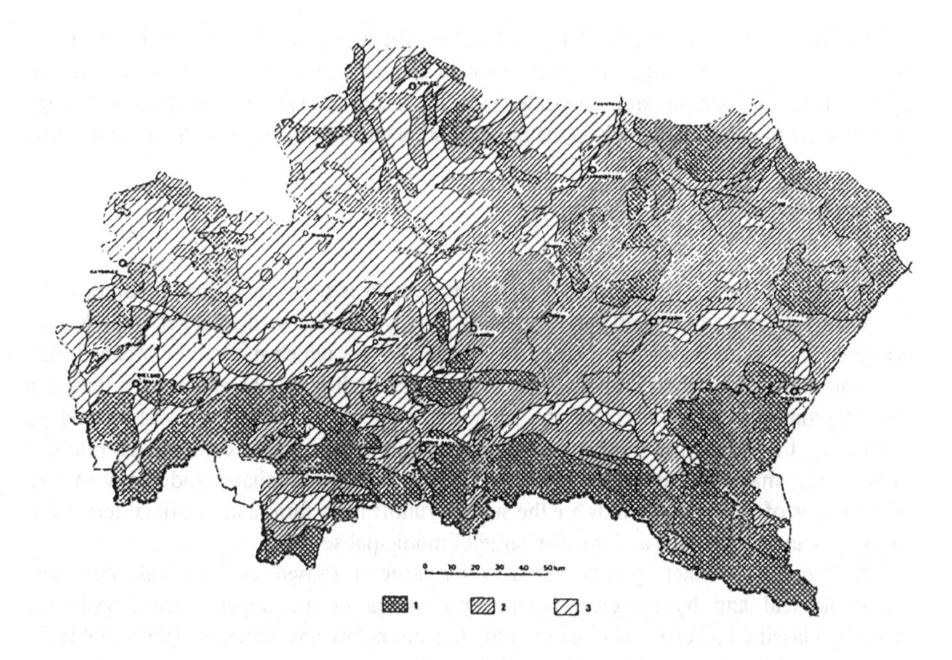

Figure 4. Anthropogenic changes in the vegetation of the upper Vistula basin.
1 - low; 2 - moderate; 3 - high [2]

The highest loads of pollutants are emitted by factories of the Silesian, Kraków and Tarnobrzeg Industrial Regions, and the lowest loads are recorded in the agricultural areas of the Nowy Sacz, Krosno and Przemysl voivodships (Table 1).

TABLE 1. Gas and dust emissions - upper Vistula basin (1)

Voivodship	SO_2	NO_2	DUST	HEAVY METALS (kg/year)		
	Thousand tons/year			Cd	Pb	Zn
Katowice	320.0	148.0	82.5	1663	6893	133404
Tarnobrzeg	83.2	25.3	16.2	-	32	25
Kraków	58.4	25.9	21.5	68	4360	221
Bielsko	13.6	5.4	6.7	-	6	22
Kielce	11.3	6.3	8.4	-	40	254
Tarnów	8.2	7.6	7.6	-	20	485
Rzeszów	5.5	2.3	2.3	2	55	4
Krosno	3.1	1.3	2.1	3	2	496
Nowy Sacz	2.0	0.8	1.3	-	31	-
Przemyœl	1.8	1.8	1.4	-	8	-

The concentrations of SO_2, NO_2 and suspended particulates recorded in the upper Vistula basin are highly variable. They reach an annual average of a few to several dozen $\mu g/m^3$. Dust deposition and concentrations of heavy metals in the dust also vary significantly [4]. Pollutant loads in the upper Vistula basin decrease from the west towards the east.

3.3.2. CONTAMINATION OF WATER

The quality of surface water in the upper Vistula basin is poor. Its chemical composition has changed very significantly over the last forty years due to discharges of industrial wastewater, mining water and municipal sewage. Waters which naturally contained high concentrations of calcium carbonate currently carry high levels of chloride- sulfate and sodium compounds [5]. Besides the high salt contents in the upper Vistula and its tributaries, the water also contains high concentrations of organic compounds, nitrogen compounds, mainly in ammonia form (NH_4^+), as well as phosphates and heavy metals. The sources of these compounds are the surface runoff from farmland (artificial fertilizers, pesticides and herbicides) and the discharge of municipal sewage.

According to water quality classification criteria (based on physical, chemical, bacteriological and hydrological parameters) waters in the upper Vistula basin are partially classified as class III - heavily polluted and below any standards (not suitable for any economic or household application) [4]. The most heavily loaded waters occur in rivers crossing the most industrialized areas (Katowice, Kraków and Tarnobrzeg voivodships) and dominantly agricultural areas (Rzeszów voivodship) (Table 2).

TABLE 2. Surface water quality in the upper Vistula basin in 1994-1995

Voivodship	Classes of water purity (% of river length)			
	I	II	III	below any standards
Katowice	0	0	0	100.0
Tarnobrzeg	0	0	0	100.0
Kraków	0	0.4	0.6	99.0
Rzeszów	0	0	1.7	98.3
Tarnów	0	0	27.0	73.0
Kielce	0	2.2	27.3	70.5
Krosno	0	18.9	27.0	53.0
Nowy Sacz	0	4.8	42.0	53.0
Przemyœl	0	1.5	46.7	51.8
Bielsko	0	25.7	33.5	40.8

There are several retention reservoirs in the upper Vistula basin, playing an important role for the municipalities, in power generation, recreation and flood control. The waters in most of these reservoirs are significantly polluted, generally classified as class II or III or even as below any standards. They are generally strongly eutrophicated, which causes bloom of algae and shallowing of the reservoirs.

Underground water in the upper Vistula basin is badly affected by both excessive exploitation and pollution. This particularly applies to the underground water in the strongly urbanized Upper Silesian and Kraków conurbations, and also to the industrialized zones of the Tarnów and Tarnobrzeg voivodships. The sources of underground water deterioration are agricultural non-point pollution, weakly sewered rural and municipal settlements and also municipal and industrial waste storage facilities.

3.3.3. SOIL CONTAMINATION

In the industrialized areas of the upper Vistula basin the soils are significantly deteriorated due to the input of toxic elements and acidification. Heavy metals reach the soil through dry and wet deposition from the atmosphere and are introduced along with fertilizers and pesticides. The input of nitrogen and sulphur compounds from the atmosphere causes soil acidification, thus increasing the availability of heavy metals to plants. Soils contaminated with heavy metals (particularly cadmium and lead) occur in urban areas (allotment gardens), suburbs and also along transportation routes in the Katowice, Kraków, Tarnobrzeg and Tarnów voivodships. The soils in the remaining areas of the basin contain toxic elements in amounts below the permissible levels.

3.3.4. LAND SURFACE ALTERATION

The relatively primitive Polish mining and raw material processing industry and other activities of man generate tremendous amounts of waste. The lack of an efficient waste management system causes a continuous input in waste quantity. In the upper Vistula basin the largest amounts of waste are accumulated in the Katowice, Kraków and Bielsko voivodships. The lowest amounts of waste are deposited in the Nowy Sacz, Krosno and Przemysl voivodships. Of the waste generated annually, 50-80% is economically utilized and a small percentage (2-7%) is neutralized. The rest of the waste is deposited in industrial and municipal refuse dumps. The number of properly managed refuse dumps is still too low. The existing dumps often do not meet technical requirements and thus are a threat to the environment. In rural areas there are many illegal "midnight-dumping" sites, usually located in creek and river valleys.

3.4. Threats to the vegetation and fauna in the upper Vistula basin

Over the last century many species of algae disappeared in Poland, along with about forty species of pteridophytes and flowering plants, every fifth vascular plant is on the endangered species list. Among nonvascular plants about 40% of lichen and slime-mould species died out or became endangered, about 25% of large fruiting body fungi species and 20% of mosses and liverworts species are threatened [6]. A similar situation is observed in particular regions of the country, including the upper Vistula basin [2,3,7]. A serious hazard for the indigenous flora is the expansion of alien species (including *Solidago serotina, S. canadensis, Polygonum cuspidatum, P. sachalinense, Heracleum sosnowskii, H. mantegazzianum, Aster ssp.*) mainly observed in river valleys. Among plant communities in the upper Vistula basin, particular attention should be focused on the forests. Their average percentage reaches 32% of the total basin's area, ranging between a dozen percent (Tarnów and Kraków) and over 50% (Krosno). The natural composition of the tree stands was significantly altered in the past, due to the activities of man. Large areas, especially those located in river valleys, have been deforested. Another factor destroying the forests of the upper Vistula basin over the last forty years is pollution. As compared to the northern parts of Poland, forests in the upper Vistula basin (part of the Malopolska and Carpathian regions) show extensive of damage.

Non-forest plant communities, particularly rushes, marshes and wet meadows, depending on the groundwater table, have undergone large scale transformation over the last fifty years, very often disappearing totally (Figure 5) [8]. The changes in management patterns caused the impoverishment and often the disappearance of species-rich meadows [9]. Changes of similar magnitude have occured in various groups of animals. In the upper Vistula basin, various types of pollution have caused quantitative and qualitative alterations in many groups of invertebrates, fish and birds (particularly aquatic and wetland birds) [3,5,10,11].

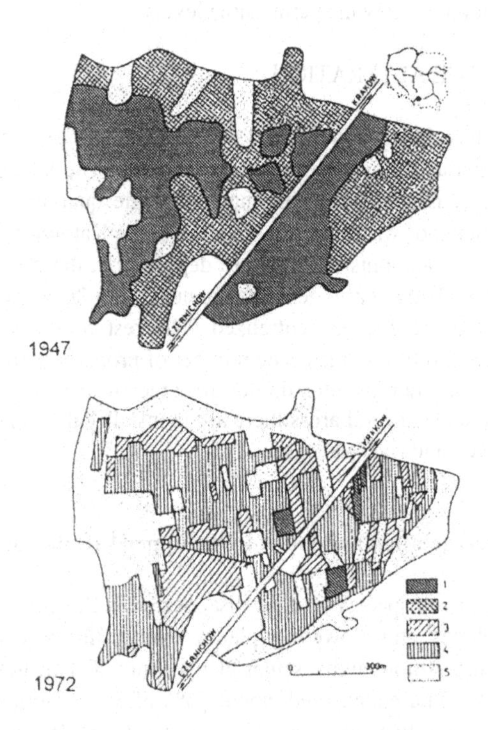

Figure 5. Meadow communities in the upper Vistula basin - Kraków region. 1 - wet meadows (typical *Molinietum coerulae*); 2 - poorer *Milinietum coerulae*; 3 - transitional communities from wet (*Molinietum*) to drier meadows (*Arrhena-theretum*); 4 - *Arrhenatheretum elatioris*; 5 - fields [8]

3.5. Examples of environmental problems in the Kraków voivodship

3.5.1. GENERAL CHARACTERISTICS OF THE VOIVODSHIP

The Kraków voivodship is one of the smallest (3 254 km^2) and most densely populated (1 241 000) areas of Poland. It is also one of the most interesting geographical regions in the country. The heart of the région is Kraków, a city of great historical and cultural tradition. Together with the salt mine in nearby Wieliczka, Kraków is on the UNESCO World Cultural Heritage List. The Kraków voivodship is strongly industrialized (ranked fifth in Poland), and its industrial production accounts for 4% of the national total. The best developed industrial sectors are metallurgy, the metallurgical industry and the chemical industry.

In 1996, forty large pollution sources emitted about 17 thousand tons of dust, which is 5% of the total emissions in Poland, and 159 thousand tons of gas. The emission levels have dropped tremendously over the last ten years (Table 3) [12].

The average annual (1996) concentration of SO_2 in Kraków voivodship (excl. Kraków city) was 24.5 μg m^{-3} (standard 32 μg m^{-3}), NO_2 was 18.5 μg m^{-3} (standard 50 μg m^{-3}), F was 1.70 μg m^{-3} (standard 1.6 μg m^{-3}), and suspended dust particles 24.5 μg m^{-3} (standard 50 μg m^{-3}). The average annual concentration of heavy metals (Cd, Pb, Cr, Ni, Cu, Zn, Fe) in suspended dust particles was much lower than standards. The average annual (1996) dust deposition for Kraków voivodship was 52 g m^{-2} (standard 200 g m^{-2}) [12].

Atmospheric precipitation in Kraków voivodship in 1996 was acidic (pH 4.50). High deposition of sulphates (1374 mg m^{-2}), calcium (1798 mg m^{-2}), chlorides (994 mg m^{-2}), nitrogen (492 mg m^{-2}) and heavy metals (ca 59 mg m^{-2}) [12].

TABLE 3. Dust and gas emissions in Kraków voivodship in 1987-1996 [12]

Year	Dust	SO_2	NO_x^-	CO	Others	Total
			thousand tons $year^{-1}$			
1987	111.1	101.7	49.3	4.05.5	34.3	590.8
1988	101.4	93.8	46.1	378.4	34.3	552.6
1989	95.6	94.3	42.6	405.4	33.1	575.4
1990	69.0	67.1	44.3	295.6	25.8	432.8
1991	44.2	57.5	31.6	165.5	19.9	272.5
1992	34.2	50.5	23.0	127.8	8.7	210.0
1993	24.1	49.6	22.4	142.7	1.4	216.1
1994	20.0	50.6	22.2	147.7	2.0	222.5
1995	20.9	60.8	27.0	145.4	3.7	236.9
1996	17.6	53.2	25.8	77.3	3.4	159.8

In 1996 about 547 hm^3 of sewage was discharged to the surface waters of the voivodship, of which only 27% was properly treated. The remaining 73% was untreated or insufficiently treated waste water [12].

TABLE 4. Surface water quality in Kraków voivodship in 1996 [12]

Type of pollution	Classes of water purity (% of river length - 509.2 km)			
	I	II	III	below standard
Organic substances	30.6	57.5	9.9	2.0
Inorganic substances	63.8	11.1	2.7	22.4
Suspension	11.4	7.6	27.8	53.2
Biogenic substances	0.0	28.0	5.7	66.3
all physicochemical parameters	0.0	7.8	13.3	78.9
hydrobiological index	0.0	42.0	56.1	1.9
sanitary status	0.0	0.4	0.6	99.0

3200 thousand tons of solid waste were generated in the Kraków voivodship in 1996. Industrial waste amounted to 90% of this figure, the remaining 10% being municipal and sanitary (hospital) waste. In recent years industrial and municipal waste has become a major problem, its volume is increasing continually. Waste is disposed in several controlled dumps, and often in an uncontrolled manner posing threats to health and the environment.

The surface water quality in Kraków region is very poor. According to the general classification there are no water courses with class purity I and II in the voivodship. Almost 100% of the water does not meet required standards (Table 4) [12].

As much as 65% of the area of the Kraków voivodship is agricultural land. Arable land is usually of good and very good soil, particularly in the north-eastern parts of the voivodship. At least 75% of this land represents very acidic ($pH<4.5$) and acidic ($pH<5.5$) soils. In terms of heavy metal content the soils of the Kraków region are mostly classified as unpolluted. Soils with moderate cadmium, zinc and lead contamination occur in small areas, mainly in western part of voivodship and within large cities [13].

The Kraków voivodship is one of the least afforested regions in the country (16% of the area). The most frequent tree species are the pine *Pinus sylvestris* (42.4% of the area), the oak *Quercus pedunculata* (14%) and the beech *Fagus sylvatica* (12%). The forests of the Kraków voivodship are currently classified as a zone of average and moderate alteration. The health condition of forests in the Kraków voivodship has improved significantly over the last three years [12,14].

The smallest national park in Poland is located in the Kraków voivodship (Ojców National Park), along with six landscape parks, one protected landscape area and twenty-four reserves. All these sites are of a high natural value.

In the past, the Kraków voivodship was classified as an environmentally threatened area and in the 70-ties and 80-ties it was considered to be an area of ecological disaster. Thanks to the consequent environmental protection policy enforced by local authorities, the condition of the environment in Kraków has improved significantly. Thanks to these activities the area is no longer referred to as an ecological disaster area [12].

3.5.2. THREATS TO THE FOREST CATCHMENTS

The forests in the Vistula basin, in its northern tributary catchments and in the lower parts of the southern tributary catchments within the Carpathian Foothills are usually small complexes surrounded by strongly fertilized fields, meadows and human settlements. The small (88-hectare) catchment of the Ratanica Stream is typical of the wide area of the Carpathian Foothills (Figure 6). Ratanica Stream is a tributary of the mountain river, the Raba, which in turn is a right hand side tributary of the Vistula [15]. This catchment is located within forty kilometers of Kraków, near a large drinking water reservoir supplying the city. For the last forty years this catchment has been exposed to moderate but chronic industrial emissions transported from distant regions (Upper Silesia-Kraków and Ostrava), local emissions from coal-fired home stoves in the surrounding villages and intense agricultural activities. The average annual pollutant concentrations in 1991-1995 were 20 mg/m^3 of SO_2, 8.7 mg/m^3 of NO_2 and 18.7 mg/m^3 of suspended particulates. They did not exceed the Polish standards. It is, however, important to note the high concentration of pollutants over short periods of time are responsible for damage to forests.

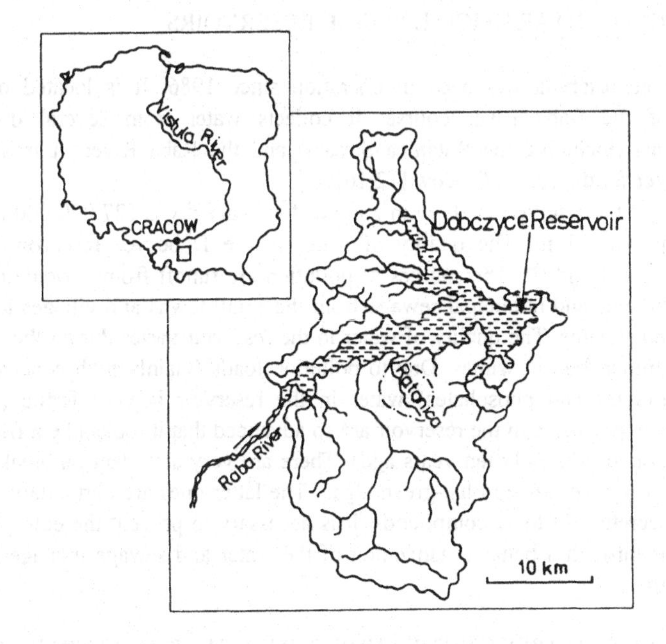

Figure 6. Location of the Ratanica catchment (Kraków region) [15]

Each hectare of the Ratanica catchment receives yearly: 26 kg of sulphur, 25 kg of nitrogen, 15 kg of calcium, about 3 kg of magnesium, about 1 kg of heavy metals and about 1.3 kg of hydrogen ions. In areas which are not exposed to anthropogenic emissions sulphur deposition does not exceed a few kilograms per hectare per year. The input of nitrogen in the Ratanica catchment exceeds the maximum values referred to as critical for forest communities, the annual demand of a growing forest is estimated to be 5-8 kg per year. The Ratanica catchment continuously accumulates hydrogen ions and heavy metals (Cd, Pb, Zn and Cu). This phenomenon is very unfavorable to the forest ecosystem. With the gradually progressing acidification processes heavy metals may become one of the environmental deterioration factors. Also nitrogen (mainly in the form of ammonia compounds) is accumulated in the catchment. Sulphur, chlorine, sodium, magnesium and calcium have a negative balance in the catchment area. The output of large amounts of alkaline metals is also an unfavorable phenomenon (Figure 7).

The Ratanica catchment, located on flysh rock formations rich in calcium carbonate, has a very high buffering capacity. Thanks to this, despite a continuous input of acid precipitation (pH about 4.3), the water in the stream is slightly alkaline (pH>7.4). Buffering processes occur in deep layers of the soil profile.

The Ratanica catchment, despite the fact that it is not located within the range of high pollutant emissions, is a threatened ecosystem. The negative consequences of this fact may already be observed, such as partial defoliation of tree stands, damage to pine needles or alteration of the species composition of lichen flora [9].

3.5.3. THREATS TO ARTIFICIAL WATER RESERVOIRS

The Dobczyce reservoir has been in operation since 1986. It is located on the 60th kilometer of the Raba River course. It collects water from several dozen small subcatchments (including the Ratanica Stream) and the Raba River.. It is a source of drinking water for the city of Cracow [12,16].

The area of the reservoir is 10.7 km^2, and its total capacity is 127 000 000 m^3, with an average depth of 11 m. The quality of water in the Dobczyce reservoir is poor. It represents class II and III. The sources of pollution are runoff from agricultural lands in the subcatchments and municipal sewage from the small towns and villages located near the rivers and streams. The quality of water in the reservoir varies during the year, being worse in summer than in winter. Due to pollution loads (mainly high concentrations of ammonia, nitrates and phosphates) water in the reservoir is very fertile (eutrophic). Eutrophication processes in the reservoir are so advanced that it looks like a fifty year-old reservoir (though built only ten years ago). There are very abundant outbreaks of algae: diatoms, *Chlorophyceae* and blue-green algae. The latter ones are particularly dangerous due to the secretion of toxic compounds. It is necessary to prevent the eutrophication of the reservoir through a better organization of the water and sewage management in the catchment area.

3.5.4. RECENT STUDIES ON THE ENVIRONMENTAL SITUATION IN KRAKÓW VOIVODSHIP

In Kraków voivodship several ecological studies concerning environmental contamination have been performed. The most comprehensive is Ecological Monitoring Programme which has been created in 1981 (see also: Collier et al. this volume). The network has got more than 70 sites (41 agricultural, 16 forest, 12 allotment gardens inside the city). In Kraków voivodship (excl. Kraków city) several parameters have been determined: dust

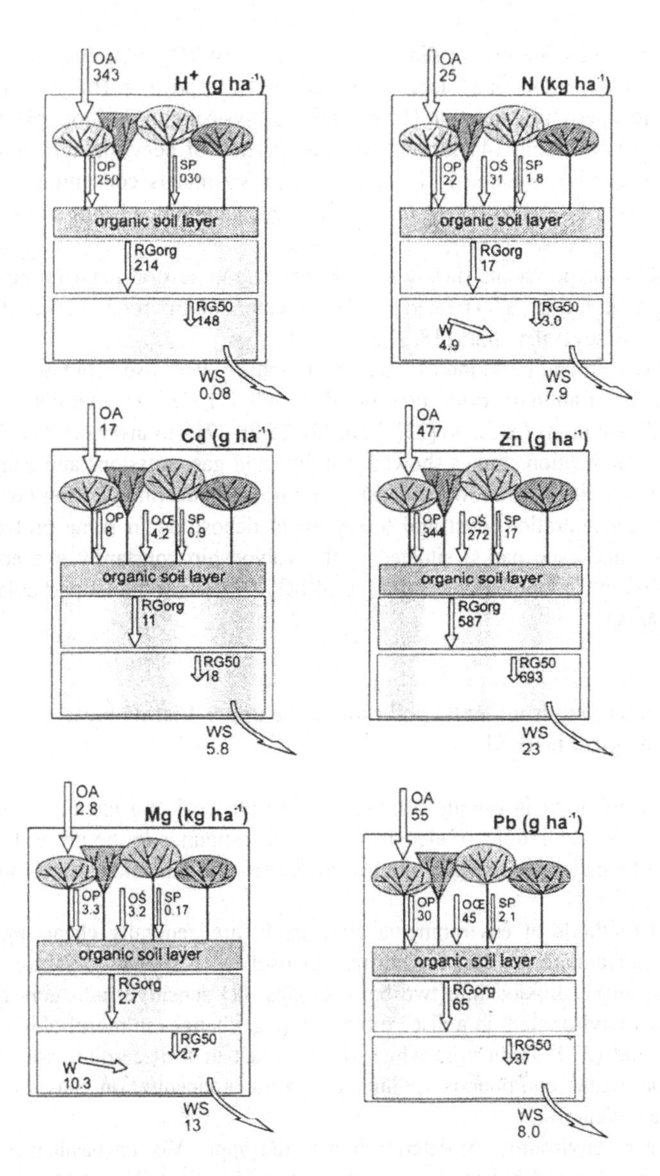

Figure 7. Yearly chemical budget of elements (H, N, Cd, Zn, Pb, Mg) in the Ratanica forest catchment. OA - bulk precipitation, OP - throughfall, OS litterfall, SP - stemflow, RGorg - soil solution below the organic horizon, RG 50 - soil solution at 50 cm depth, W - weathering, WS - output with streamwater (9)

and heavy metal deposition (40 localities), concentration of SO_2, NO_2, suspended particles (incl. heavy metals) in the air (7 localities), chemical composition (biogens, pollutants) and pH of atmospheric precipitation (11 localities), physical composition, pH and heavy metal concentration in soils (44 localities), concentration of heavy metals in vegetables (lettuce, carrot, parsley, potatoes) (44 localities), heavy metals concentration in livers, kidneys and blood of roe deer (*Caprelous caprelous* L.) and hare (*Lepus europaeus* Pall.) [12,17].

Concentration of pollutants (heavy metals, sulphur) in several plant groups (mosses, lichens, pine needles, tree bark) treated as a bioindicators of environmental contamination were estimated by several authors [18,19,20,21,22].

Level of heavy metal pollution of Kraków voivodship was also estimated with a help of various groups of animals. Concentration of Cd, Pb, Zn, Fe was determined in various tissues of small mammals (rodents) [23], birds [24,25,26,27] and invertebrates [28,29].

All these bioindication studies showed that dust and gas emissions and concentration of pollutants in the air in Kraków voivodship significantly declined during last 10 years. However SO_2 concentration, dust and heavy metal deposition in some protected areas (national park, landscape parks) situated in the voivodship constantly exceed accepted Polish standards (mean annual concentration of SO_2 - 11 $\mu g\ m^{-3}$, dust particulates 40 $\mu g\ m^{-3}$) for such areas.

3.6. Estimation of environmental pollution in the upper Vistula basin using biological methods

Environmental pollution is usually estimated using physical and chemical methods by determining the concentrations of elements or their compounds in the air, water and soil. They can also be estimated using biological methods - bioindication using bioindicators [30,31,32].

Biological methods of environmental assessment are generally cheap and have an other major advantage - the organisms themselves respond to toxic pollution. Bioindicators can be divided into two basic groups: (1) sensitive indicators reacting to changes in their environment in a discernible way (e.g. changes in morphology or species composition) and (2) bioindicators which do not react in a discernible way but which accumulate particular compounds (pollutants) whose concentration may be measured (accumulative indicators).

The extent of environmental deterioration in the upper Vistula basin was estimated using these two groups of indicators. The parameters investigated were (1) the frequency and abundance of lichens (the most pollutant-sensitive group of plants), (2) heavy metal concentration in the moss *Pleurozium schreberi*, and (3) sulphur concentration in the needles of the pine *Pinus sylvestris*.

Lichenological studies carried out in various areas of the upper Vistula basin over the last forty years demonstrated a significant impoverishment in the lichen flora. Many epiphytic and ground species sensitive to pollution died out, the number and abundance of other species decreased significantly, and the lichen thallus underwent various degrees of deterioration. Based on a ten-degree biological scale for epiphytic lichens [33,34] lichen-indicator maps were drawn for several areas of the Vistula basin (Czarna Wiselka and Biala Wiselka streams) [7], the Kraków voivodship [34] (Figure 8) and the Przemysl voivodship [35]. The lichens in studied areas represent toxicophilic species or species strongly resistant (biological scale 2 and 3) and resistant (scale 4) to pollution. Species which are more sensitive to pollution cover in the upper Vistula basin small areas only.

The moss *Pleurozium schreberi*, a species accumulating large amounts of heavy metals is being used in the European monitoring program (incl. Poland), to estimate the extent of environmental contamination by these elements [36,32]. The levels of analyzed metals (Cd, Cr, Zn and others) in the moss specimens collected in the upper Vistula basin are high (higher than in central and northern Poland and many Central Europe countries). These levels correspond to the distribution of emission sources in the Vistula basin (Figures 10, 9).

The level of sulphur in pine needles is considered to be a sensitive indicator of sulphur dioxide contamination [15,37]. The level of sulphur in pine needles collected in the upper Vistula basin is higher than normal. The highest concentrations were found in material collected in the western part of the basin (Figure 11).

3.7. Activities aimed at improving the environmental condition of the upper Vistula basin

The upper Vistula basin is an important and valuable part of the country from natural, historical and economic points of view. It should receive particular attention and financial resources to protect its natural environment against deterioration caused by industrialization. Immediate and consistent activities are necessary, in particular:

* the radical reduction of pollutant emissions (achieved after basic restructuring of the economy)
* the implementation of an adequate sewage water management policy, by the construction building waste water treatment plants and technically adequate waste landfills
* the further development of reclamation of deteriorated areas (especially mining areas and industrial waste dumps)
* an increase of environmental awareness in the society through a proper educational program
* an increase in funds available for environmental protection.

References

1. GUS, *Maly rocznik statystyczny (GUS, small yearly statistic book).* (1996) Warszawa, 505 pp.
2. Dynowska, I. and Maciejewski, M. (eds.) (1991a) *Dorzecze górnej Wisly (Upper Vistula Basin)* I, PWN, Warszawa-Kraków, 341 pp.

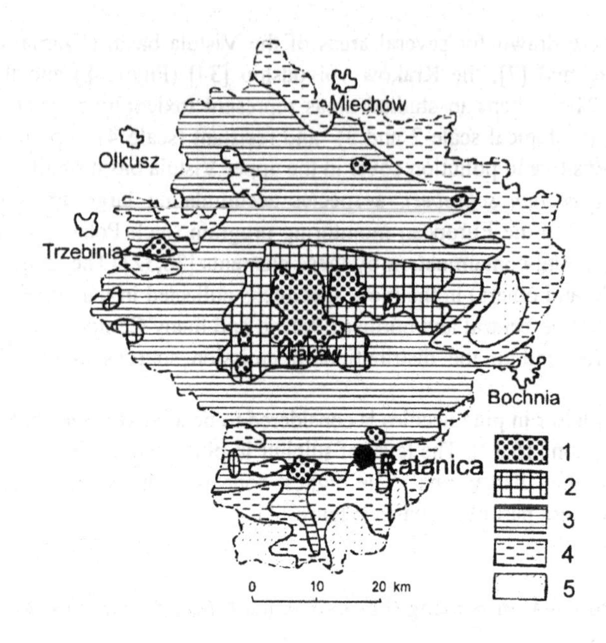

Figure 8. Zones of environmental degradation in Kraków region based on lichen sensitivity; Degree of degradation: 1-extremely high; 2-very high; 3-high; 4-moderate; 5-relatively low [15]

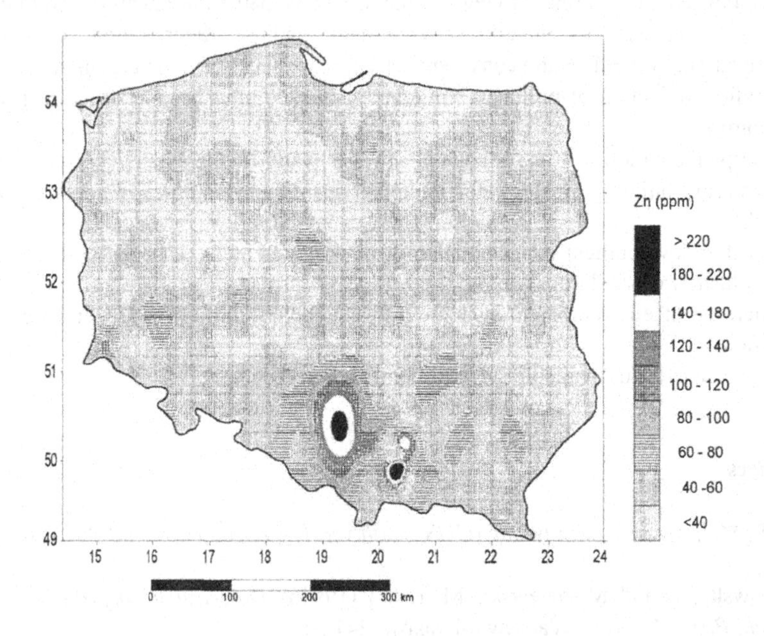

Figure 9. Zones of heavy metal (Zn) pollution in Poland (1995) based on zinc concentration in moss (*Pleurozium schreberi*) [10]

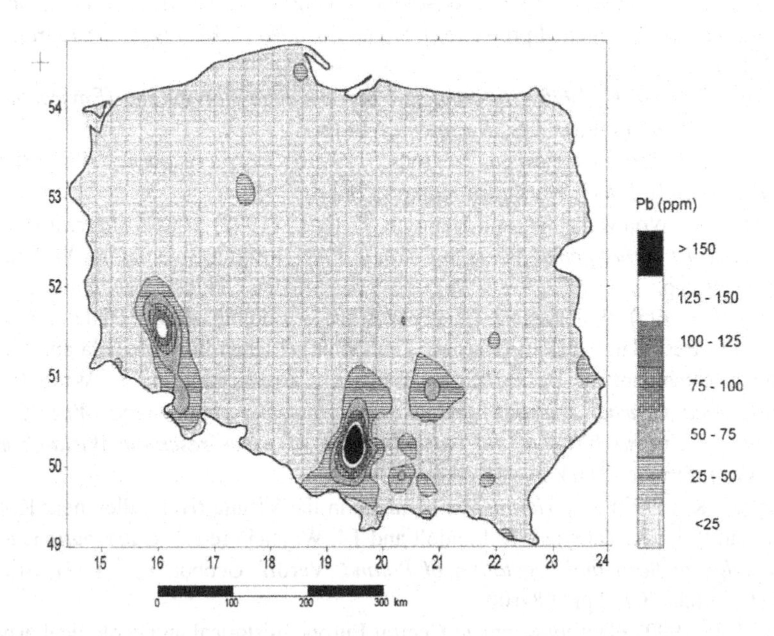

Figure 10. Zones of heavy metal (Pb) pollution in Poland (1995) based on lead concentration in moss (*Pleurozium schreberi*) [10]

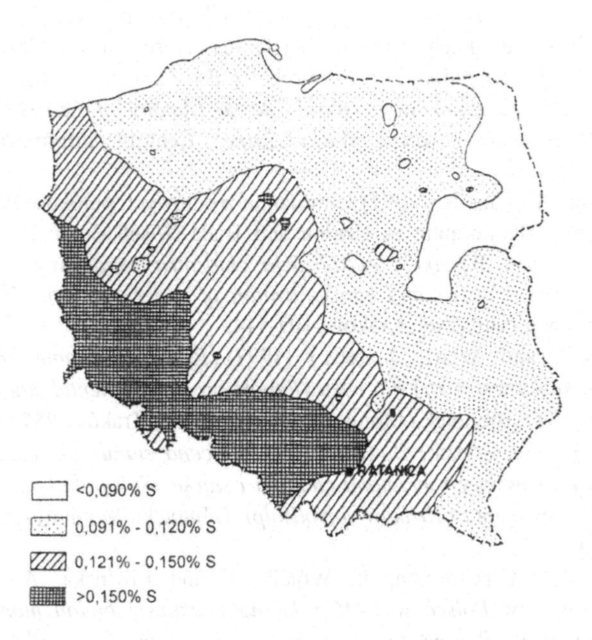

Figure 11. Zones of sulphur pollution in Poland based on sulphur concentrations in Scots pine (*Pinus sylvestris*) needles [2]

3. Kleczkowski, A. (ed.) (1989) Zagrozenie i ochrona srodowiska przyrodniczego (Threats and environmental protection). *Studia Osrodka Dokumentacji Fizjograficznej* 17, 1-405.
4. *Ochrona Srodowiska 1996 - Informacje i opracowania statystyczne (Environmental Protection 1996).* (1996) GUS, Warszawa, 514 pp.
5. Dynowska, I. and Maciejewski, M. (eds.) (1991b) *Dorzecze górnej Wisly (Upper Vistula Basin)* II, PWN, Warszawa-Kraków, 282 pp.
6. Zarzycki, K., Wojewoda, W. and Heinrich, Z. (eds.) (1992) *Lista roslin zagrozonych w Polsce (List of threatened plants in Poland),* PAN, Instytut Botaniki im. W. Szafera, Kraków, 98 pp.
7. Kiszka, J. (1995) Waloryzacja flory porostów w obrebie zlewni Bialej i Czarnej Wiselki w Beskidzie Slaskim (Karpaty Zachodnie) (Lichen flora in Biala and Czarna Wiselka catchment in Beskid Slaski-Western Carpathians), in: S. Wróbel (ed.) *Zakwaszenie Czarnej Wiselki i eutrofizacja zbiornika zaporowego Wisla-Czarne (Acidity of Czarna Wiselka and eutrophication of water reservoir Wisla-Czarne).* Centrum Informacji Naukowej, Kraków, 33-45.
8. Zarzycki, K. (1992) The *Molinietum ceruleae* in the Vistula river valley near Kostrze (S. Poland), in: K. Zarzycki, E. Landolt and J.J. Wójcicki (eds.), *Contributions to the knowledge of flora and vegetation of Poland,* Veröff. Geobot. Inst. ETH, Stiftung Rübel, Zürich, 107, 2 pp. 98-100.
9. Kornas, J. (1990) Plant invasions in Central Europe: historical and ecological aspects, in: F. Di Carstri, A.J. Hansen and M. Debuscho (eds.). *Biological Invasions in Europe and Mediterranean Basin.* Kluwer Academic Publ., Dordrecht. 19-36.
10. Szczesny, B. (1995) Bezkregowce bentosowe zródlowych potoków Wisly (Czarnej i Bialej Wiselki) w warunkach niskiego pH (Bentos invertebrates in Czarna and Biala Wiselka in low pH condition), in: S. Wróbel (ed.) *Zakwaszenie Czarnej Wiselki i eutrofizacja zbiornika zaporowego Wisla-Czarne (Acidity of Czarna Wiselka and eutrophication of water reservoir Wisla-Czarne)*, Centrum Informacji Naukowej, Kraków, 97-106.
11. Szczesny, B. and Kukula, K. (1995) Ichtiofauna zródlowych potoków Wisly (Czarnej i Bialej Wiselki) (Ichtiofauna in Czarna and Biala Wiselka), in: S. Wróbel (ed.) *Zakwaszenie Czarnej Wiselki i eutrofizacja zbiornika zaporowego Wisla-Czarne (Acidity of Czarna Wiselka and eutrophication of water reservoir Wisla-Czarne)*, Centrum Informacji Naukowej, Kraków, 107-114.
12. Turzanski, K.P. and Wertz, J. (eds.) (1997) *Raport o stanie srodowiska w województwie krakowskim w 1996 roku (Report on environmental status of Kraków voivodship in 1996)*, Biblioteka Monitoringu Srodowiska, Kraków, 287 pp.
13. Turzanski, K.P. and Godzik, B. (eds.) (1996) *Ocena stanu zanieczyszczenia gleb województwa krakowskiego metalami ciezkimi i siarka (Soil pollution (heavy metals and sulphur) assessment in Kraków voivodship)*, Biblioteka Monitoringu Srodowiska, Kraków, 66 pp.
14. Wawrzoniak, J., Malachowska, J., Wójcik, J. and Liwinska, A. (1996) *Stan uszkodzenia lasów w Polsce w 1995 roku na podstawie badan monitoringowych (Forest damages in Poland in 1995 based on monitoring studies)*, Biblioteka Monitoringu Srodowiska, Warszawa, 190 pp.

15. Grodzinska, K. and Laskowski, R. (eds.) (1996) *Ocena stanu srodowiska i procesów zachodzacych w lasach zlewni potoku Ratanica (Pogórze Wielickie, Polska poludniowa) (Environmental assessment and biogeochemistry of a moderately polluted Ratanica catchment (Southern Poland))*. Biblioteka Monitoringu Srodowiska, Warszawa, 143 pp.
16. Mazurkiewicz, G. (1996) Zbiornik dobczycki - zródlem wody pitnej dla Krakowa (Dobczyce reservoir-drinking water source for Kraków). *Aura* **8**, 12-14.
17. Grodzinska, K. (ed.) (1996) *Monitoring ekologiczny województwa krakowskiego w latach 1993-1995 (Ecological monitoring in Kraków voivodship in 1993-1995)*, Biblioteka Monitoringu Srodowiska, Kraków, 1-48.
18. Makomaska, M. (1978) Heavy metal contamination of pine-wood in the Niepolomice Forest (Southern Poland), *Bull. Pol. Sci. Ser. Sci. Biol. Cl. II*, **26**, 679-685.
19. Grodzinska, K. (ed.) (1980) *Acidification of forest environment (Niepolomice Forest) caused by SO_2 emissions from steel mills.* Final report on investigations for the period July 1, 1976 - June 30, 1980. Grant No FG-PO-355 (JB-22). Project No PL-FS-75, Institute of Botany, Polish Academy of Sciences, Cracow, 1-143.
20. Grodzinski, W., Weiner, J. and Maycock, P.F. (eds.) (1984) Forest ecosystems in industrial regions. Studies on the cycling of energy, nutrients and pollutants in the Niepolomice Forest, Southern Poland. *Ecological Studies* **49**, 1-277. Springer Verlag.
21. Godzik, B. and Szarek, G. (1993) Heavy metals in mosses from the Niepolomice Forest, southern Poland - changes in 1975-1992, *Fragm. Flor. Geobot.* **38**, 199-208.
22. Grodzinska, K., Szarek, G., Godzik, B., Braniewski, S. and Chrzanowska, E. (1994) Mapping air pollution in Poland by measuring heavy metal concentration in mosses, in: J. Solon, E. Roo-Zielinska and A. Bytnerowicz (eds.), *Climate and Atmospheric Deposition Studies in Forests, Conference Papers* **19**, IG SO PAS, Warszawa, 197-209.
23. Sawicka-Kapusta, K. and Zakrzewska, M. (1996) Heavy metal concentrations in small mammals from Poland. *Proceedings of Third International Symposium and Exhibition on Environmental Contamination in Central and Eastern Europe*, September 10-13, Warsaw 1996, Florida State University, Tallahassee, Florida, USA, 933-935.
24. Swiergosz, R. (1991) Cadmium, lead, zinc and iron levels in the tissues of pheasant *Phasianus colchicus* from Southern Poland, in: B. Bobek, K. Perzanowski and W. Regelin (eds.) *Global trends in wildlife management*, Trans. 18th IUGB Congress, Kraków 1987. Swiat Press, Kraków-Warszawa, 433-438.
25. Nyholm, N.E.J., Sawicka-Kapusta, K., Swiergosz, R. and Laczewska, B. (1995) Effects of environmental pollution on breeding populations of birds in southern Poland, *Water, Air and Soil Pollution* **85**, 829-834.
26. Sawicka-Kapusta, K., Górecki, A., Bryl, B. and Moszczynska A. (1996) Heavy metal concentration in the tissues of common partridge (*Perdix perdix* L.) from southern Poland. *J. Wildl. Research* **1**, 62-64.
27. Dmowski, K. (1997) Biomonitoring with the use of Magpie *Pica pica* feathers: heavy metal pollution in the vicinity of zinc smelters and national parks in Poland, *Acta Ornithologica* **32**, 15-23.

28. Grodzinska, K., Godzik, B., Darowska, E. and Pawlowska, B. (1987) Contamination of heavy metals in trophic chains of Niepolomice Forest, S Poland, *Ekol. pol.* **35**, 327-344.
29. Laskowski, R. and Maryanski, M. (1993) Heavy metals in epigeic fauna: thropic level and physiological hypotheses. *Bull. Environ. Contam. Toxicol.* **50**, 232-240.
30. Spellerberg, J.F. (1991) *Monitoring ecological change*, Cambridge Univ. Press, Cambridge-New York-Port Chester-Melbourne-Sydney, 334 pp.
31. Markert, B. (1993) *Plants as biomonitors. Indicators for heavy metals in the terrestrial environment.* Weinheim-New York-Basel-Cambridge, 644 pp.
32. Grodzinska, K., Szarek-Lukaszewska, G., Godzik, B., Braniewski, S., Budziakowska, E., Chrzanowska, E., Pawlowska, B. and Zielonka T. (1997) *Ocena skazenia srodowiska Polski metalami ciezkimi przy uzyciu mchów jako wskazników (Assessment of heavy metal pollution in Poland using the moss monitoring method).* Biblioteka Monitoringu Srodowiska, Warszawa, 83 pp.
33. Hawksworth, D.L. and Rose F. (1970) Qualitative scale for estimating sulphur dioxide air pollution in England and Wales using epiphytic lichens, *Nature* **227**, 145-148.
34. Kiszka, J. (1990) Lichenoindykacja obszaru województwa krakowskiego (Lichen indication in Kraków voivodship). *Studia Osrodka Dokumentacji Fizjograficznej* **18**, 201-212.
35. Kiszka, J. (1992) Lichen indication in the Przemysl District (S.E. Poland), *Veröff. Geobot. Inst. ETH, Stiftung Rübel, Zürich*, **107**, 287-291.
36. Rühling, A. (ed.) (1994) Atmospheric heavy metal deposition in Europe - estimations based on moss analysis, *NORD 1994*, **9**, 1-53.
37. Dmuchowski, W. and Bytnerowicz, A. (1995) Monitoring environmental pollution in Poland by chemical analysis of Scots pine (*Pinus sylvestris* L.) needles. *Environ. Pollut.* **87**, 87-104.

4. ECOLOGICAL PROBLEMS OF THE KATOWICE ADMINISTRATIVE DISTRICT

GODZIK S., KUBIESA P., STASZEWSKI T. & SZDZUJ J.
Institute for Ecology of Industrial Areas
6 Kossutha St. 40-833 Katowice
Poland

Abstract

The Katowice Administrative District (KAD), with its Upper Silesia Industrial Region (USIR), is the oldest and largest industrial region in Poland. The population density is about five times the average for the whole country. Due to deposits of coal and ores of lead and zinc, and power generation, a heavy industry has developed here. As a result, all elements of the environment are of low quality as compared to other parts of Poland. The highest concentrations of both gaseous and non-gaseous pollutants, and the most severe impact to the environment are found in the central part of the KAD, the USIR. This report contains a brief account of the characteristics of the District, and the results of recent investigations and measurement programs of the status of the environment. Reviewed here are reports on biomarkers used for identification of pollutants, and their impact on the environment.

Keywords: Katowice Administrative District, ecological problems, heavy metals, organic pollutants, air pollution, biomonitoring.

4.1. Introduction

The area of the Katowice Administrative District is 6650 km^2. It comprises 2% of the total area, and 10% of the population of Poland. It is the largest and oldest industrial region in Poland. Due to long lasting industrial activities, the quality of the environment and the structure of major land use categories it differs from that of all the other 48 administrative districts of Poland.

In 1984 only 300 copies of the first official document of the Central Statistical Office (GUS) on the state of the environment of 27 Areas of Ecological Hazards in Poland were published [1]. Currently, reports on the environment quality for the whole country are published annually [2]. State Inspection for the Environment (PIOS), coordinating monitoring activities country wide, is also the publisher of reports on specific problems of the environment. Reports on the air quality in Poland are published each year. Each of

D. B. Peakall et al. (eds.),
***Biomarkers: A Pragmatic Basis for Remediation of Severe Pollution in Eastern Europe,* 49–73.**

administrative districts in Poland reports on the state of the environment of the district on a biannual basis. Data for the KAD for the years 1995 and 1996, obtained by different institutions, and presented in this contribution have been collected by the District Inspector for the Environment [3]. No authors of chapters nor sources of data are possible to identify, unfortunately.

As already mentioned, the contemporary state of the environment of this region is caused by the industry and urbanisation. This region is the major if not the sole source of hard coal, cadmium, and to a large extent also lead and zinc, for the whole country.

All power stations of a capacity above 6500 MW, located in this region are burning hard coal. They are the largest consumers of coal and the largest single sources of air pollutants. Individual householders and district heating units are another group of users. Processing of coal, coke production and distillation of hydrocarbons have been the most serious sources of organic air pollutants. One of the specific problems of the environment of this region is subsidance as a result of underground mining activities, and dumping of solid industrial wastes (mining, metallurgy, coal burning).

The political and economic situation after 1989 has caused significant changes in the production and emission of most of environmental pollutants [2,4]. However, because of methods used for collection of data, amounts of emitted air pollutants, solid wastes and waste waters given by the GUS are probably underestimates. The reason is that statistics does not include emissions from smaller enterprises, the number of which has markedly increased since 1989. Simultaneously an improvement in monitoring network has taken place. More data from monitoring activities are available, and they are published by PIOS [5,6].

According to the best knowledge of the authors, air pollutants are the major source of environmental problems, when assessing the whole area of KAD. Locally, more serious problems may be due to other causes, e.g. coal mining (subsidance and/or solid wastes dumping, alterations of water regime etc.), noise from industrial facilities and traffic. Changes are less dynamic if compared to air pollution. However, effects caused by accumulation of pollutants or structural changes of the landscape are long lasting and they are more difficult to correct.

4.2. Land use

The structure of the land use categories for the KAD in comparison to the whole country is presented in Table 1. The category of land use referred to as "other" consists of technical infrastructure (roads, industrial areas, etc.), waste land due to subsiding (and other industrial activities), dumping sites, etc. Estimations of areas already degraded and being at risk of degradation, are presented in Table 2. The most degraded environment is in the central part of the District: the USIR, where the largest number of industrial facilities is located and where the population density is the highest. Areas to the north and south of the central part, are more rich in forest and/or agriculture areas (also Figures 1 and 2).

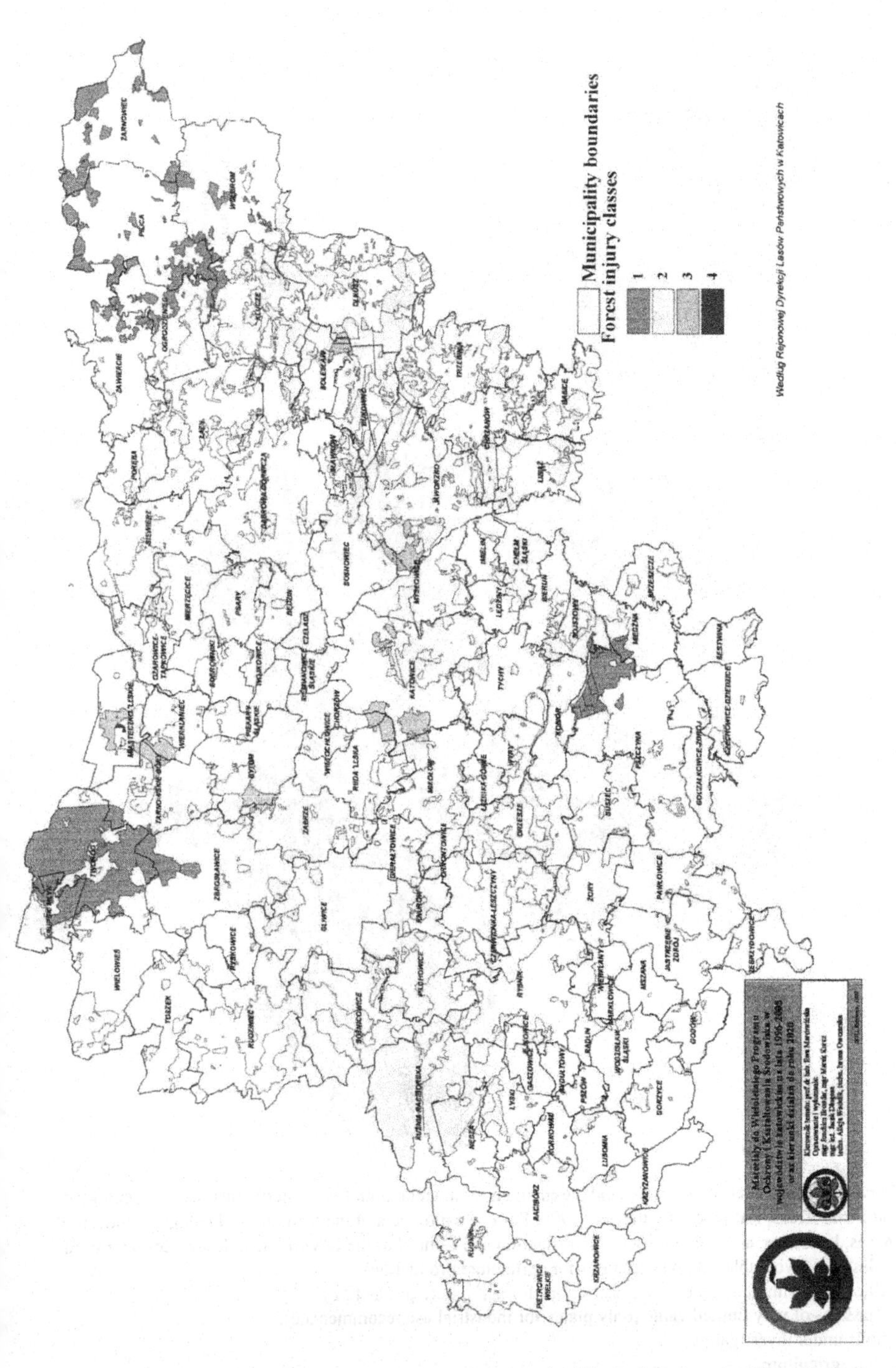

Figure 1. Forest injury classes in the Katowice Administrative District based on data from the Regional State Forest Administration in Katowice (1997).
1 - 4 - Forest injury classes according to UN ECE classification 1 - slight - 4 - dying forest;

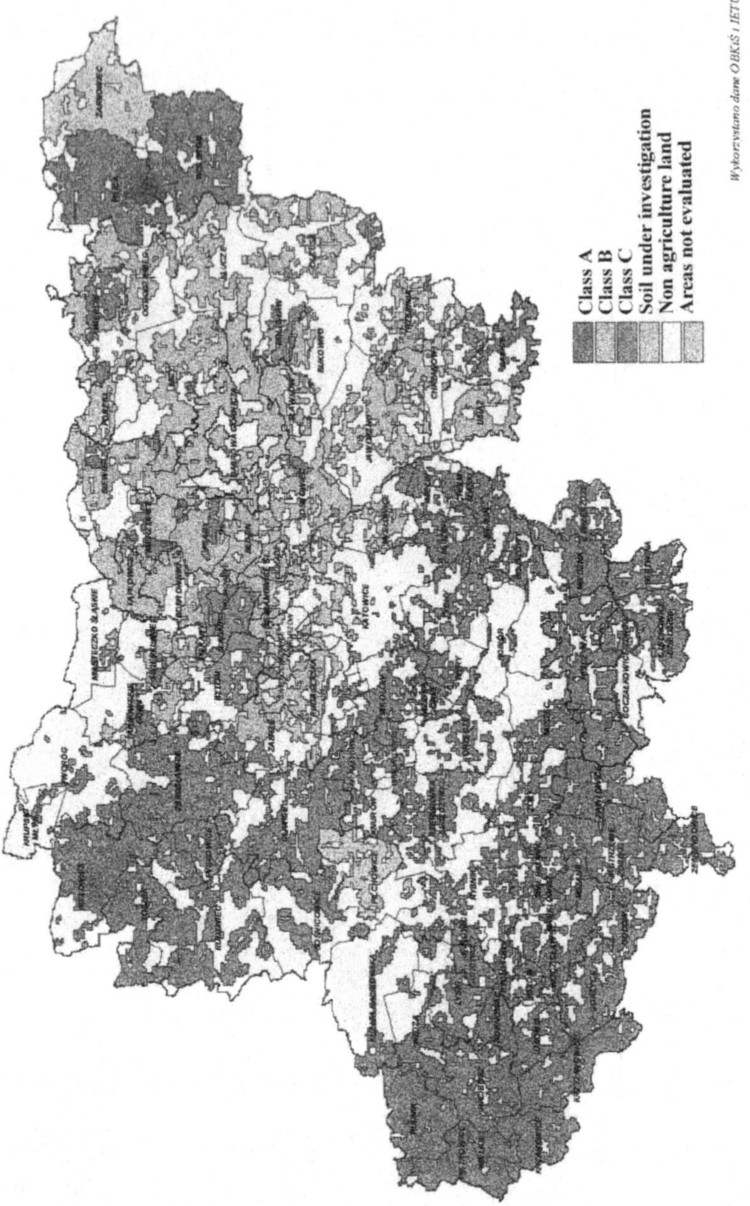

Figure 2. Categorisation of arable land based on environmental pollution (criteria: Institute for Ecology of Industrial Areas, Katowice) for the year 1996. For evaluation data of the Institute for Ecology of Industrial Areas, Katowice, and Laboratory for Investigation and Control of the Environment, Katowice, were used.

1. Class A - favourable (of no restrictions for agriculture production);
2. Class B - of limited value (some restrictions for agriculture production);
3. Class C - of very limited value (only plants for industrial use recommended);
4. Soils under investigation;
5. Non agriculture land;
6. Areas not evaluated.

TABLE 1. Comparison of land use categories in Poland and the Katowice Administrative District in percent of the area [2, 9*].

	Land use category					
	Arable		Forest		Other	
Area of	Min - Max	Average	Min - Max	Average	Min - Max	Average
Poland	-	59.1	-	28.2	-	12.7
KAD	10.0 90.0*	48.5	0 77.1*	29.6	4.0 90.0*	22.1

TABLE 2. Estimation of land degradation and land exposed for degradation in Poland and the Katowice Administrative District [2].

	Degree of degradation				Exposed for degradation	
	Very Large and Large		Medium and Low			
Area of	103 ha	%	103 ha	%	103 ha	%
Poland	151.6	0.5	689.1	2.2	3976.7	12.7
KAD	75.9	11.4	297.5	44.7	207.5	31.2

For most land use categories, differences between the KAD and the whole country are large. Using Landsat images from the year 1992, and CORINE (CORdination d'INformation Environmentale) Land Cover programme launched in 1985 by the Commission of European Communities, the land use of the KAD has been assessed. A 5 ha grid has been used and 27 out of 44 land use categories identified [7].

The comparison of land use categories between years 1970 (base line values) and 1986 performed for the 14 largest cities of the Katowice district has shown an increase of the category "other", from 9% to 306% (three times), with the average of 31% for the same cities [8]. No similar comparisons has been made more recently.

A discrepancy between the statistics based on information from local authorities and/or industries, and remote sensing data, confirmed by ground control, has been found [9]. One of the possible reasons for these discrepancies, is that changes in the deterioration of land surface are very fast. This is especially true for subsidance and/or uncontrolled dumping of solid wastes. About 50% of all industrial solid wastes of Poland is located in the KAD. The amount of industrial solid wastes in the Katowice District for the year 1995 was assessed to be close to 800 million tons [2,3,9].

Dumping of solid industrial wastes, from coal and ores mining, ores enriching and processing, are leading to leaching of elements to the environment and ground water, although no precise data are available . Iron and steel production also results in solid waste emission. Of importance are dumping sites for ash from coal fired power stations. If not protected, they may be a significant source of secondary emission.

Because some of the dumping sites for solid wastes are old, documentation is not available. Data for possible contamination of ground water, are scarce.

Subsidance due to lead and zinc mining in the area of Olkusz and coal mining in other parts of KAD, leads to distortion of the water regime. In the Olkusz area the water table is decreasing. In the other parts the water table is increasing, and ponds with no effluent are formed.

The comparison of data presented by Ciolkosz and Gronet [7] with earlier results have not been performed. This may not be possible because of different criteria for distinction of land use categories, techniques of data collection and size resolution used for evaluations.

The increasing role of local administrations (municipalities) for the environment is the source of some concern about future development. A recent decisions of concern is the localisation of new industrial investments made by some municipalities which are preferring less disturbed areas. The more polluted sites or former industrial areas are used for new investments only very rarely. New regulation are needed for protecting agriculture or forested land use categories.

4.3. Air

Air quality in the region depends on emissions of air pollutants from local sources and long range transport. The largest single sources of air pollutants are coal fired power stations. Next are the other industrial enterprises, e.g. metallurgy (iron and non-ferrous metals). District heating and heating of individual houses are a significant source of pollution, especially during the heating season: from October to April. Of increasing importance is emission from cars, one of the fastest growing sources [2, 3].

Monitoring of air quality in KAD started in 1975 and since then has been operated by the Sanitary and Epidemiological Station in Katowice (Sanitary Network). Non-automatic volumetric methods for measurements are used at 10 locations for formaldehyde and 30 locations for SO_2., NO_2 and NH_3. No information on quality assurance is given. The comparison of air pollution levels for 1990, 1995 and 1996 obtained by the Sanitary Network, are presented in Table 3.

A Regional Environmental Network has been established in 1993 (10, Fig. 3). Automatically recorded are concentration of SO_2, NO, NO_2, TSP (Total Suspended Particulates) at 8 locations in the central part of the region and two locations on the outskirts. Carbon monoxide and ozone are measured at 8 and 4 locations, respectively. Data from each monitoring station are transferred to a centre, where they are stored and interpreted. Local radio, TV and daily newspapers are making data public. The Sanitary and Regional networks are run independently. Data for the same area (not necessary the same locality) shows some differences, as seen for the years 1995 and 1996 in Table 3. For the few components measured in both networks, maximal values from the Regional Network, are lower , except for SO_2 (Table 3). Unrealistically low are yearly mean concentration of SO_2 2-5 $\mu g/m^3$ determined by Sanitary Network in the southern part of the district, and the concentrations of CO_2 in the range of 465 to 1980 ppm seem to be too high (Table 3). Independent measurements of SO_2 performed at exactly the same

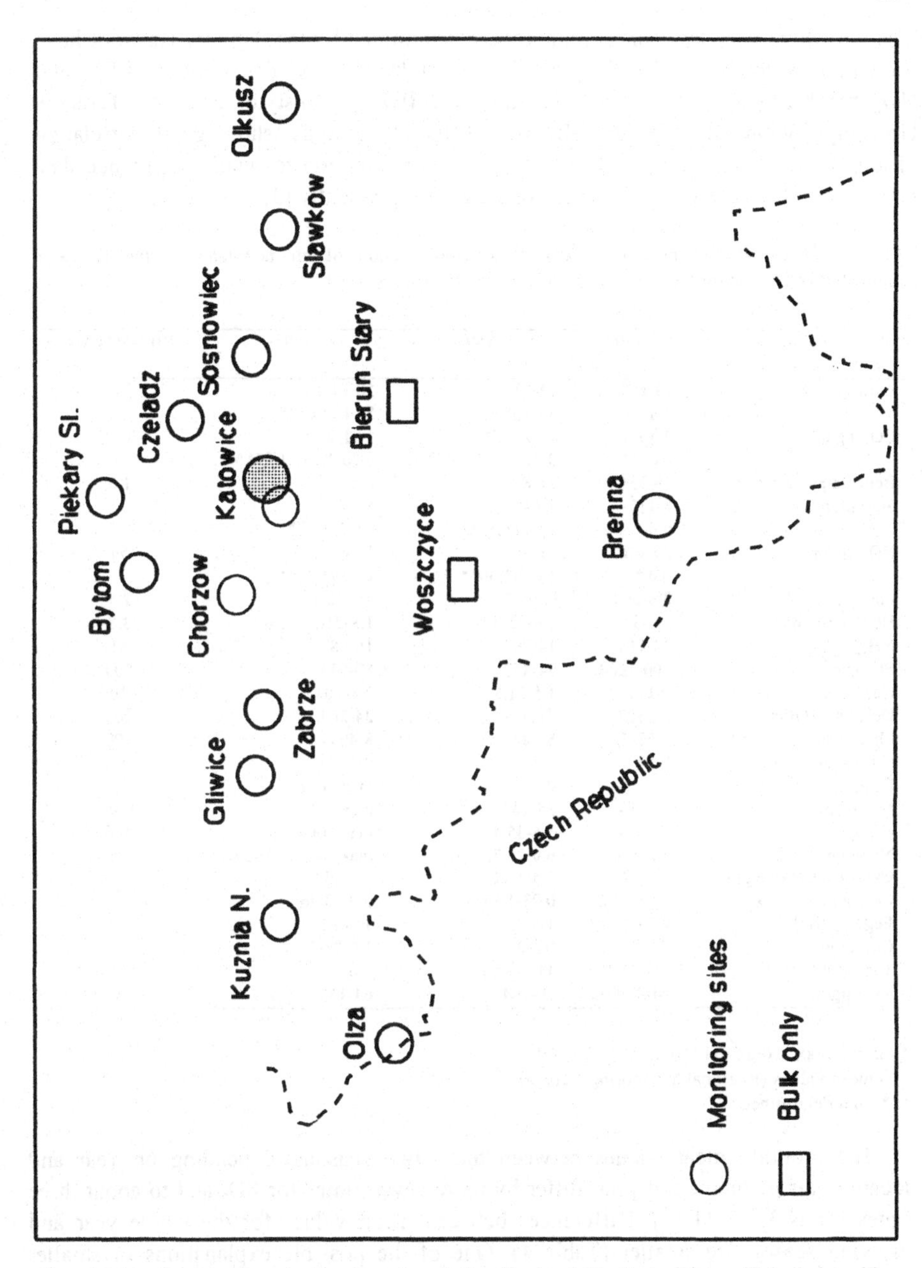

Figure 3. Localisation of monitoring sites: air pollution (Regional Monitoring Network), bulk deposition and biomonitoring (Institute for Ecology of Industrial Area, Katowice).

location where very low concentration were determined by the Sanitary Network have shown a concentration of 32.6 μg/m³ (Table 4, Olza,(11)). Concentrations of SO_2 and NO_2 for Katowice, (Katowice and Katowice IETU = Instytut Ekologii Terenów Uprzemyslowionych), measured by two different methods, show good correlation (Table 4). It seems, to be necessary to pay more attention to quality assurance data presented in both networks which are operated independently [3].

TABLE 3. Minimal and maximal yearly mean concentration of air pollutants in the Katowice Administrative District for the years 1990, 1995 and 1996 [3].

Pollutant	1990	1995	1996	Permitted conc. a^{-1}
Aerosols (PM10) μg/m³	54-259	54-134	58-146	50
	n.d.	30-72*;56**	52-82*;66**	
CO; mg/m³	4-15.8	1.59-2.52	1.91-2.65	0.12
	n.d***	0.71-1.16*;0.91**	0.76-1.30*;1.03**	
Benzo-a-pyren; ng/m³	19-134	27-85	n.d.	1.0
SO_2; μg/m³	8-113	2-61	5-67	32
	n.d.*	26-77*;52**	42-78*;57**	
NO_2; μg/m³	29-136	11-65	13-82	50
	n.d.*	19-39*;30**	25-39*;33**	
Phenol; μg/m³	3.8-26.6	0.2-9.9	2.3-12.1	2.5
HCHO; μg/m³	3.5-34.1	2.6-22.0	1.8-21.5	3.8
NH_3; μg/m³	14-114	14-76	16-98	51
Pb; ng/m³	100-3200	74-453	87-533	200
Cd; ng/m³	2-119	1.5-21.5	2.3-103.7	10
Dust fall; g/m²/a	26-507	21-355	24-263	200
Pb deposition; mg/m²/a	2-2822	8-624	8-491	100
Cd deposition; mg/m²/a	0-87.5	0-56	0-110.6	10
F^-; μg/m³	-	0.71-1.37	max.-1.28	1.6
Mn; ng/m³	-	24-143	max.-88	1000
Cr; ng/m³	-	5.6-35.4	max.-14.9	400
Ni; ng/m³	5.2-15	6.0-26.7	max.-16.2	25
CxHy (as CH4); ng/m³	5.05-7.39	2.4-3.11	2.99-3.96	-
CO_2; g/m³	1.08-1.22	0.93-1.15	0.97-3.96	-
Fe_2O_3; μg/m³	2.44-25.12	1.6-7.1	1.2-6.1	-
Co; μg/m³	-	0.8-2.5	1.0-2.5	-
Tar; μg/m³	6.5-27.2	11.5-22.6	n.d.	-
TSP; μg/m³	54-259	60-144	64-156	-

* - min. - max. (Regional Monitoring Network);
** - mean values (Regional Monitoring Network).
*** - not determined

The monthly mean values between these two seasons, depending on year and location, air pollutant and year, differ by up to seven times for SO_2 and to about three times for NO_2 [3,11,12]. Differences between mean values for the whole year and growing season are smaller (Table 4). One of the possible explanations of smaller variation in mean concentration of NO_2 in comparison to SO_2 between seasons is the increasing role of NOx emission from cars. During winter months only some of the cars are driven.

In the most areas of KAD the yearly mean concentrations of SO_2, TSP, and CO exceed the national standards. Permitted daily values for CO and ozone are exceeded quite frequently. Short term concentrations (30 minutes), except in the case of ozone are not exceeding national standard [3,6].

TABLE 4. Concentration of sulphur dioxide and nitrogen dioxide - yearly mean values, and sulphur dioxide, nitrogen dioxide and ozone during the growing season in 1995 [11].

Location	Yearly mean values		Mean values for the growing season		
	SO_2 [μg/m³]	NO_2 [μg/m³]	SO_2 [μg/m³]	NO_2 [μg/m³]	O_3 [μg/m³]
Czelad	55.50	26.63	19.78	17.47	-
Chorzów	63.64	36.63	29.15	28.47	-
Gliwice	38.67	28.72	13.29	22.63	-
Katowice	43.51	39.06	22.09	34.21	37.80
Ku nia Nieborowicka	29.35	23.96	14.55	18.18	44.77
Piekary Slaskie	51.37	35.78	24.91	25.53	-
Slaskie	45.18	19.88	23.77	21.42	-
Katowice (IETU)	45.00	39.45	22.59	37.56	39.48
Brenna (IETU)	29.72	5.20	22.78	3.79	74.23
Olza (IETU)	32.63*	14.48*	23.73**	10.43**	-

* - mean values for June - December
** - mean values for the time June - September

The most likely source of organic compounds were and are installation of coal processing: coke production and distillery of its products. Another source of organic compounds is coal burning for heating of flats or individual houses. Coal is the cheapest source of energy. No incinerator for wastes has been built so far.

Concentration of some organic compounds (formaldehyde, phenol, benzo (a) pyrene (B(a)P), sum of hydrocarbons), and ether extract from particles collected on filters are also measured. Results are presented in Table 3. No violation of the recommended values for sum of hydrocarbons was found for 1996. The concentration of phenol has exceeded the national standard for air quality at 22 from 23 measuring stations. Formaldehyde exceeded the standard at 7 from 10 stations. BaP has exceeded standards at all locations [3].

The emissions inventory for Volatile Organic Compounds (VOC) are made using the CORINE Air methodology. A report on air quality for the year 1995 in Poland does not contain any information about organic air pollutants [6].

4.4. Bulk deposition

Characterisation of bulk deposition collected at 14 locations was performed for the years 1995 and 1996. Except for Bieruñ and Woszczyce, concentration of gaseous air pollutants were also measured (Regional Monitoring Network and IETU). Location of monitoring sites are seen in Fig. 2. It has been found, that pH at all sampling sites is acidic (below 5.6) and differences between locations shown in Fig. 2 do not exceed 0.4 pH unit. The lowest mean pH (4.5 and 4.12) values were found in locations outside USIR (Bieruñ and Brenna, respectively [11,12]. Data from the year 1968 for the USIR show, that the pH was in the range from 7.2 to 8.1, but outside the area varied from 4.5 to 5.5 [13]. Comparison of data for these two periods for the USIR confirms that improvement in control of emission of particulate matter has taken place.

Mean values of sulphur and nitrogen loads (nitrates and ammonium), calculated for the whole area (14 locations) are in same range between 22 and 24 kg ha^{-1} yr^{-1}, depending on year [11,12]. These values exceed the critical loads for soil or vegetation type, accepted (recommended) under the UN ECE (Economic Commission for Europe) Convention on Long Range Transboundary Air Pollution.

Heavy metals, lead and cadmium, are mostly deposited in soluble form [11,12,14]. Depositions of total lead and total cadmium have not exceeded recommended values (100 mg m^2 yr^{-1}, and 10 mg m^2 yr^{-1} for lead and cadmium, respectively) according to data from the Regional Monitoring Network [11,12,14]. According to the Sanitary Network the recommended values were exceeded at about 2% to 8% of sampling sites [3].

Some of the reasons for improvement in air quality in respect of contamination by heavy metals, are the decreased use of leaded fuel and improved control of emissions of particulate matter from industrial sources. Since 1995 it is required that each car registered for the first time in Poland must have a catalytic converter. No organic compounds were determined in bulk deposition.

4.5. Soil

Most soils in the KAD are sensitive to acidification (sandy soils with low buffering capacity), but with naturally low content of heavy metals. Elevated or even very high concentrations of heavy metals are the result of deposition of metals from long lasting industrial activity. Soils located on calcareous rocks (Jura), areas of Olkusz, Bytom and Tarnowskie Góry, contain high concentration of heavy metals, because of outcropping beds of ores deposits.

Soil pollution by organic compounds, according to information available, seems to be of little importance if larger areas are under consideration. Sites known to be polluted by organic compounds are around coal processing plants and distilleries of its products (tar). Improvements in installations and technology, and closing of old units has resulted in significant reductions of emission and deposition of organic compounds from cokeries. Knowledge about the extend of pollution is scarce.

Use of herbicides and other pesticides in agriculture and horticulture is significantly less than in west European countries. Persistent compounds, like DDT, have been removed from the market and are not produced in Poland. Contamination of soil by organic compounds is not mentioned in the report on the quality of soil, plants, agriculture products and food, published recently [15].

Concentrations of lead, cadmium and zinc in soils of the KAD is presented in Figures 4-6 and Table 5. Evaluation of suitability of land for agriculture purposes, based on environmental pollution, heavy metals in particular, is shown in Figure 2. Zones of elevated concentrations of Zn, Pb and Cd, are quite well defined. The highest, extreme concentrations are found around the zinc and lead smelters, which have operated for a very long time, with no or very poor control of emissions. The best examples are areas of Bytom and Szopienice, east of Katowice, where the oldest smelters were placed. Two other smelters, located in Miasteczko Slaskie - north of Bytom, and Bukowno - west of Olkusz, have been established later. The increase in heavy metals concentration around one of the smelter for ten years period has been presented by Karweta [16]. First samples of soil were taken before smelting processes started. Samples of surface layer soil of arable land, were taken from the same site. The concentration of Zn has increased from 162 mg/kg in 1966 to 1240 mg/kg of soil in 1976. The concentration of lead increased from 91 to 850 mg/kg. The increase in concentration of Cd was much smaller: from 1.6 to 4.0 mg/kg. Concentration of Zn and Pb in 1976 in litter from the same area were about ten times higher if compared to arable land [16]. A great variability in lead content in soil was found in the area before the smelter was put in operation - from 91 to 3605 ppm [17].

TABLE 5. Concentration of Cd, Pb and Zn in soil of arable land (0-20 cm) [after 21, modified].

Element	Area	Concentration range, mg/kg;	Mean, geometrical, mg/kg	Degree of contamination, % of the area	
				low	high
Cd.	Katowice	0.10 - 49.73	1.06	70.4	29.6
	Poland	0.01 - 49.73	0.22	98.2	1.8
Pb	Katowice	10.0 - 1722.7	50.90	83.2	16.8
	Poland	0.1 - 1722.7	13.80	99.5	0.5
Zn	Katowice	5.0 - 2837.5	116.20	74.0	26.0
	Poland	0.5 - 2837.5	32.70	98.6	1.4

One sample in the Katowice Administrative District represents 700 ha and for Poland 800 ha of land.

The concentration of heavy metals in the vicinity of the zinc and lead smelters has a strong vertical profile: the highest concentrations are in surface soil layer as shown in the case of the smelter in Szopienice [18]. The highest concentrations found were the following (ppm): 34,000 for Zn, 3800 for Pb, and 290 for Cd. These extreme concentrations were found in the same soil sample taken from the layer 0-10 cm, 400 m

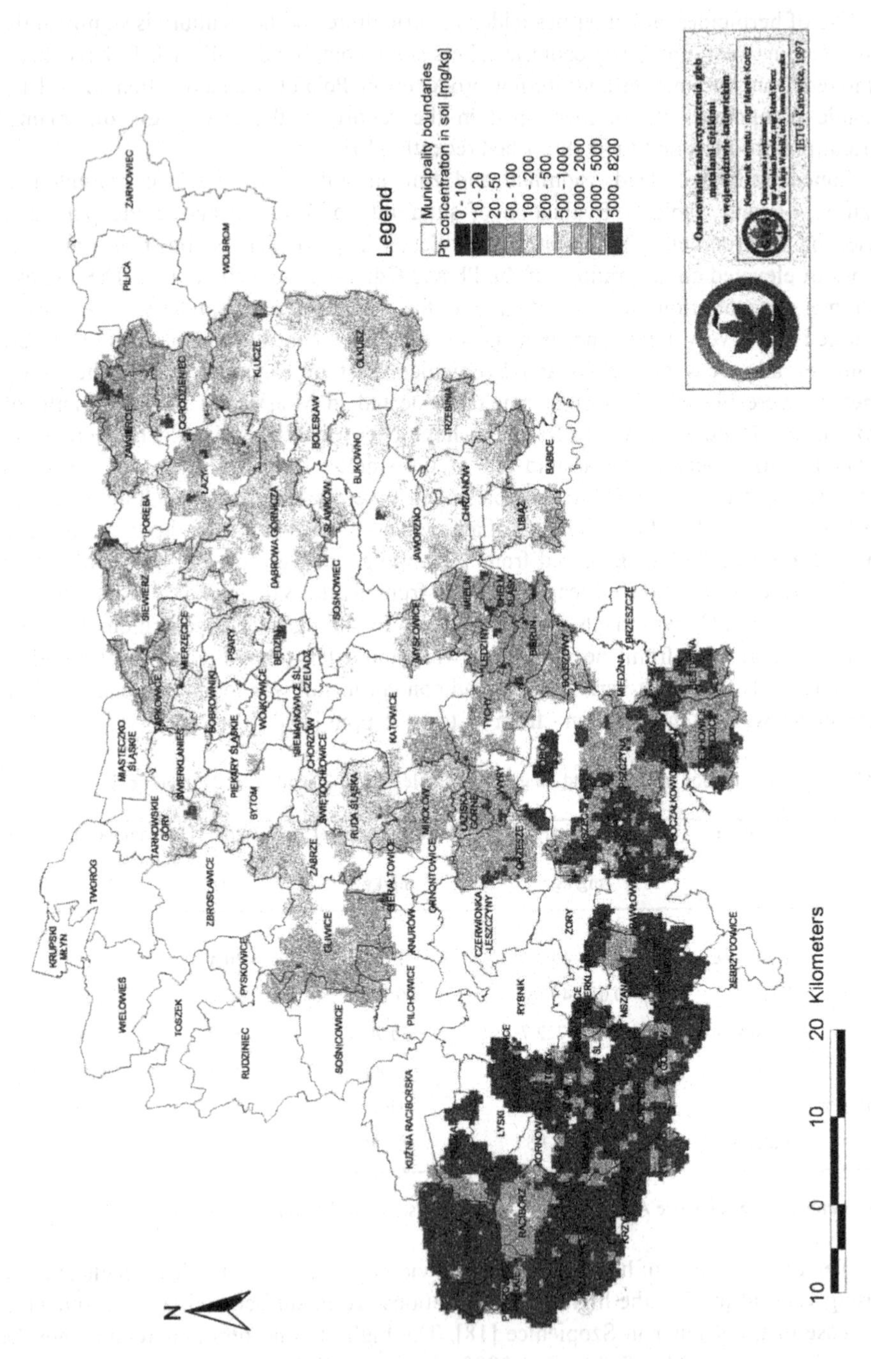

Figure 4. Estimation of soil contamination by heavy metals of the Katowice Administrative District Concentration of lead in soil (mg/kg dry weight).
Empty boxes - no estimations performed.

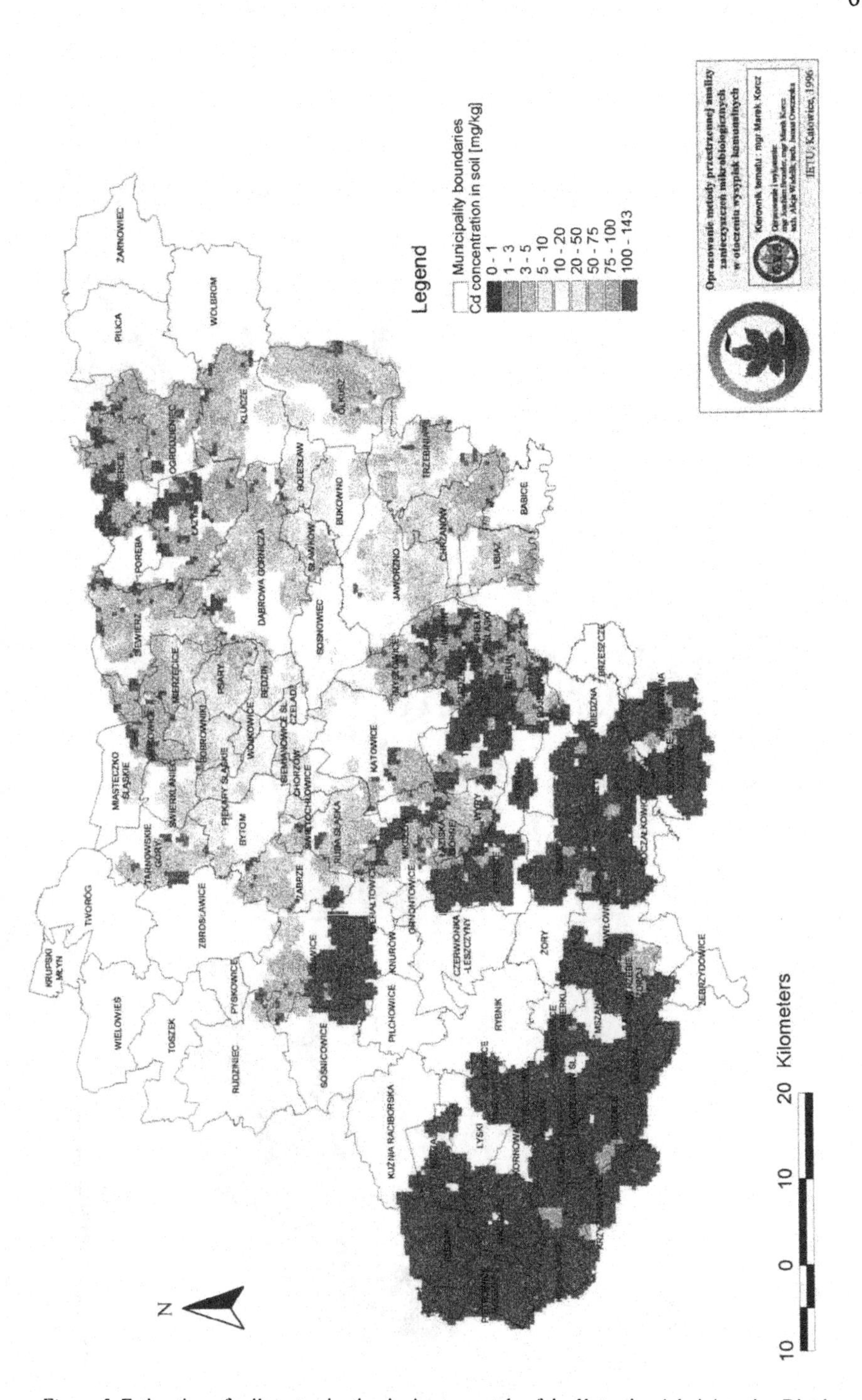

Figure 5. Estimation of soil contamination by heavy metals of the Katowice Administrative District. Concentration of cadmium in soil (mg/kg dry weight).

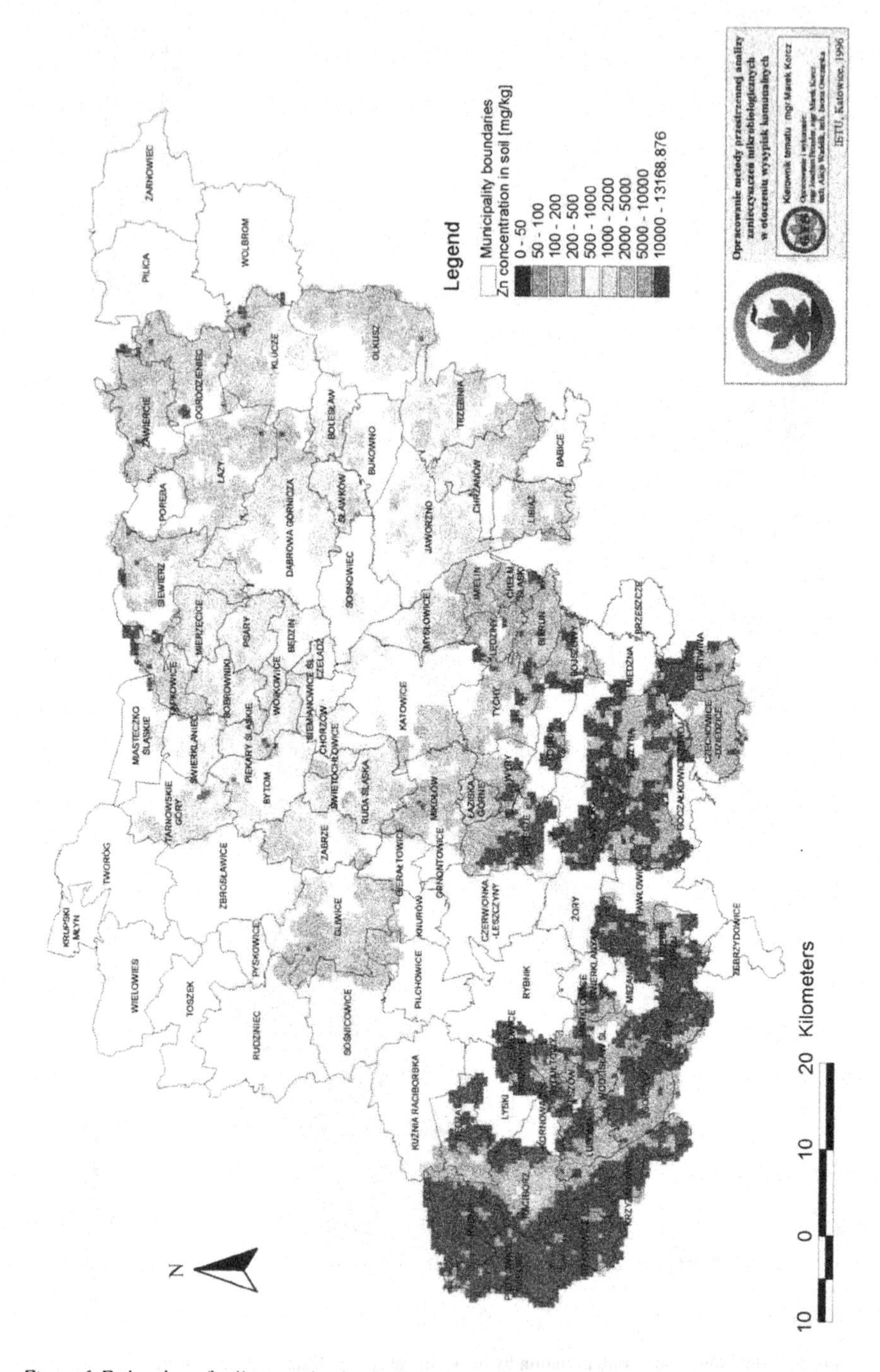

Figure 6. Estimation of soil contamination by heavy metals of the Katowice Administrative District. Concentration of zinc in soil (mg/kg dry weight).

from the smelter. Concentrations of Zn, Pb and Cd in a sample from the layer 11-20 cm were: 600, 610 and 50 ppm, respectively. The average values, from the same distance, but different exposure, were several times lower [18]. Soil samples in this study were taken from unmanaged areas, and in the area adjacent to the smelter, bare land. Concentrations of heavy metals in cultivated soils, are significantly lower [17,19,20,21]. All smelters have improved the control of emission, or have been closed. Places of very high concentration of zinc, lead and cadmium in the soil were also found in several locations, where old smelters have been operated, but abandoned for several tens of years, and are now forgotten.

The largest smelters, still in operation, are in the areas of Miasteczko Slaskie (north of Bytom) and Bukowno (close to Olkusz) in the east of the district. Areas of Tarnowskie Góry, Bytom and Olkusz are the oldest places in Poland where ores were mined and processed for silver. The concentration of lead suggested as permissible are 50 mg/kg for sandy soil and 100 mg/kg for loamy soils. Concentrations for cadmium are the same for both soil types: 3 mg/kg.

4.6. Surface waters

No natural lakes are found within the KAD. Water reservoirs, for drinking water supply for the Silesian conurbation are located on the outskirts of the KAD, or outside of it. The quality of surface waters, mainly rivers, is monitored at 169 places [3].

According to characteristics of water in rivers in 1996, 89.6% of rivers, belonging to watersheds of Wisla (Vistula) or Odra (Oder), are beyond classification (Class I to III). In 1991 the rivers beyond classification were 96.6%. These numbers are misleading, because of the criteria used for the assessment. The quality of water is scored according to the worst single parameter: physical, chemical or bacteriological. If this approach of assessing is not used, the water quality of rivers should be scored higher.

The major sources of water pollution in rivers are waste water that is not biologically treated from both municipal and industrial sources. A somewhat local problem is saline waters from coal mines, discharged to both Wisla and Odra rivers, with predominance of Wisla. Concentrations of nitrogen and phosphorus in reservoirs for drinking water cause danger for summer "blooming", mostly of Cyanophyceae (blue green algae). No systematic hydrobiological characteristics of the surface waters are performed.

Concentrations of polyaromatic hydrocarbons, petroleum products and expressed as DOC (dissolved organic carbon), have been described as generally low, with the highest values in rivers in the central part of the district [3].

Surface waters are the major source of drinking water to the Silesian conurbation, contributing about 80%. The remainder comes from underground water reservoirs. The largest water reservoirs are located outside the main industrial area: Goczalkowice, Czaniec, Kozlowa Góra and Maczki. Quality of the drinking water is controlled for chemical and bacteriological parameters, and fulfil requirements set by the Ministry of Health. Concentrations of organic compounds are low (e.g. trihaloethanes), below standard values. In villages, water is coming mostly from individual sources (wells - ground water), is usually used without any treatment, and are not checked on a regular basis for their quality. In 60% of controlled wells, the quality of water did not meet standards, mostly bacteriological [3,9].

4.7. Biological monitoring

Lack of analysers for measuring concentrations of air pollutants and inadequate infrastructure were the reasons for using biological methods to identify the pollutant(s), and determine the extent of adverse effect (impact) to the environment and inhabitants of the region. Most common were observations of visible foliar injury of vegetation and chemical analysis of sampled plant material. In the surroundings of large industrial facilities, like iron sintering and iron processing plants, zinc and lead smelters, coal processing plants, small experimental plots were established to determine the extent of injury and effects. In these experimental plots agricultural plants were grown. Yields of crops and chemical analysis were the criteria used for determining the effects [22,23,24]. An overview of these results has already been published [25,26]. Field sampled plants were used for biological and chemical analysis. For diagnosing early stage of air pollution effects, activity of enzymes and structure of chloroplasts and surface of foliage were investigated using transmission and scanning electron microscopes [27,28,29,30]. The basis for investigations at lower levels of biological organisation was an assumption, that detection of changes caused by environmental pollutants should be possible before any visible foliar injury do occur. Investigations on structure of plant and animal associations started later, and will be discussed later.

The historical priority of using biomarkers in the Katowice region must be given to Reuss, who in 1893 published a report on visible injuries of spruce needles (*Picea abies*) in a forest around a single source of air pollutants in the Katowice area [31]. Forest injury and its expansion from a regional to national and international scale, have contributed very much to increasing interest of the society to the quality of the environment.

Foliar injury has remained the most common criterion for diagnosing air pollution levels. Defoliation (or crown transparency) is also used for the forest health monitoring within the UN ECE Convention on Long Range Transboundary Air Pollution. It is the only country - wide monitoring system using the same methodology and criteria. According to these data, no uninjured forests in the KAD exist [2,9,32,33]. The highest injury classes III and IV, have been identified in the surroundings of large industrial installations when outside the centre of the district, or around of it (Fig. 1). For poor forest health in Poland, air pollutants are mainly responsible. The improved air quality was until 1994 not followed by improvement in forest health status. A reverse trend has been reported for 1995 [33].

It was assumed, that for early diagnosis (early response) to stress, examination should be performed and changes found at lower levels of biological organisation. The greatest expectation for success were in examination of biophysical (electrical parameters of plant cells), biochemical (enzyme activity) or physiological (water loss and transpiration) parameters. Changes in electrical parameters of cell of *Nitellopsis obtusa* and *Lychnothammus barbatus* exposed to heavy metals, the increase of polyphenol oxidase of homogenates of *Aesculus hippocastanum* leaves and the increase of uncontrolled water loss of *Aesculus hippocastanum* leaves have been found [27,28,30]. Changes were found between control or reference samples and from polluted areas, also with no visible foliar injury. Adaptation to and/or recovery from stress factors are making interpretations of results difficult. Structural alterations, representing irreversible changes, were seen as more promising for diagnostic purposes. To be able to find changes, microscopical methods were used for examining structures on the subcellular level (chloroplasts) or organ levels (foliar surface structure) [29,34,35,36,37]. Differences between samples from reference and polluted sites were found. However, it is technically difficult to examine a large enough number of samples to reduce risks for misinterpretations.

Foliar injury, yield of potatoes, alfalfa and pods of bean have been used by Warteresiewicz [24] as criteria for assessment the impact of two lead and zinc smelters (one of which was the Szopienice smelter). Yield of potatoes and vegetables was used earlier for the same purposes [22,23]. The yield of several plant species has been used as a routine criterion in investigations on effects of large sources of air pollution. An overview of these results has been already published [25,26]. In most of these studies, cultivars available on the market and unified soil in "microplots" have been used.

Concentration of elements of air pollutants, in plant material were determined more often since mid sixties. The concentration of fluorine in leaves of several plant species (lupine - *Lupinus luteus* L, beet - *Beta vulgaris* L, parsley - *Petroselinum sativum* Hoffm.), and two bean (*Phaseolus* sp.) species, growing in surrounding of enamel factory have been determined [38]. A similar approach has been used by Paluch and Karweta [17] for assessing the increase in zinc and lead concentrations in plants (plantain - *Plantago lanceolata*, rye - *Secale cereale*, clover - *Trifolium repens*, pine - *Pinus silvestris*) after one year of the smelter operation in Miasteczko Slaskie. Leaves of beet (mangel) in 13 municipalities of the KAD were analysed for content of lead and cadmium [39,40]. The highest concentration of lead: 49.5 ppm (average: 18.1 ppm) was found in samples from Bukowno. Concentrations of cadmium: 51.9 ppm (average: 23 ppm) were the highest in a location between Bytom and Tarnowskie Góry, area of lead and zinc ores mining in the past [39]. Concentrations of lead and cadmium in forage plants from around the smelter in Bukowno was the highest in leaves and varied from 24.5 ppm Pb in straw to 50.6 ppm Pb in corn leaves [39]. Concentrations of cadmium varied between 1.5 ppm in straw to 8.8 ppm in beet leaves [39].

The large variation in response to pollutants within a population of organisms, and adaptation processes to local environment are the reasons for using selected and previously tested plant material. Existence of genotypes of *Lollium perenne* within the Silesia region has been confirmed experimentally, and in field studies for *Arabidopsis thaliana* [41,42].

For diagnosing the existence of photochemical oxidants, ozone in particular, the tobacco Bell W3, and several other pre-tested plant species and cultivars are used. A bean (*Phaseolus vulgaris* cv. Lit) has been found to be most sensitive [11,43]. For diagnosing the possible contamination of vegetables by heavy metals taken up from the soil or deposited from the air, clover (*Trifolium repens* L., *T. subtereanum*), lettuce (*Lactuca sativa* L.) and radish (*Raphanus sativa* L.) have been used. Experiments performed for three years at 14 locations within the KAD have shown the decreasing importance of lead contamination in a direct way - as deposited from the air. In the case of cadmium, the current situation is more complicated. Concentrations of Cd exceeding the recommended values for food and forage were found also in Brenna, which has no local sources of air pollutants, and a content of this element in soil below recommended values [11,43]. Instead of naturally growing "wild", plant species or cultivars have been selected for certain properties: sensitivity to single pollutant, e.g. sulphur dioxide, ozone, or serving as "accumulators" of environmental pollutants. The advantage of using selected plants, grown in standardised and local soil, exposed in different locations is the possibility for comparison of results. It is also possible to assess the route of entry of pollutants to plants: a direct way from the air, or indirect from the soil.

Plant communities have been analysed in areas adjacent to large single air pollution sources, e.g. "Miasteczko Slaskie" Zinc Smelter, "Katowice" Steel Mill, cokery in Zdzieszowice, or of whole areas [44,45,46]. Long lasting air pollution has caused changes in the structure of plant communities - disappearance of species more sensitive or site specific in favour of synanthropic and/or new for these sites. *Calamagrostis epigeios* became one of dominant plant species of ground flora, also in forests.

Concentrations of the heavy metals lead and cadmium in animals were also used to assess the contamination of the environment [47, and Migula and others this volume]. Highest concentrations of lead, cadmium and also thallium, in feeders, liver or kidney of some birds (e.g. magpie - *Pica pica*) from surroundings of lead and zinc smelters in Bukowno, Miasteczko Slaskie and Szopienice, are reported by Dmowski (this volume). Concentrations reported by Dmowski are very often the highest ever presented for any environmental samples from surroundings of the same smelters, (except soil). The concentration of thallium in the kidney of magpie is higher by about three orders of magnitude if compared to other environmental samples from the same areas [48].

Associations of aphids (*Aphis sp.*) on birch (*Betula sp.*) were tested for their usefulness to characterise the role of heavy metals and sulphur dioxide over the KAD and adjacent areas [49]. Data have shown, that lower numbers of species and individuals were found in areas where chemical factories (refineries, chemical synthesis) are located. However, no chemical analysis of the air and leaves of birch were performed. Honey bees, and their products were also used as biomonitoring tools [47,50, Migula and others, this volume].

Reduced intervillous space and low cytochrome oxidase activity in syncytiotrophoblasts of villi of human placenta have been found in samples taken from women living in surroundings of large industrial air pollution sources or in urban areas [51,52]. These parameters may influence the pregnancy, birthweight, neonatal mortality, and health of infants and children [53,54,55]. Low birthweight, and higher neonatal mortality were found more often in cases when the mothers of children were born and grew up in polluted areas [54]. A decreased concentration of osteocalcin was found in the blood of children with concentrations of lead above 20 $\mu g/100\ cm^3$ blood [56].

To assess the exposure of human population to fluorine, magnesium and fluorides in urine, (Mg, Na, K) ATPase, ATP, lactate and glucose in blood, were determined [57,58]. Reduced gluthatione and delta aminolevulic acid (ALA) in blood, and porphobilinogen in urine of humans exposed to lead were determined [59]. According to the authors, relations were found between concentrations of contaminants and parameters used as criteria of effect.

Significantly higher lead and cadmium concentration were found in hairs from school children in Miasteczko Slaskie, Piekary Slaskie and Bukowno if compared to a similar group of children from the Kraków area [60].

Use of corn (*Zea mays*), pea (*Pisum sativum*) and oat (*Avena sativa*) for characterising bulk deposition has been tested by Paluch [13]. Reductions of germination rate and root growth were found in samples from the USIR (13). Samples of snow and bulk deposition have also been tested for toxicity, morphological anomalies and teretogenicity [61]. As test organisms *Paramecium primaurelia*, *Tubifex sp.* and *Rana temporaria* were used. Depending on sampling site, acute toxicity to *Paramecium* and *Tubifex* and *Rana* was determined. In samples from surrounding of cokeries, changes in the course of *Rana temporaria* ontogenisis have occurred [61]. Similar analysis was performed for PAH fractions of three, four and five rings from extracts of particulate matter collected using high volume samplers. The most active was the five ring PAH fraction with a high content of three and four ring compounds [62]. Several other tests, using plant species, were also used for evaluation of the same organic compound fractions [41]. The Ames test, number of micronuclei and chromosomal aberration of bone marrow, were also performed with crude extracts and fraction of filter sampled organic compounds [63]. All tests have shown effects. Number of DNA adducts (lymphocytes of peripheral blood) of people employed in a cokery and in the population leaving nearby does not differ markedly, but in both groups are higher than in rural areas of east Poland [62].

For soil characterisation by biological methods, decomposition of organic matter, characteristics of micro-organisms in experimentally introduced organic matter, soil associations of Diptera larvae, and Protozoa have been studied [64,65,66,67].

To the best of the authors knowledge, no test was or is used for routine analysis of environmental (air) pollutants.

4.8. Conclusion

Contamination of and disturbance in the environment of the region under consideration has a long lasting history. An improvement of air and surface water quality has been observed in recent years. However, national standards and/or international recommendations for several compounds are exceeded.

Of greatest concern were gaseous air pollutants - mainly sulphur dioxide. Sulphur dioxide has been recognised as the major cause of forest injury in this region. More recently some attention has been given to ozone as a major component of photochemical smog. Knowledge about organic pollutants of the air, water, soil and effects to organisms is poor.

Major sources of organic compounds such as PAHs are cokeries and processing of tar products. A rapid increase in number of cars and the extent of their use will change the ratio between sources of emission of air pollutants.

Because of shortage in equipment for measuring concentration of air pollutants using analysers, biological methods have been applied.

Over the years, a number of attempts have been made and methods developed for evaluating the response of organisms, populations, associations or ecosystems, to environmental pollutants. Deterioration of forests in this region was the "driving force" for environmental protection movement.

Heavy metals, lead and cadmium, were related mostly to human health, although plants and animals were most often used for assessing the possible risk to the population. Children have been taken as a group of high risk.

The improved control of emissions from industrial sources, and to a lesser extent from municipal sources, has resulted in decreasing concentrations of gaseous air pollutants and deposition of dust. Observations of vegetation both natural for this region and test plants, are suggesting that sulphur dioxide is no more causing visible foliar injury. The only gaseous air pollutant causing foliar injury is ozone of photochemical origin. An increased role of photooxidants and their precursors to the environment, the general population included, can be expected.

A decreasing share of leaded petrol in the market, the reduction of the number of lead and zinc smelters, and improvements in emission control are resulting in reduction of the contamination of plants. Experiments carried out in 1994-1996 are confirming these predictions. More attention must be given to heavy metals already deposited in the soil, in particular those that are poorly buffered and of low pH. Increased uptake of heavy metals by plants from the acidic soils has been confirmed under field and experimental conditions.

Analysing the data obtained, it can be concluded that the results of biophysical and biochemical methods are difficult to interpret, because of processes of recovery and adaptation. Easier for interpretation are irreversible changes, e. g. structural changes on subcellular or organs (leaf, needle) levels. However, because of technical reasons (availability of electron microscopes: - transmission or scanning), the size of sample and number of investigated objects is limited.

It should be suggested that routine procedures are introduced for assessing the presence (and levels) of organic pollutants, which have been identified in other countries. Implementation also sophisticated methods (biomarkers) is possible, because several laboratories located in the KAD are well equipped, and have skilled teams.

4.9. Recommendations

1. Identify the extent of occurrence of organic compounds, of high potential biological impact, to the environment and human beings.
2. Introduce a simple, non invasive approach to assess both the most detrimental factors (even locally) and the most widespread factors - for the largest area and the general population.

References

1. GUS (Glówny Urzad Statystyczny - Central Statistical Office) (1984) *Regional Documents. Areas of Ecological Hazard in Poland* Warszawa. (in Polish).
2. GUS (Glówny Urzad Statystyczny - Central Statistical Office) (1997) *Environment 199.* Warszawa. (in Polish).
3. Jarzêbski, L. (Ed.) (1997) *Report on the state of the environment in the KAD for years 1995 - 1996.* Biblioteka Monitoringu Serodowiska, Katowice (in Polish).
4. *National program of natural environment protection till the year 2010* (1988) Draft. Ministerstwo Ochrony Serodowiska, Zasobów Naturalnych i Leœnictwa, Warszawa (in Polish).
5. Ciolkowska-Dygas, L. (Ed. in chief) (1996) *Air pollution in Poland in 1994.* Biblioteka Monitoringu Serodowiska. Warszawa. (in Polish).
6. Ciolkowska Dygas, L. (Ed. in chief) (1997) *Air pollution in Poland in 1995.* Biblioteka Monitoringu Serodowiska. Warszawa. (in Polish).
7. Ciolkowska, A. and Gronet, R. (1997) *Use of teledetection in assessment of the environmental devastation extend in Katowice Region.* Archiv. Environ. Protection 23, 3-4, 159-167 (in Polish).
8. Dabrowska L., Gronet, R. and Wrona, A. (1990) Land use in KAD. in S. Godzik (ed.), *Hazards and state of the environment of the Cracow - Silesia region*, 32-38. CPBP 04.10 SGGW AR Warszawa, nr 62; (in Polish).
9. Urzad Statystyczny w Katowicach (Statistical Office in Katowice). (1996) *Statistical yearbook of the Katowice Administrative District* XIX. Katowice.
10. Stawiany, W. (1993) Design and management of the regional air pollution monitoring system) in *Biuletyn Sluzby Monitoringu Powietrza*, nr 1, Wojewoda Katowicki (in Polish).
11. Godzik, S.; Staszewski, T.; Kubiesa, P. and Szdzuj, J. (1996) and (1997) *Biological Monitoring of the Katowice Administrative District).* Report and 546/DN/95. Instytut Ekologii Terenów Uprzemyslowionych, Katowice (in Polish).
12. Godzik, S., Kubiesa, P., Staszewski, T. and Szdzuj, J. (1997) Ecological monitoring of the Katowice Province in Livia Navon (Proceedings coordinator) *Symposium on Environmental Contamination in Central and Eastern Europe*, 30 - 32. Warsaw.
13. Paluch, J. (1968) Die Moeglichkeit der Anwendung von Pflanzentesten zur Beurteilung des Luftverunreinigungsgrades in J. Paluch and S. Godzik (eds.). Proceedings VI Internationale Arbeitstagung Forstlicher Rauchschadens-achverstaendiger. Katowice, 219 - 234.
14. Staszewski, T., Szdzuj, J., Godzik, S., Kubiesa, P. and Stawiany, W. (1997) *Time dependent fluctuation (changes) of mineral nutrients concentration in bulk deposition and soil solution in Upper Silesia Industrial Region* J. Conf. Abstracts vol. 2(2), 308. Cambridge Publications 1997.
15. Michna, W. (Ed.) (1997) *Report on monitoring of quality of soils, plants, agriculture products and food in 1996)* Biblioteka Monitoringu Serodowiska Warszawa (in Polish).

16. Karweta, S. (1980) Effects of industrial air pollution on soil and plants contamination by heavy metals in the Upper Silesia Industrial Region in H. Janczewski (ed.). *Proceedings of the Conference on Effects of Air Pollution on the Environment* 314-324. NOT Warszawa (in Polish).
17. Paluch, J. and Karweta, J. (1968) Die akkumulierung von Zink und Blei im Boden und in Pflanzen in J. Paluch and S. Godzik (eds.). *Proceedings VI Internationale Arbeitstagung Forstlicher Rauchschadensachverstaendiger,* Katowice. 127 - 138.
18. Greszta, J., and Godzik, S. (1968) *Influence of zinc metallurgy on soils,* Roczniki Gleboznawcze **20**, 195 - 213 (in Polish; summ. English).
19. Gzyl, J.and Nowińska, Z. (1985) *Agricultural management of areas adjacent to the Smelter - Mining Complex "Boleslaw" in Bukowno.* Raport. Instytut Ksztaltowania Œrodowiska Oddzial w Katowicach. (in Polish).
20. Kucharski, R., Marchwińska, E., Gzyl, J., Wala, B., Matula, F., and Strzêpek, P. (1994) *Assessment of contaminated agriculture land for growth of edible and forage plants* PAN Oddzial w Katowicach. Katowice (in Polish).
21. Terelak, H., Stuczyński, T., Motowocka, T., and Terelak, M. (1997) *Content of Cd, Cu, Ni, Pb, Zn and S in soils of Katowice Voivodship as compared to soils of Poland Archiv. Environm. Protection* **23**, 3-4,167 -180 (in Polish, summ. English).
22. Szalonek, I., and Warteresiewicz, M. (1966) *Growth and yield of potatoes under industrial air pollutio.* Biul. Zakl. Badań Naukowych GOP PAN. 8,59 - 74 (in Polish, summ. English).
23. Szalonek, I., Warteresiewicz, M. (1966) *Effects of air pollution on some vegetable species* Biul. Zakl. Badań Naukowych GOP PAN. 8, 75 - 84 (in Polish, summ. English).
24. Warteresiewicz, M. (1968) Einfluss der Luftverunreinigung auf Pflanzen in der Naehe einiger Zinkhuetten in J. Paluch, S. Godzik (eds.). *Proceedings VI Internationale Arbeitstagung Forstlicher Rauchschadensachverstaendiger.* 185-195. Katowice.
25. Godzik, S. (1990) Impact of air pollution on agriculture and horticulture in Poland 196-214 in W. Grodziński, E.B. Cowling, A. Breymeyer (eds.). *Ecological risks. Perspectives from Poland and the United States.* National Academy Press. Washington D.C.
26. Godzik, S. and Krupa, S.V. (1982) Effects of sulphur dioxide on the growth and yield of agricultural and horticultural crops in M.H. Unsworth and D.P. Ormrod (eds.). *Effects of gaseous air pollution in agriculture and horticulture* 247-265. Butterworth Scientific London.
27. Godzik, S. (1967) *Polyphenol oxidase activity in vegetation injured by industrial air pollution.* Biul. Zakl. Badań Nauk. GOP PAN 10, 103-113.
28. Godzik, S. and Piskornik, Z. (1966) Transpiration of *Aesculus hippocastanum* L. leaves from areas of various air pollution. *Bull. Acad. Pol. Sci.* **14**, 181-184.
29. Staszewski, T., Godzik, S., and Poborski, P. (1994) Physico chemical characteristic of pine needles surface exposed to air pollution sources in K.E. Percy, J.N. Cape, R. Jagles and C.J. Simpson (eds.) *Air pollutants and the leaf cuticle.* 341-350. NATO ASI Series (G), Vol. 36, Springer Verlag.

30. Spiewla, E., Godzik, S. and Jaskowska, A. (1983) Effects of heavy metal ions on some electrical parameters of plant cells. *Archiv. Environm. Protection* 1-2, 57-64.
31. Reuss, C. (1893). Rauchbeschaedigung in dem von Thiele-Winkler'schen Forstreviere Myslowitz-Katowitz. Goslar.
32. Godzik, S. and Sienkiewicz, K. (1990) Air pollution and forest health in Central Europe: Poland, Czechoslovakia and the German Democratic Republic 155-170. in W. Grodziński, E.B. Cowling, and A. Breymeyer (eds.). *Ecological risks. Perspectives from Poland and the United States*. National Academy Press. Washington D.C.
33. Wawrzoniak, J., Malachowska, J., Wójcik, J. and Liwińska, A. (1997) *Forest injury in Poland in 1995 - based on monitoring data* Biblioteka Monitoringu Serodowiska, Warszawa. (in Polish, summ. English).
34. Godzik, S.and Halbwachs, G. (1986) Structural alterations of *Aesculus hippocastanum* L. leaf surface by air pollutants *Z. Pflanzenkrank. Pflanzensch.* **93** (6): 590-594.
35. Godzik, S. and Knabe, W. (1972) Vergleichende elektronenmikroskopische Untersuchungen der Feinstruktur von Chloroplasten einiger Pinus-Arten aus den Industriegebieten an der Ruhr und in Oberschlesien *Proceedings Third Internatl. Clean Air Congr.* A 164 - 170.
36. Godzik, S. and Sassen, M.M.A. (1974) Einwirkung von SO_2 auf die Feinstruktur der Chloroplasten von *Phaseolus vulgaris*. *Phytopathologische Zeitschrift* **79**, 155 -159.
37. Godzik, S. and Sassen, M.M.A. (1978) A scanning electron microscope examination of *Aesculus hippocastanum* L. leaves from control and air polluted areas *Environ. Pollut.* **17**, 13 - 18.
38. Szalonek, I. (1968) Die durch Fluor hervorgerufene Luftverunreinigung und die Pflanzenbeschaedigung in der Naehe eines Emallienwerkes in J. Paluch and S. Godzik (eds.). *Proceedings VI Internationale Arbeitstagung Forstlicher Rauchschadensachverstaendiger* 117-125. Katowice.
39. Marchwińska, E. and Kucharski, R. (eds.) (1995) *Monitoring of edible and forage plants contamination in the Katowice Administrative District*. PIOS, Biblioteka Monitoringu Serodowiska. Katowice (in Polish).
40. Piesak, Z. (1991) Lead and cadmium content in the mangel leaves in some regions of Katowice District *Archiv. Environm. Protection* **2**, 111-115 (in Polish; summ. English).
41. Ashmore, M.R., Bell, J.N.B. and Godzik, S. (1984) Differential tolerance to SO_2, NO_2 and their mixture in *Lolium perenne* L. populations from southern Poland *Bull. Pol. Acad. Sci.* **32**, 9-10, 339-345.
42. Kilian, A. and Malusyński, M. (1990) Comparison of natural and model plant populations) in S. Godzik (ed.). *Hazards and state of the environment of the Cracow - Silesia region.* 86 - 90. CPBP 04.10 SGGW AR Warszawa, nr 62. (in Polish).
43. Godzik, S., Kubiesa, P., Szdzuj, J., Staszewski, T., and Lukasik, W. (1996) *Use of plants for assessment of ozone effects to the environment.* Report research project

4 S401 014 05. Instytut Ekologii Terenów Uprzemys'owionych (IETU) Katowice (in Polish).

44. Przybylski, T. (1990) Degradation of the forest environment of industrialised regions of south Poland in S. Godzik (ed.). *Hazards and state of the environment of the Cracow - Silesia region* 70 - 85. CPBP 04.10 SGGW AR Warszawa, nr 62 (in Polish).
45. Rostański, K. (1990) Impact of human activities on vascular flora of industrial region on the example of the USIR and surrounding areas in S. Godzik (ed.). *Hazards and state of the environment of the Cracow - Silesia region* 58-85. CPBP 04.10 SGGW AR Warszawa, nr 62 (in Polish).
46. Jêdrzejko, K. (1990) Bryological problems of the Silesia - Krakow upland in S. Godzik (ed.). *Hazards and state of the environment of the Cracow - Silesia region.* 39-57. CPBP 04.10 SGGW AR Warszawa, nr 62. (in Polish).
47. Migula, P., Dolezych, B., Kielan, Z., Laszczyca, P., and Howaniec, M. (1990) Response of animals on environmental stresses in industrial areas in S. Godzik (ed.) *Hazards and state of the environment of the Cracow - Silesia region.* 108-129. CPBP 04.10 SGGW AR Warszawa, nr 62.
48. Zielonka, U. (1996) *Methods of determination of Mn Tl and S in environmental samples using the ICP technique* Report 122/DV/96. Instytut Ekologii Terenów Uprzemyslowionych, Katowice (in Polish).
49. Czylok, A., Herczyk, A., Klimaszewski, S. M. and Wojciechowski, W. (1990) Associations of Aphids of birch (*Betula pendula*) as a marker of environmental quality in areas of industrial influence of the Silesia and Kraków agglomerations in S. Godzik (ed.) *Hazards and state of the environment of the Cracow - Silesia region.* 130-135. CPBP 04.10 SGGW AR Warszawa, nr 62. (in Polish).
50. Migula, P., and Jethon, Z. (1990) The resistance of aphids and moths to air pollution *Archiv. Environm. Protection* 3-4, 141-156.
51. Niweliński, J., Zamorska, L., Kaczmarski, F. and Pawlicki, R. (1990) Enzyme histochemistry and microstructure of the human placenta as indicators of environmental pollution *Archiv. Environm. Protection.* 3-4/1990. 53-59.
52. Niweliński, J., and Zamorska, L. (1990) Monitoring of environmental pollution on the base of human placenta investigations. Results for south Poland for the period 1986 - 1990) in S. Godzik (ed.). *Hazards and state of the environment of the Cracow - Silesia region.* 149-156. CPBP 04.10 SGGW AR Warszawa, nr 62, (in Polish).
53. Hager Malecka, B., Seliwa, F., Lukas, W., Frydrych, J., and Franiczek, W. (1991) Assessment of state of health in children exposed to zinc smelter emission *Archiv. Environm. Protection* **2/91**, 21-27.
54. Norska-Borówka, I., Bursa, J., Bober-Dubik, E., and Genge, E. (1990) Environmental pollution, living conditions of parents and infant mortality in S. Godzik (ed.) *Hazards and state of the environment of the Cracow - Silesia region.* 157-162. CPBP 04.10 SGGW AR Warszawa, nr 62. (in Polish).
55. Norska-Borówka, I., Bursa, J., Rzempoluch, J., and Wawryk, R. (1990) The influence of environmental factors on pregnant women and infants health in Silesia *Archiv. Environm. Protection* **3-4**, 45 - 51.

56. Seliwa, F., Lukas, W., Schneiberg, B. and Ziora, K., 1997 Osteocalcin and some parameters of calcium metabolism serum concentration in children with lead intoxication *Archiv. Environm. Protection* **23**, 3-4, 209-214 (in Polish, summ. English).
57. Ignacak, J., Gumińska, M. and Stachurska, M.B. (1990) Fluoride and magnesium in urine of humans living in polluted environment. Studies *in vitro* and *in vivo* *Archiv. Environm. Protection.* **3-4**, 123 -128.
58. Kêdryna, T., Marchut, E. and Gumińska, M. (1990) ATP, lactate and glucose in blood of humans living in polluted area *Archiv. Environm. Protection.* **3-4**, 129-134.
59. Ignacak, J., Brandys, J., Danek, M. and Moniczewski, A. (1990b) Biochemical effects of lead in humans living in the polluted environment *Archiv. Environm. Protection.* **3-4**, 135-140.
60. Zachwieja, Z., Schlagel-Zawadzka, M., Ziêba, K. (1990) Concentration of lead and cadmium in hair of children - marker of environmental pollution in S. Godzik (ed.) *Hazards and state of the environment of the Krakow - Silesia region.* 163-175. CPBP 04.10 SGGW AR Warszawa, nr 62. (in Polish).
61. Jordan, M., Komala, Z. and Przybos, E. (1990) Changes in ontogensis of selected animal species from polluted areas *Archiv Environm. Protection* 3-4, 157 - 168.
62. Przybos, E., Komala, Z., Luks-Betlej, K. and Bodzek, D. (1990) Effects of polyaromatic hydrocarbons fractions extracted from dust in Silesia on animal organisms, *Archiv. Environm. Protection.* **3-4**, 169-184 (in Polish, summ. English).
63. Chorazy, M. (1990) Biological hazard by air pollution from the Katowice Administrative District in S. Godzik and P. Wójcik (ds.) 1990. *Ecological conditions of health and life of the Polish society.* 1-12. Suplement. SGGW AR Seria programu CPBP 04.10, nr 13. (in Polish).
64. Kulesza, A. and Sztrantowicz, H. (1987) Decomposition of different plant materials and cellulose in forest soil of the Rybnik Coal Mining Region *Polish Ecological Studies* **13**, no 1, 9-27.
65. Paplińska, E. (1984) Density, biomass and qualitative structure of soil Diptera larvae communities in an industrial landscape *Polish Ecological Studies* **10** no 1-2, 93-110.
66. Paplińska, E. (1987) Preliminary quantitative and qualitative estimation of soil associations of Diptera larvae occurring in forest ecosystems and in woodlots in the vicinity of Knurów (Silesia) *Polish Ecological Studies* **13**, no 1, 95-111.
67. Sztrantowicz, H. (1984) Soil Protozoa as indicators of environment degradation by industry *Polish Ecological Studies* **10**, no 1-2, 67-92.

5. VALIDATION OF SELECTED BIOMARKERS IN INVERTEBRATES FROM THE POLLUTED SILESIAN REGION

P. MIGULA, M. AUGUSTYNIAK, P. LASZCZYCA and G. WILCZEK
Department of Human & Animal Physiology
University of Silesia,
Bankowa 9, 40-007 Katowice, Poland

Abstract

A number of technological changes has led to an improvement in the condition of the heavily polluted region of Upper Silesia. Nevertheless, organisms living in this area are exposed to a combination of toxic chemicals at levels exceeding those permissible by the environmental law. In this report some research activities from our laboratory, concentrated on application of biomonitoring and biomarkers in terrestrial invertebrates from this region to monitoring programmes, are presented. Their advantages, validation and constraints are discussed. Non-specific biomarkers, such as metabolism rates of intact animals and general indices of energetic processes - adenine nucleotide concentrations, and the adenylate energy charge (AEC) - were studied in various groups of insects and spiders. These biomarkers are also affected by many non-chemical factors. It is difficult to separate toxic effects from natural factors since time-related assays with the same organisms often produced significantly different results.

The detoxification enzymes (mixed function oxidases, glutathione transferases or carboxylesterases) can show exposure to a series of xenobiotics and endogenous compounds. Also the validity of free radical scavengers, such as superoxide dismutase or catalase, was tested in animals from a pollution gradient and in laboratory experiments dose-effect relationships were assessed.

Biomarkers as additional tools in environmental risk assessment will help in solving the problem of environmental management. Existing data from the Silesian Region are still too scanty for their proper use and can serve only as additional information in taking environmental decisions.

Keywords. Upper Silesia; biomarkers, adenylate energy charge, free radicals, detoxification enzymes, insects, spiders.

D. B. Peakall et al. (eds.),
***Biomarkers: A Pragmatic Basis for Remediation of Severe Pollution in Eastern Europe*, 75–90.**

5.1 Introduction

Depledge and Fossi [1] proposed three stages of investigations allowing extrapolation from a biomarker response to the ecosystem level: (1) identification of ecosystems at risk; (2) identification of critical species and target populations in a limited range of species occupying different trophic levels and ecological niches and (3) predicting the likely impact of chemical pollutants including measurements of a suite biochemical, physiological or behavioural biomarkers. The first stage is obvious, since in the region of Upper Silesia practically all ecosystems are at risk. The problem was how to find where the level of impact might cause irreversible biological functioning.

The area of Upper Silesia is badly affected by man-made pollutants, due to concentrated heavy industry and inadequately functioning systems of environmental protection against emissions of toxic compounds to the soil, water and air. Disturbances to natural conditions may lead to losses of compensatory mechanisms of species and communities [2,3,4]. Despite recent technological incentives, closure of major polluters in this region and stronger restrictions forced by a new environmental law, the monitoring services still show that concentrations of many toxic compounds seriously exceed the levels permissible by law. Heavy metals and organic compounds (such as aromatic hydrocarbons, polychlorinated biphenyls, polychlorinated dibenzodioxins) in particular, remain in some sites at very high levels [5,6]. Although regional pollution, in terms of its effects on plants and animals, has been regarded as the abiotic environmental stress [7]. these pollutants do reach biotic elements. Organisms living in this area are exposed to a combination of various chemicals, showing additive toxic effects, but examples of harmful effects of pollutants were also indicated as dependent on the physiological state of organisms [3,8]. Taking also into consideration distinct changes of natural factors, which may influence organisms, one may accept the hypothesis of Nuorteva [9] on the multiple stress experienced in this region

The invertebrate species of special importance for the functioning of terrestrial ecosystems under multistress from industrial pollution, such as earthworms, slugs small soil-dwelling microarthropods, social, carnivorous or herbivorous insects and spiders have been well identified [9,10,11,12,13,14,15,16,17]. The majority of our studies were carried out on insects, since they play an essential role in natural forests and managed plantations as the dominant components of many food webs, and their sensitivity to pollutants may also affect other elements of an ecosystem [18]. These studies were concentrated on the application of biomonitoring and biomarkers to monitoring programmes. Some of molecular biomarkers for invertebrates mentioned above will be reported below. Their validity for the natural conditions was based on laboratory dose-effects models tested in a measurable gradient of selected stress factors.

5.2. Biomonitoring

Effects of pollutants on organisms are difficult to predict when only changes in concentration of harmful compounds in abiotic environment are monitored. The degree of pollution-induced stress for organisms living in Upper Silesian ecosystems was assessed through monitoring of some persistent compounds, mostly heavy metals. Comparative studies did not establish that the accumulation of metals along the food chains is proportional to the concentration gradient in the environment or in consumed food [3]. Body burdens with metals were measured, however, in a limited number of species from the described area. There is limited data on vertebrates (mostly passerine birds, small rodents or frogs) [3,19,20,21,22,23,24], Some data are available for invertebrates, such as earthworms, terrestrial molluscs, insects and spiders [3,17,25].

Looking for an "early warning system" we used honey bees for *in situ* monitoring, together with sources of their food and products [26]. Honey bees can concentrate elevated levels of trace metals while foraging the area, which extends to about four kilometres. Their life span, behaviour and visited areas are easy to recognise. Moreover, they are active throughout the spring-autumn period, making them easy for sampling and their food and products (honey, royal jelly, propolis) are identified. We used them to study bioconcentration of metals in practically the whole Upper Silesian Industrial Region and even beyond in the Czestochowa-Krakow Upland [26,27,28,29]. Our description of risk associated with the industrial pollution was based on the cumulative index of metal concentrations in produced honey, royal jelly, pollen, propolis, nectar and bees of various castes. The index based on sampling during the years 1986-95, from more than 50 sites, varied between 0.4 and 0.8, allowing us to construct a map of hazard taking into account a possible distance of flight of the honey bee foragers from sampled hives [28,29]. It was difficult, however to predict from these data the degree of hazard as the function of both exposure and toxicity. In other words, we were not able to transfer these data into functional possibilities [30], thus it was necessary to use biomarkers.

5.3. Biomarkers of energetic status of target organisms

As we have been working in the region where animals are exposed mainly to a mixture of pollutants, biomarkers of a low specificity seemed the most suitable. Our studies were concentrated on assessing pollutant effects on general indices of energetic processes, by measurements of metabolism rates and adenine nucleotide concentrations and calculations of the adenylate energy charge index - AEC [15,26,31,32]. The concentration of high-energy phosphates (ATP, ADP, AMP) in animal tissues, and calculated from their content the adenylate energy charge value, may reflect a metabolic imbalance of an organism exposed to various environmental stressors [31]. Determinations were made for the representatives of insects from variously contaminated sites and the dose-effects relationships between metabolism rate (or adenylate concentration) and concentration of applied pesticides or heavy metals in consumed food were tested in laboratory conditions.

5.3.1. OXYGEN CONSUMPTION

A valuable physiological non-destructive biomarker in environmental assessment of intact organism may be the oxygen uptake [2]. Most of the metabolic processes in insects are strictly oxidative, with the Respiratory Quotient between 0.8 and 1.0; thus measurements of oxygen uptake looked very promising. However, there are many difficulties in transfer of the results to environmental status. Some factors, such as a temperature range, compensatory mechanisms, effects of physiological and developmental stages of an organism were discussed in detail elsewhere [2,16,26,33]. For the interpretation of data from direct measurements of respiratory rates in animals taken from their natural conditions in variously contaminated sites [20], some guidance was obtained by comparison with laboratory measurements of metabolism rates in animals exposed to various toxic compounds [2,3,33]. For example, from studies on earthworms [2,34,35] we know that the oxygen uptake by *Eisenia foetida* exposed to lead was more pronounced at higher temperatures, but only at 15°C correlated with metal body burdens and metal contents of soil (Fig 1). Similar temperature-related

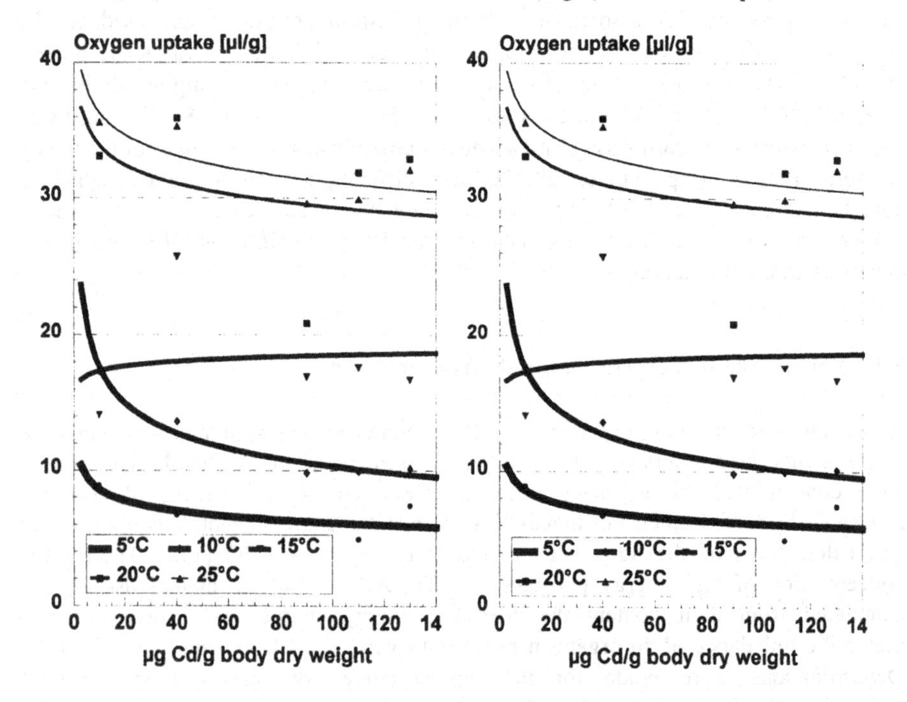

Figure 1. Oxygen uptake by earthworms, *Eisenia foetida* against cadmium concentration at different temperatures.

effects were obtained for *Lumbricus terrestris* living in soil mixed with the dust taken from the zinc smelter collectors [35]. Concentration effects were more pronounced at higher temperatures, showing that heavy metal 'accumulators' respond more markedly to increased temperature than species presumed to be good metal regulators [36].

5.3.2. ADENYLATE NUCLEOTIDES AND ADENYLATE ENERGY CHARGE.

Insects from uncontaminated or moderately contaminated areas usually showed a higher level of both the AEC and adenylate pool size than those from highly contaminated areas [4,26]. Only in grasshoppers (*Chorthippus brunneus*) from heavily polluted site did the AEC exceed the values calculated for insects from more favourable areas. The strategy of insects was to maintain the AEC above the value of 0.8, notwithstanding anthropogenic stress, but the ATP level and the total phosphoadenylate pool may decrease proportionally to a concentration load. Energetic strategy was species-dependent, as it has been shown in some Lepidopterans [4]. Also honey bee workers, taken from various polluted areas indicated a maintenance of the AEC (changes from 0.81 to 0.85; $p > 0.05$) with a significant reduction of both the pool of adenylates and ATP levels (Fig. 2) [26].

The red wood ants (*Formica polyctena*) from Silesia, experimentally poisoned in the field with cadmium, showed slightly lower concentrations of total phosphoadenylates than the closely related "Finnish" ants *Formica aquilonia,* similarly exposed to known amounts of added cadmium, but the AEC index was always above 0.8. The strategy of *F. polyctena* ants differs from that of *F. aquilonia*, since they maintain narrow limits of the AEC, due to a higher variability of the phosphoadenylate pool size [15]. This strategy needs more effective systems for energy restoration and more energy allocated for detoxification processes [26]. Measurements done in spring of the year following the poisoning period showed nearly complete recovery to the initial values, but high levels of cadmium in their tissues remained (Fig. 3).

From studies on *Araneid* and *Linyphid* spiders we concluded that energetic biomarkers should also be used carefully, bearing in mind differences in hunting activity, type of web production, developmental processes or sexual differences [17,25,37]. Examined species may exhibit various strategies to maintain optimal AEC level, depending on the intensity of stressing factors influencing their biotopes. Spiders, taken from a site highly polluted by heavy metals, enlarged the adenine nucleotides pool, which did not necessarily lead to increased energy expenditure. In areas with dominating organic and gaseous pollutants the *Araneids* showed high aerobic activity, but the *Linyphids* showed increased anaerobic activity and pool of adenylates while the AEC index was strongly reduced, even below 0.7 [25,37].

5.4. Detoxification enzymes

Induction of various detoxification enzymes such as mixed function oxidases, glutathione transferases (GSTs) or carboxylesterases (ESTs) might also be used as general biomarkers showing unspecific effects of exposure to a series of xenobiotics and endogenous compounds [38,39]. The GSTs, existing in multiple molecular forms, are able to metabolise various pesticides, allelochemicals, heavy metals and other compounds toxic to the organisms [4]. We also tested the responses of free radical scavengers, such as superoxide dismutase (SOD) or catalase (CAT) as biomarkers

[17,41]. SOD and CAT serve all aerobic organisms as a major defence against oxygen radical toxicity [42]. SOD catalyses the dismutation of superoxide radical to hydrogen peroxide, which is subsequently decomposed to molecular oxygen and water by CAT. Heavy metals and other industrial pollutants as well, may accelerate a synthesis of secondary metabolites, which in turn are leading to adverse effects in their consumers [43].

A flavoprotein - NADPH-cytochrome reductase, assayed in phytophagous insects sampled along a pollution gradient (heavy metals, hydrocarbons, polychlorinated biphenyls, dioxins) was positively correlated with the level of pollutants [3,4,16]. Patterns of detoxification for the same species (*Pieris brassicae, Lepidoptera*) from variously contaminated areas were different when exposed under laboratory conditions to excessive amounts of cadmium and lead in their diet, with a stimulation in insects from a moderately polluted site, but a fall in activity rates was observed in insects sampled from a heavily polluted site [16].

Time-related changes of selected detoxification enzymes in the red wood ants treated under natural conditions with a measurable excess of cadmium, showed different patterns of enzyme recovery. A significant reduction from the initial value was found with GST. Superoxide dismutase was stimulated, while carboxylesterases remain unchanged (Fig. 4), in contrast to a study with ants in a Finnish forest where opposite effects were shown [44,45].

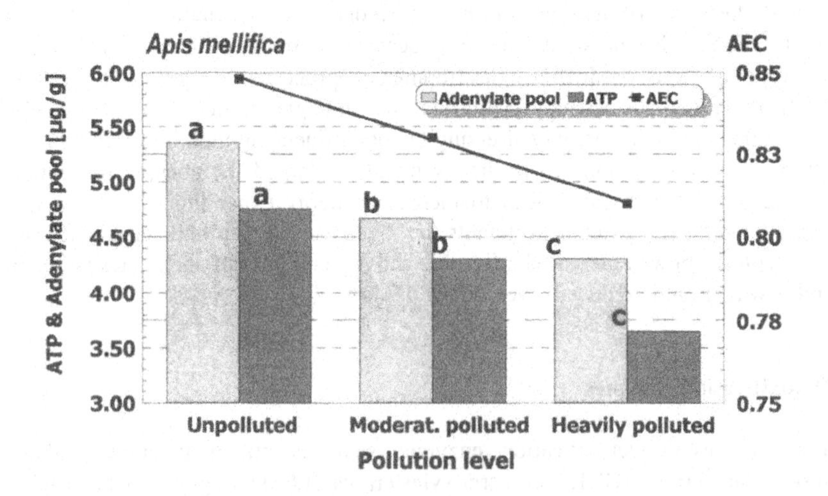

Figure 2. Adenylate nucleotides and the AEC index for honey bee workers from variously polluted areas in the Upper Silesia. The letters, a, b and c indicate heterogeneous groups in relation to the pollution level at $p < 0.05$.

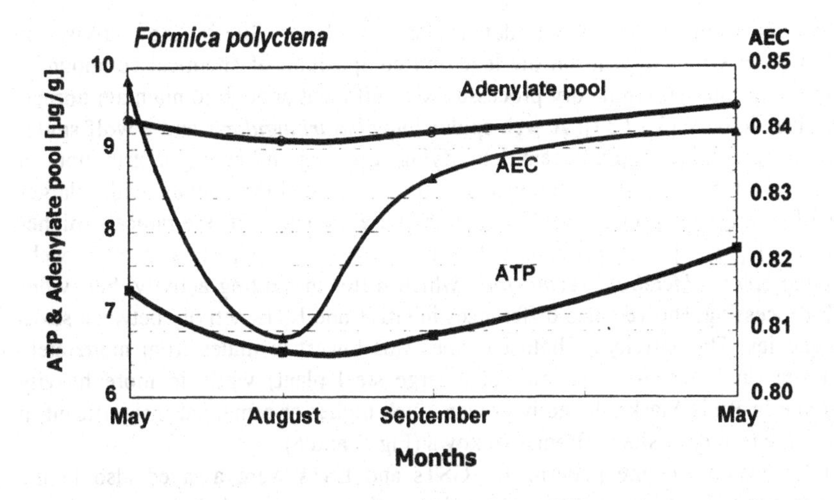

Figure 3. Time-related changes in the total adenine nucleotide pool, ATP and Adenylate Energy Charge (AEC) in red wood ants from Zarzecze (Southern Poland) exposed to excess of cadmium (300 µg $CdCl_2$/g honey; 150 mg/ant family) in natural conditions.

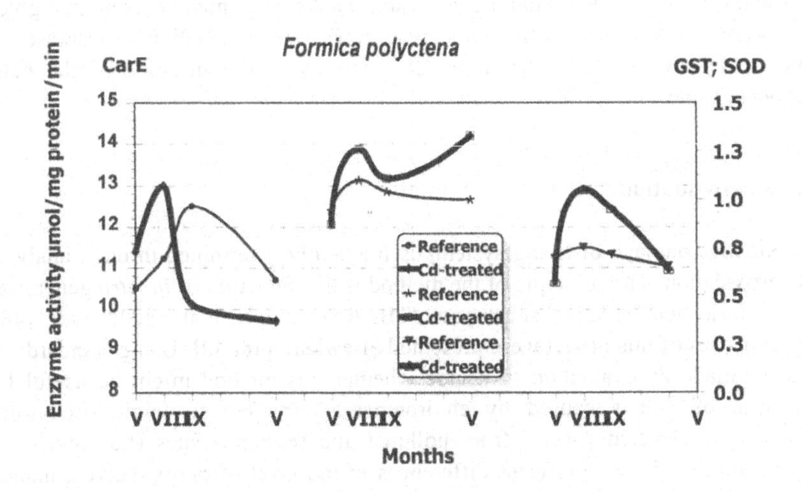

Figure 4. Time-related changes in activity of carboxylesterases, glutathione S-transferase and superoxide dismutase in red wood ants, untreated and exposed to excess of cadmium (300 µg $CdCl_2$/g honey; 150 mg/family) in natural conditions (from Zarzecze, Southern Poland).

In spiders from industrially polluted sites of Upper Silesia the activity of detoxification enzymes depend on the load and composition of chemical compounds. Generally, their internal regulatory processes were efficient enough to maintain normal physiological activity [3,17,37]. A web spider *Linyphia triangularis* and a wolf spider *Pardosa palustris* have used different detoxifying strategies and energy distribution. In comparison with the wolf spider the web spider accumulated less metals, with a larger increase of conjugation processes (GSTs) and better elimination of free radicals (higher SOD activity) [17,37]. Comparisons between *Linyphia triangularis* and other web-constructing spider, *Metellina segmentata,* which differ in hunting activity, behaviour and web-processing, showed also differences in GSTs and ESTs activity between sexes in both species. The activity of both enzymes was lowest in males from moderately polluted site (in Losien - in vicinity of a large steel plant) while in more heavily polluted site in Ruda Slaska the activity was much higher or remained insignificant in relation to the reference site at Brenna-Bukowa (Fig. 5 and 6).

Gender-related enzyme patterns for GSTs and ESTs were assayed also in the intestine and gonads of grasshoppers (*Chorthippus brunnes*) sampled along a pollution gradient in five plots from Silesia [46]. Earlier studies showed that local populations of this species from more polluted areas were better equipped with compensatory mechanisms against heavy metals in their diet than insects from a reference site [47]. Also dose-related effects of cadmium were confirmed in laboratory experiments. Significant correlations were confirmed between cadmium concentration in the consumed grass and the intestine of both sexes as well as between esterase activity in the testes and GST in female sexual organs (Table 1). As far as metal content in a given organ is concerned a positive correlation was stated only in relation to esterases in testes but a negative one in relation to GST activity in the intestine of the male grasshoppers (Table 1).

5.5. Lipid peroxidation

The peroxidative damage of living systems is frequently determined through analyses of lipid peroxidation. The principle of the method is the induction of *in vitro* generation of thiobarbituric acid reactive substances (TBARS), mainly malondialdehyde [48]. Kinetic properties of this model were presented elsewhere [48, 49]. Using standardised methods we made an evaluation to decide whether this method might be useful for determination of effects caused by environmental stressors to living organisms. Examinations of selected species from polluted and reference sites (Sosnowiec vs. Rudziniec) did not show significant differences in the level of peroxidative damages (Fig. 7). Laboratory testing on earthworms and slugs for the effects of various chemical compounds on lipid peroxidation in earthworms and slugs confirmed that lipid peroxidation is useful only at high levels of industrial or agricultural pollutants. A decrease of TBARS was organ-dependent and the most sensitive organ was the intestine (Fig. 8 and 9).

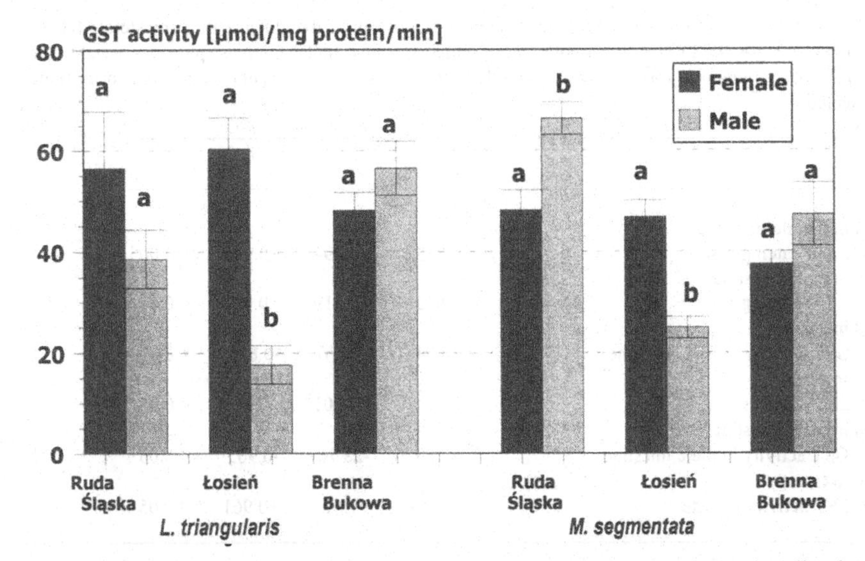

Figure 5. Glutathione S-transferase activity in two species of spiders from variously polluted sites of the Silesian Region. Different letters denote gender-related significance of means ($p < 0.05$).

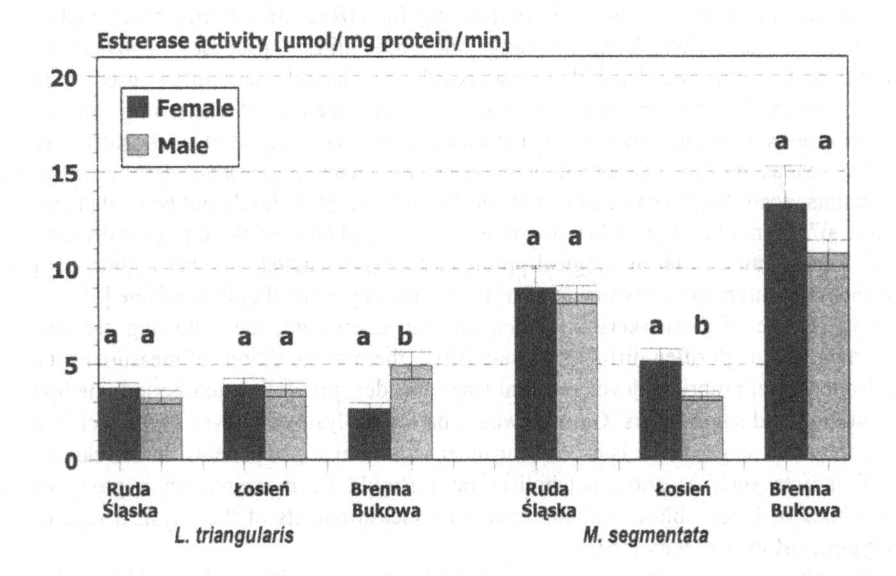

Figure 6. Esterase activity in two species of spiders from variously polluted sites of the Silesian Region. Different letters denote gender-related significance of means ($p < 0.05$).

TABLE 1. Coefficients of correlation and coefficients of regression of the glutathione S-transferase (GST) and carboxylesterase (EST) activity (μmol/min/mg protein) against metal burdens (μg/g dry weight) for grasshoppers *Chorthippus brunneus*. Regression follows the model: Y = a ±bX; only significant correlations are presented.

	a	b	r	p
Cd in female intestine *vs.* Cd in eaten grass	0.753	+0.094	+0.995	< 0.005
Cd in male intestine *vs.* Cd in eaten grass	0.653	+0.219	+0.991	< 0.01
Cd in grass *vs.* GST activity in fem. intestine	30.04	-0.276	-0.831	< 0.02
Cd in grass *vs.* EST in testes	0.07	+0.0012	+0.96	< 0.05
Cd in male intestine *vs.* GST activity in male intestine	57.89	-28.76	-0.965	< 0.05
Cd in testes *vs.* EST activity in testes	-0.259	+1.693	+0.961	< 0.05

5.6. Validation and constraints in the use of proposed biomarkers as a "warning system"

Biochemical measures are useful in monitoring for effects before they reach higher biological organisation level. Among the biomarkers proposed above, the detoxification enzymes have been successfully adapted and often used [see 1,4,30,39,50,51,52 for literature review]. The parameters characterising energy expenditure or energetic state of an individual have been criticised in the literature [16,50]. Measurements of metabolism rates in intact organisms, despite many constraints, according to van Gestel and van Brummelen [53] should not be treated as a „classical" biomarker but rather as "bioindicators", characteristic for an individual level. This is rather a terminological problem, and is discussed by other authors [54] and such measurements are considered as biomarkers by Peakall in this volume [55].

For the use of biomarkers in ecological risk assessment, the following are also important (although often difficult to establish) - thermal conditions of measurements, a selection of appropriate developmental stage, gender, preacclimation period, method of handling and some others. Optimal would be the analysis of a given biomarker in a developmental stage which is crucial for other community components. In the case of social insects such as ants, metabolism rates should be measured on a group of individuals in favourable conditions, involving measurements of the physical load or studying working capacity [56].

The advantage of using measurements of adenine nucleotide pool and AEC is that we are able to state whether toxicants caused changes which are irreversible for an organism. This will work in heavily polluted sites. For less polluted sites there is a strong dependence on metabolic strategy and this will be valid only for species where adenine nucleotides play a major role in transport of energy.

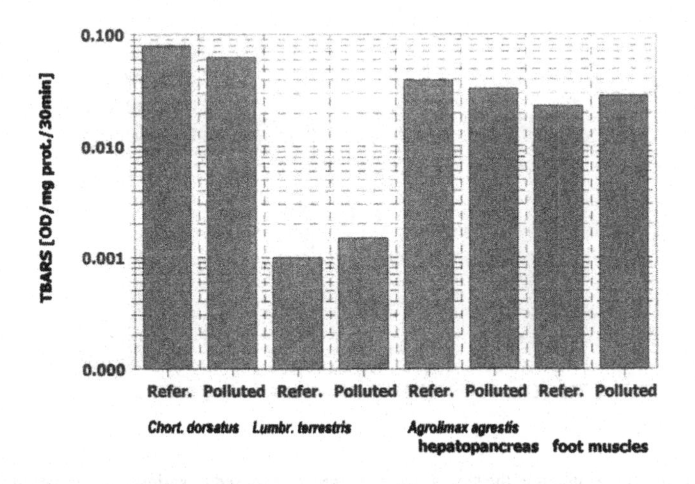

Figure 7. A comparison of lipid peroxidation in three species of invertebrates (a grasshopper, an earthworm and a slug) from unpolluted and polluted site in Upper Silesia.

The availability of metabolic energy through measurements of nucleotide concentrations should be measured in the whole body, since all preparatory techniques take time and thus affect a real pattern of nucleotides in examined organs. A disadvantage is that all collected samples for adenylate measurements should be deep-frozen in liquid nitrogen immediately after sampling. Honey bees for such measurements should be selected from apiaries kept in similar conditions, representing the same race of insects, in a comparable time period, without artificial feeding and from colonies free of pathogens, specially *Varroa jacobsoni*. Also the weather conditions during sampling should be similar. It is important to know whether foraging insects are returning or leaving the hive.

A number of limiting factors make difficult a proper interpretation of changes in enzyme activity which should be related to changes of other biological factors, as well as to changes at higher levels of biological organisation [53,57]. Measurements repeated with the same model organisms in consecutive seasons under similar conditions are often giving significantly different results [3,37].

The usefulness of changes in lipid peroxidation as a biomarker depends not only on a proper selection of species in danger, but also on selection of sensitive organs for analyses. From our studies this seems to be the intestine - the first target organ for many toxic compounds [49].

There is no doubt that biomarkers, as new tools in environmental risk assessment, will help us in solving problems of environmental management. We should always take into account a great individual variability in a population, which is why it may not respond homogeneously to a chemical stressor. As yet it is impossible to determine what proportion of a population and which individuals must be adversely affected by pollutants before decline in the population ensues [43]. Little attention has been paid to genotypically or phenotypically determined interpopulation differences in susceptibility

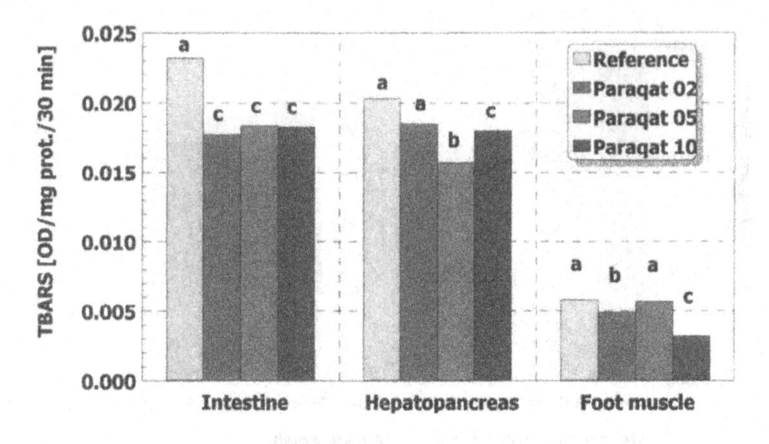

Figure 8. Lipid peroxidation in the slug *Agrolimax agrestis* exposed to various concentrations of the bipiridyl herbicide paraqat. Different letters denote gender-related significance of mean TBARS in relation to paraqat treatment within the organs ($p < 0.05$).

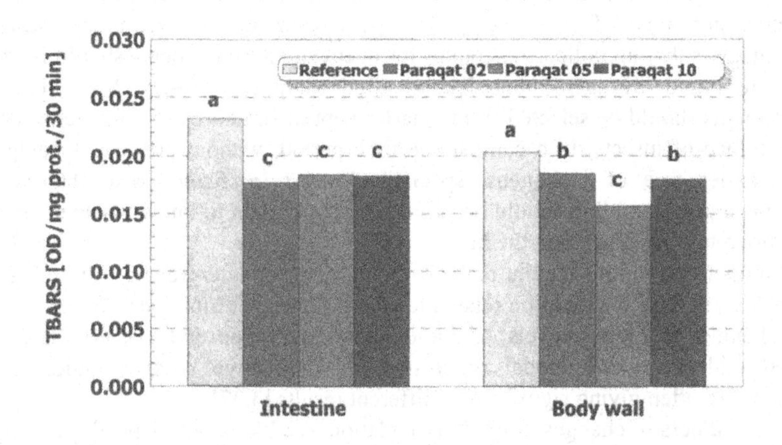

Figure 9. Lipid peroxidation in the earthworm *Lumbricus terrestris* exposed to various concentrations of the bipiridyl herbicide paraqat. Different letters denote gender-related significance of mean TBARS in relation to paraqat treatment within the organs ($p < 0.05$).

to pollutants and to biogeographic differences in pollutant toxicity, and persistence of biomarker response with time.

For biomarkers we need to know their normal levels in various ambient situations and to know how environmental factors and biotic relations may influence the responses to pollutants. We are still not sure to what extent we can transfer the data from dose-effect relationships established in controlled experiments to the field conditions and how valid are correlations from the field studies between a biomarker response and internal amounts of contaminants [50,52]. This step is needed before we can know to what extent various laboratory tests are capable of predicting likely exposure or effects of pollutants on ecosystems in danger.

References

1. Depledge, M.H. and Fossi, M.C. (1994) The role of biomarkers in environmental assessment (2). Invertebrates. *Ecotoxicology* **3**, 161-172.
2. Migula, P. (1989) Bioenergetic indices as indicators of environmental contamination in insects. In: Goel, S.C. (ed.) *Nutritional Ecology of Insects and Environment,* Muzzafarnagar, India, pp.157-166.
3. Migula, P., Dolezych, B., Kielan, Z., Laszczyca, P. and Howaniec, M. (1990) (Responses to stresses by organisms from industrially contaminated regions) In: Godzik, S. (ed.) *Srodowisko Przyrodnicze Regionów Szczególnego Zagrozenia.* Publ. SGGW, Warszawa **62**, 108-29.
4. Migula, P. (1997) Molecular and physiological biomarkers in insects as the tools in environmental hazard assessment. *Acta Phytopatol. Entomol. Hung.* **32** (1-2), 231-243.
5. Wojewodzki Urzad Statystyczny (1994) (*Environmental protection in Katowice woivodship in the years 1990-1993*). WUS OP Katowice.
6. Godzik, S., Kubiesa, P., Staszewski, T. and Szdzuj J (1997) Ecological problems of the Katowice administrative district.. (this volume).
7. Heinrichs, E.A. (1988) *Plant stress-insect interactions*, Wiley, New York.
8. Dolezych, B. (1994) Influence of cadmium on the absorption. motor and enzymatic functions of the alimentary tract of rats. *Prace Nauk. Univ. Slask.* **1402**, 1-80. Katowice
9. Nuorteva, P. (1990): *Metal distribution patterns and forest decline*. Publ. Dept Environ. Conservation, Univ. of Helsinki Press, Helsinki, pp.1-77.
10. Koricheva, J., Laappalainen, J. and Haukijoja, E. (1995): Ant predation of *Eriocrania miners* in a polluted area. *Entomol. Exp. Appl.* **75**, 75-82.
11. Bengtsson, G. and Tranvik L. (1989) Critical metal concentrations for forest soil invertebrates. *Water Air Soil Pollution* **47**, 381-417.
12. Neuhauser, E. F., Cukic, Z.V., Malecki M.R., Loehr, R.C. and Durkin, P.R. (1993) Bioconcentration and biokinetics of heavy metals in the earthworm. *Environ. Pollution* **89**, 3, 293-301.
13. Maavara, V., Martin, A. J., Oja, A. and Nuorteva P. (1994): Sampling of different social categories of red wood ants (Formica s.str.) for biomonitoring. In: Merkert,

B. (ed.) *Environmental sampling for trace analysis*. VCH, Weinheim, New York **26**, 465-489.

14. Van Straalen, N. M. and Donker, M. (1994): Heavy metal adaptation in terrestrial arthropods - physiological and genetic aspects. *Proc. Exper. & Appl. Entomol.*, NEV. Amsterdam **5**, 3-17.
15. Migula, P. and Glowacka, E. (1996): Heavy metals as stressing factors in the red wood ants (*Formica polyctena*) from industrially polluted forests. *Fresenius' J. Anal. Chem.* **354**, 5-6, 653-659.
16. Migula, P. (1996): Constraints in the use of bioindicators and biomarkers in ecotoxicology. In: Van Straalen, N. and Krivolutsky, D.A. (eds). *Bioindicator systems for soil pollution.* Kluwer Acad. Publ., Dodrecht, Netherlands **3**, 17-31.
17. Wilczek, G. and Migula, P. (1996): Metal body burdens and detoxifying enzymes in spiders from industrially polluted areas. *Fresenius' J. Anal Chem.* **354**, 5-6, 643-647.
18. Koricheva, J. (1994) *Air pollution and Eriocrania miners: observed interactions and possible mechanisms.* Reports Dept. of Biology, Univ. of Turku No. **44**.
19. Dmowski, K., Karolewski M.A. (1979) Cumulation of zinc, cadmium and lead in invertebrates and in some vertebrates according to the degree of an area contamination. *Ekol. pol.* **27**, 333-349.
20. Pytasz, M., Migula, P. and Krawczyk A. (1980). (The effect of industrial pollution on the metabolism rate in several animal species from the ironwork „Katowice" region) *Prac. Nauk. Uniw. Slask.* Katowice **348**, 81-103.
21. Dmowski, K. (1993) Lead and cadmium contamination of passerine birds (Starlings) during their migration through a zinc smelter area. *Acta Ornithol.* **28**, 1, 1-9.
22. Dmowski, K., Kozakiewicz, M. And Kozakiewicz, A. (1995) Ecological effects of heavy metal pollution (Pb, Cd, Zn) on small mammal populations and communities. *Bull. Pol Acad. Sci. Biol. Sci.* **43**, 1, 1-10.
23. Sawicka-Kapusta, K., Laczewska, B. and Kowalska, A. (1995) Heavy metal concentration in Great Tit nestlings from polluted forest in southern Poland. In: Wilken, R.D., Forstner, U and Knochel, A (eds): *Heavy Metals in the Environment.* Proc. Intern. Confer. **1**, 244-247.
24. Dmowski, K. (1997) Biomonitoring with the use of Magpie *Pica pica* feathers: heavy metal pollution in the vicinity of zinc smelters and national parks in Poland. *Acta Ornithol.* **32**, 1, 15-23.
25. Marczyk, G., Migula, P. and Trzcionka, E. (1993) Physiological responses of spiders to environmental pollution in the Silesian Region (Southern Poland). *Sci. Total Environment,* Suppl. **2**, 1315-1322.
26. Migula, P., Biñkowska, K., Kafel, A., Kêdziorski, A. and Nakonieczny, M. (1989) Heavy metal contents and adenylate energy charge in insects from industrialized region as indices of environmental stress. In: Bohac, J., Ruzicka, V. (eds) *Bioindicatiores Deteriorisationis Regionis*, Ceske Budejovice, **2**, 340-349.
27. Migula, P. (ed) (1990) (*Instructions for bee-keeping in polluted regions*). Univ. Silesia, Katowice, pp. 1-83.

28. Migula, P. (ed) (1993) (*Hazards for the bee-keeping from heavy metals in Krakow-Czestochowa Upland*). Zarzad Zespolu Jurajskich Parków Krajobrazowych, Dabrowa Górnicza 1993, pp.1-23.
29. Migula, P. (ed) (1995) (*Hazards for the bee-keeping from heavy metals in Krakow-Czestochowa Upland*). Zarzad Zespolu Jurajskich Parków Krajobrazowych, Dabrowa Górnicza 1995, pp.1-47.
30. Walker, C.H., Hopkin, S.P., Sibly, R.M. and Peakall, D.B. (1996) *Principles of Ecotoxicology*. Taylor & Francis , London
31. Atkinson, D.E. (1977) *Cellular energy metabolism and its regulation*. Academic Press, New York.
32. Migula, P., Kafel, A., Kêdziorski, A. and Nakonieczny, M. (1989) Combined and separate effects of heavy metals on energy budget and metal balances in Acheta domesticus. *Uttar Pradesh J. Zool.* **9**, 2, 140-149.
33. Migula, P. and Jethon, Z. (1990) The resistance of aphids and moths to air pollution, *Arch. Ochrony Srodowiska* **3-4**, 141-156.
34. Migula, P., Baczkowski, G. and Wielgus-Serafiñska, E. (1977) Respiratory metabolism and haemoglobin concentration in an earthworm - *Eisenia foetida* Savigny 1826, under exposure of lead intoxication. Prac. Nauk. Uniw. Slask., Katowice **4**, 64-78.
35. Jethon, Z., Migula, P., Luszczak, G. and Kêdzierska, M. (1987) The effect of soil contamination by dust containing heavy metals on the respiratory metabolism of *Lumbricus terrestris* L., *Biologia*, Bratislava **42**, 537-546.
36. Janssen, M.P.M and Bergema, W.F. (1991) The effect of temperature on cadmium kinetics and oxygen consumption in soil arthropods, *Environ. Toxicol. Chem.* **10**, 1493-1501.
37. Wilczek, G., Majkus Z., Migula, P., Bednarska, K. and Swierczek, E. (1997) Heavy metals and detoxifying enzymes ibn spiders from coal and metallurgic dumps near Ostrava (Czech Republik). *Proc. 16th Europ. Coll. Arachnol.* Siedlce, pp.317-328.
38. Terriere, L.C. (1984.): Induction of detoxication enzymes in insects. *Annu. Rev. Entomol.* **29**, 71.
39. Peakall, D.B. (1992): *Animal Biomarkers as Pollution Indicators*. London, Chapman & Hall.
40. Cohen, E. (1986) Glutathione S-transferase activity and its induction in several strains of *Tribolium castaneum*. *Entomol. exp. appl.* **41**, 39-44.
41. Chrascina, M., Kafel, A. and Migula, P. (1996) Patterns of detoxifying enzymes in larval stage *Smerinthus ocellatus* L. exposed to cadmium, tocopherol or quercetin. *Stud. soc. Sci.* Toruñ **4**, 3, 31-37.
42. Brattsten, L.B. (1988) Enzymic adaptations in leaf-feeding insects to host-plant allelochemicals. *J. Chem. Ecol.* **14**, 1919-1939.
43. Van Straalen, N. M. (1994): Biodiversity of ecotoxicological responses in animals. *Netherl. J. Zool.* **44**, 112-129.
44. Migula, P., Nuorteva, S-L. & P., Glowacka, E. and Oja, A. (1993): Physiological disturbances in ants (*Formica aquilonia*) to excess of cadmium and mercury in a Finnish forest. *Sci. Total Environment*, Suppl. 1993, **2**, 1305-1314.

45. Migula, P., Glowacka, E., Nuorteva, S_L., Nuorteva, P. and Tulisalo, E. (1997). Time related effects of intoxication with cadmium and mercury in the red wood ant (*Formica aquilonia*). *Ecotoxicology* **6**, 3, 1-14.
46. Augustyniak M and Migula, P. (1996). Patterns of glutathione S-transferase activity as a biomarker of exposure to industrial pollution in the grasshopper *Chorthippus brunneus* (Thunberg) *Stud. soc. Sci.* Toruñ, **4,** 3, 10-15.
47. Migula, P. and Binkowska, K. (1993) Feeding strategies of grasshoppers (*Chorthippus* sp.) to heavy metal contaminated plants. *Sci. Total Environment*, Suppl. **1993**, 1071-1083.
48. Laszczyca, P., Kawka-Serweciñska, E., Witas, I., Dolezych, B. and Migula, P. (1995) Iron ascorbate-stimulated lipid peroxidation in vitro. Why the method is controversial? *Gen. Physiol. Biophys.* **14**, 3-18.
49. Laszczyca, P., Kawka-Serweciñska, E., Witas, I., Dolezych, B., Falkus, B., Mekail, A., Ziólkowska, B., Madej, P. and Migula, P. (1996) Lipid peroxidation and activity of antioxidative enzymes in the rat model of ozone therapy. *Mat. Med. Pol.* **4**, 155-160.
50. Lagadic, L., Caquet, T. and Ramade, F. (1994): The role of biomarkers in environmental assessment (5). Invertebrate populations and communities. *Ecotoxicology* **3**, 193-208.
51. Depledge, M.H. (1993) The rational basis for the use of biomarkers as ecological tools. In: M.C. Fossi and C. Leonzio (eds.) *Nondestructive biomarkers in vertebrates*, Lewis Publishers, Boca Raton Fl., pp. 261-285.
52. Peakall, D.B.and Walker, C.H. (1994): The role of biomarkers in environmental assessment (3). Vertebrates. *Ecotoxicology* **3**, 173-179.
53. Van Gestel, C.A.M. and van Brummelen, T.C. (1996.) Incorporation of the biomarker concept in ecotoxicology calls for a redefinition of terms. *Ecotoxicology* **5**, 217-225
54. Peakall, D.B. and Walker, C.H.. (1996) Comment on van Gestel and van Brummelen. *Ecotoxicology* **5,** 27.
55. Peakall, D.B. Use of biomarkers in hazard assessment (Chapter 9 this volume)
56. Nielsen, M.G., Jensen, T. F. and Holm-Jensen, I. (1982) Effect of load carriage on the respiratory metabolism of running ants of *Camponotus herculeanus* (Formicidae). *Oikos* **39,** 137-142.
57. Weeks, J.M. (1995): The value of biomarkers for ecological risk assesment: academic toy or legislative tools? *Appl. Soil Ecol.* **2,** 215-216.

6. GENETIC AND ONCOGENIC CONSEQUENCES OF CHEMICAL AND RADIATION POLLUTION OF THE KUZBASS AND ALTAI REGIONS

EKATERINA N. ILYINSKIKH, NICOLAI N. ILYINSKIKH and IRINA N. ILYINSKIKH
Siberian Medical University, 634050 Tomsk-50, a/ya 808, Russia

Abstract

We have assessed frequencies of micronucleated lymphocytes in 12307 individuals living in 7 towns in the south of the Kuzbass and Altai regions. Among the towns the majority of individuals with significantly high frequencies of micronucleated lymphocytes were detected in towns adjacent to the Semipalatinsk atomic testing ground (SATG). Pollution of the environment was also caused by the activity of metallurgical plants and coal industry in this region. The greatest genome instability was found in the individuals born in the period of intensive testing on the SATG (from 1949 to 1962). Moreover, we have determined that the residents of the towns adjacent to the SATG have significantly high levels of antibodies to potentially oncogenic Epstein-Barr virus. The considerable Epstein-Barr virus contamination among the residents in the radiation polluted zone around the SATG was supposed to be caused by immunodeficiency disorders in these individuals and induce high frequencies of micronucleated cells.

Keywords: Kuzbass and Altai regions, Semipalatinsk atomic testing ground, Micronucleus test, Epstein-Barr virus

6.1. Introduction

The Semipalatinsk atomic testing ground (SATG) is situated on the left bank of the Irtysh river. During the time of its existence there have been about 470 nuclear explosions, including 87 air and 26 ground ones. The first and "the most polluting" explosions was carried out in 1949.

For a number of reasons, it was not realised for a long that there was a considerable increase in the number of people with various pathological changes in the areas close to the SATG [1]. Interpretation of the data on explosions of nuclear devices carried out from 1949-1962 allowed us to show that the greater part of radioactive products formed at the moment of testing would fall on the densely populated territory of the Kuzbass

D. B. Peakall et al. (eds.),
Biomarkers: A Pragmatic Basis for Remediation of Severe Pollution in Eastern Europe, 91–99.

and Altai regions, on the areas of intensive agriculture and naturally, would thus enter humans *via* food, drink and inhalation. During the whole period of nuclear testing on the SATG the radioactive products of 50 explosions had been spreading with the predominant winds in the direction of the Kuzbass region. For 44 settlements of 9 districts of the regions the estimate of the reconstructed radiation doses received by the population from the explosion in 1949 exceeds 0.25 gray (Gy)/per year, i.e. it falls outside the internationally established limit of applicability of the "effective dose" criterion. In 1992, our research team surveyed residents of 2 settlements of the Mezdurechensk district of the Kuzbass region. We detected significantly high frequencies of micronucleated erythrocytes in peripheral blood of the Kuzbass region residents (1,2). Shevchenko *et al.* (3) established that some residents of the Mezdurechensk district have high frequencies of chromosome aberrations in blood lymphocytes. The last atomic explosions in the atmosphere on SATG were carried out in 1962. Since the 1970s the background radiation levels have equalled the natural radiation (approximately 0.1 centrigray (cGy)/per year) so that these high frequencies of micronucleated blood cells in the residents are not explicable in terms of radiation effects. However, there is known to be a high infectious virus morbidity among the Kuzbass and Altai region residents. Moreover, the morbidity is principally associated with potentially oncogenic Epstein-Barr virus (4,5). We think that the above cytogenetic instability in the individuals residing in this area may be due to the contamination of local population by potentially oncogenic viruses.

The aim of the present investigation is the comparative assessment between frequencies of micronucleated lymphocytes in peripheral blood and levels of Epstein-Barr virus contamination among the Kuzbass and Altai regions residents living at different distances from the SATG.

6.2. Materials and Methods

We examined 12307 people living in 7 towns situated in the south of the Kuzbass and Altai region covering the area over 1,800,000 sq.km (Figure 1). We examined the population of the following towns

1. Rubtsovsk - 2560 individuals
2. Barnaul - 1196 individuals
3. Mezdurechensk - 2348 individuals
4. Novokusnetsk - 2486 individuals
5. Miski - 1248 individuals
6. Osinniki - 1189 individuals
7. Biyisk (control) - 1280 individuals

We carried out the micronucleus test to screen the levels of cytogenetically altered cells. The micronucleus test, in contrast to the chromosome analysis, allows one to estimate the levels of cytogenetically altered cells in a majority of the population during a comparatively short period of time (6). There have been attempts to evaluate radiation doses for humans, using analysis of the level of micronucleated lymphocytes (7-9).

Blood was drawn and the smears for the micronucleus test were made by a standard technique (7) at a local blood transfusion station. The frequency of micronucleated lymphocytes was determined from 1,000 cells per individual. For control we used the blood samples from inhabitants of the same age at the Biyisk settlement located outside the radiation pollution zone.

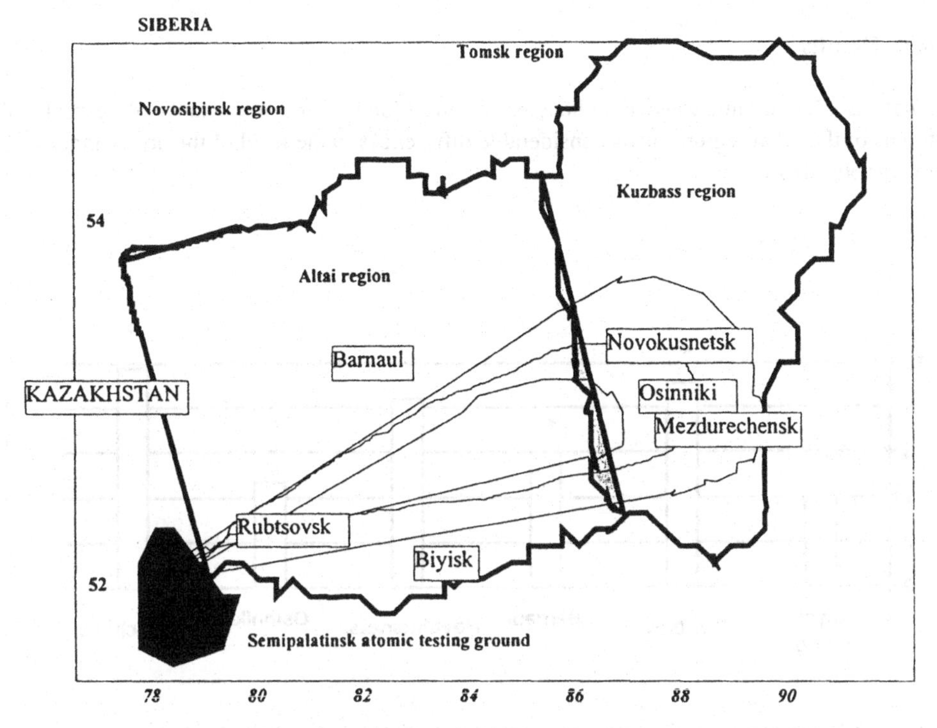

Figure 1. The main direction of the movement of radioactive clouds from the explosions on the Semipalatinsk atomic testing ground and the location of the examined towns on the territory of the Kuzbass and Altai region

In each case, an individual examined had to fill in a questionnaire by means of which we formed groups of people according the district of their residence, place of birth, sex, smoking habit, year of birth, anamnesis of diseases, the presence of children with abnormalities in the family, still-born children, the incidence of cancer, X-ray examinations taken and the length of residence in the locality.

Besides the cytogenetic analysis we also determined the following indices in peripheral blood of the 1360 residents of the Kuzbass and Altai regions: first, the level of serum IgG to the EBV capsid antigen ; second, the level of virus specific IgG to the EBV capsid antigen; third, the level of specific IgG to the complex of the early antigens of the EBV. Methods of analysis were standard (5,10). In the statistical analysis of the results, we took into account suggestions described by Hart and Engber-Pedersen [11]. Statistical analyses of the data were made by Student's "t" test. We used procedures available in the SAS statistical package [12]. The statistical inferences were based on the level of significance p<0.01.

6.3. Results

Analysis of micronucleated lymphocytes in the blood of the inhabitants of several towns of the Altai region shows considerable differences in the level of the above index in various areas.

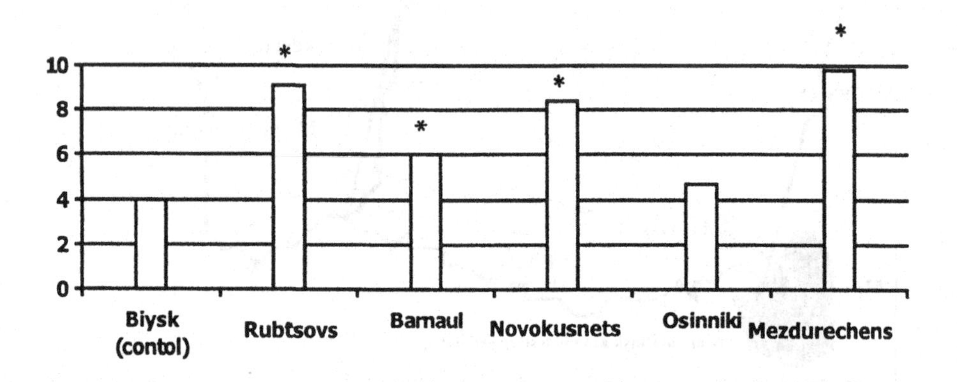

Figure 2. The frequencies of micronucleated lymphocytes in the blood of residents of various towns in the south of the Kuzbass and Altai regions.
*- markers the significantly high frequencies of micronucleated lymphocytes as compare with control individuals

A particularly high level of micronucleated lymphocytes was observed in the people living in Rubtsovsk , and Mezdurechensk (Figure 2).

The results of the analysis are presented in Figure 3. In the control settlement of Biyisk a value of the index was 3.9+_0.6%. In 5 towns (out of 7 examined) the level of the index was much higher them the control one (p<0.01).

It was established that in the migrants who came to live in Mezdurechensk after 1962 or the individuals born by the parents who did not live there from 1949 to 1962, the frequencies of micronucleated lymphocytes were significantly lower than in the indigenous population. We did not notice significant differences in frequencies of micronucleated lymphocytes between males and females (see Table 1). Moreover, the same tendency was detected between a group of individuals whose relatives suffered from cancer and a group without cancer patients in their pedigrees. Nevertheless, the smoking males with cancer pedigrees have the highest frequency of micronucleated lymphocytes (12.8+_1.4%). Among the non-smoking males without cancer patients in their pedigrees this index was 6.4±1.0% ($p<0.01$).

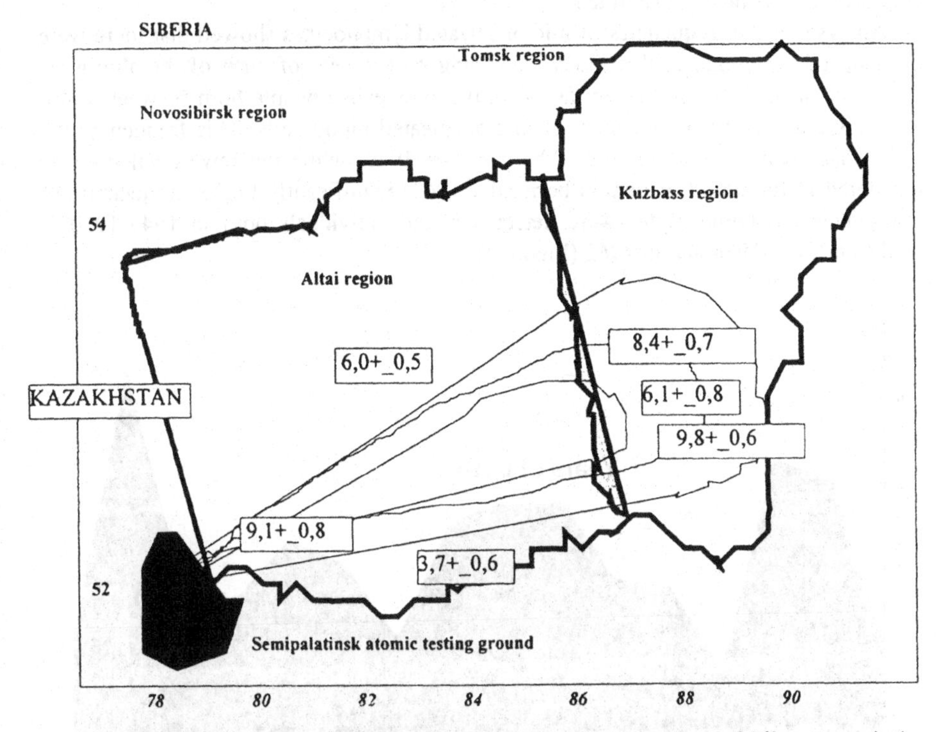

Figure 3. The frequencies of micronucleated lymphocytes in the blood of residents of different towns in the Kuzbass and Altai regions.

TABLE 1. The frequencies of micronucleated lymphocytes in the blood of the donors living in the Mezdurechensk settlement of the Kuzbass region (in %).

Groups of examined donors	Micronucleated lymphocytes +_SE	
Control group (residents of Biyisk)	3.9+_0.6	
All examined residents of Mezdurechensk	9.2+_1.2	P<0.01
Males	10.5+_1.1	P<0.01
Females	9.4+_1.2	P<0.01
Smoking	12.6+_1.6	P<0.01
Non-smoking	7.8+_1.1	P<0.01
Residents with cancer patients in their pedigrees	9.8+_1.0	P<0.01
Residents without cancer patients in their pedigrees	7.0+_0.9	P<0.01
Indigenous population	13.2+_1.3	P<0.01
Migrants who arrived after 1962	3.4+_0.4	P>0.05

However correlations between the presence of abnormal and still-born children, X-ray examinations and the frequencies of micronucleated blood lymphocytes were not detected in the residents examined.

Analysis of the frequencies of micronucleated lymphocytes showed that there were pronounced variations of this index depending on the year of birth of the aboriginal donor (i.e. an individual born in the locality). Indigenous people born between 1949-1962 have a very high frequency of micronucleated blood cells. This frequency was somewhat lower in the individuals born before 1949, while the lowest values were registered in the blood of donors born after 1962. Significantly higher frequencies of micronucleated lymphocytes were detected in the individuals born in 1949-1950 as well as in 1954-1956 and in 1962 (Figure 4).

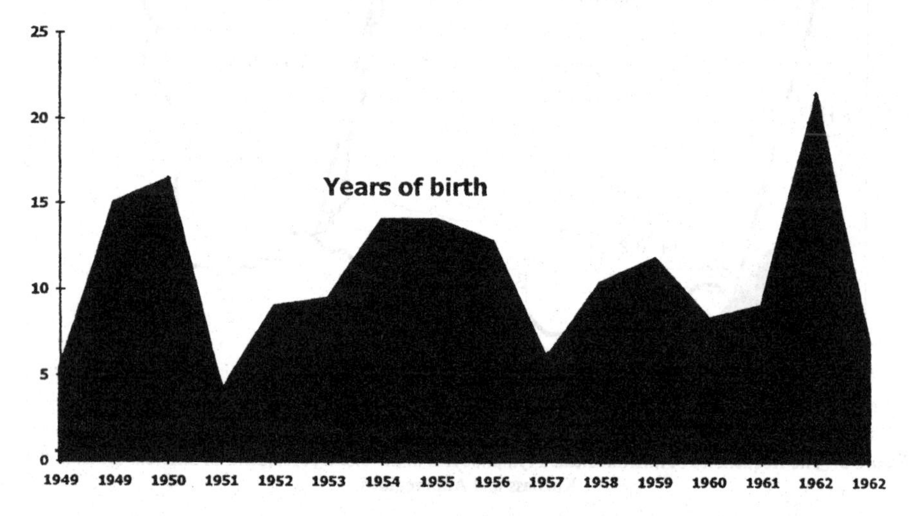

Figure 4. The frequencies of micronucleated lymphocytes in the aboriginals of Mezdurechensk depending on the year of birth.

A similar pattern was observed in the analysis of the frequencies of lymphocytes with micronuclei in the residents of other towns in which we found a significant increase in the above index .

We analysed the frequency of small (less than 3 microns) and large micronuclei in the blood of the inhabitants of Biyisk and Mezdurechensk. In the control settlement of Burla the ratio of the two kinds of the micronucleus was, practically, equal and close to the proportion of 1:1. In Mezdurechensk the frequency of small micronuclei was stongly predominant reaching 89.7%.

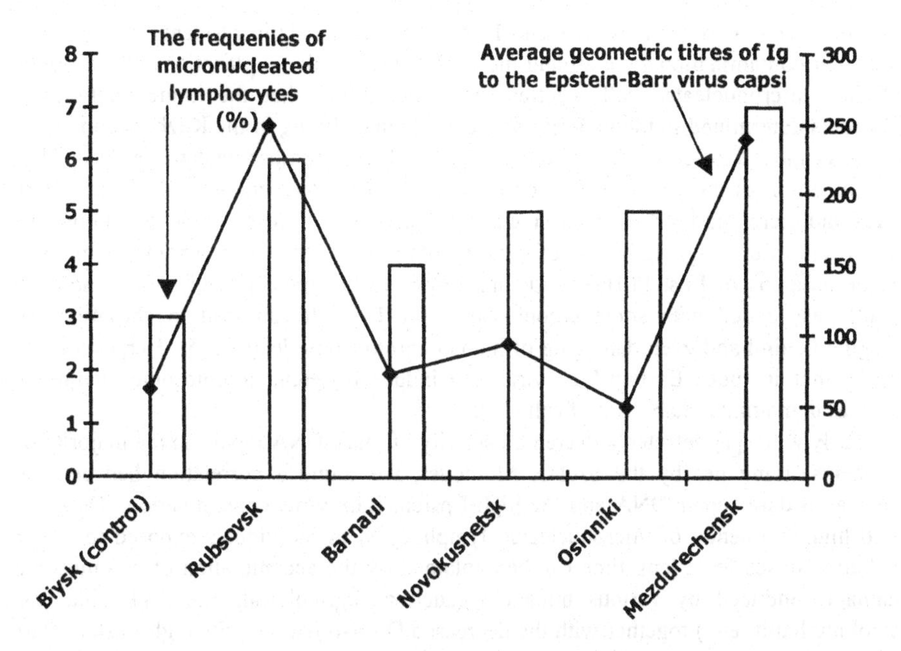

Figure 5. The frequencies of micronucleated lymphocytes in blood and average titres of antibodies to the antigens of Epstein-Barr virus in people residing in the various towns of the Kuzbass and Altai regions

Analysis of the antibody titres to the EBV have demonstrated that all the studied sera were EBV- positive, i.e. all the individuals examined residing in Rubtsovsk and Mezdurechensk were infected with the above virus. The average geometric titres of the antibodies to all studied antigens were much higher compared with those of the healthy population of Biyisk (control) the differences being statistically significant (Figure 5).

The comparison of the indices of virus specific immunity in different age groups showed in all cases for IgG to EBV capsid antigen a tendency for an increase in the antibody level. We also observed the same tendency for high titres in other individuals.

Significantly high frequencies of micronucleated cells were found only in individuals with high titres of IgG to EBV capsid antigen of the EBV . The correlation coefficient in the above case was equal to +0.89 ($p<0.01$).

6.4. Discussion

The results of the present investigation demonstrate that despite a long period of time (more than 30 years) after the last explosion in the atmosphere on the SATG we observe high levels of micronucleated blood cells in the individuals living in towns of the Kuzbass and Altai regions. Early investigations [1] found immunodeficiency disorders such as reductions of T and B lymphocytes levels in residents living in the radiation contamination zone around the SATG. One of the explanations of the origin of the micronucleated cells centres is the possibility that the immunodeficiency disorders determined in the majority of the individuals living in the Kuzbass and Altai regions can lead to some viral infections which induce chromosome damages [13-15].

Our findings show a significant correlation between some antigens of Epstein-Barr virus and increased levels of micronucleated lymphocytes. Micronuclei are known to be a result of the lagging behind of some chromosomes (or their fragments) in mitosis. Some antigens of Epstein-Barr virus are known to be found in cell nucleus and are tightly associated with some chromosomes [16-18]. It can lead to chromosomes lagging behind and even damaging of human chromosome locuses [5]. Herpes viruses group that includes Epstein-Barr virus can induce different chromosome alterations such as damaging metabolism of cell [10].

Early work [1] detected a decreased activity of dark DNA repair in the majority of residents living nearby the SATG. Moreover, we found a correlation between the decreased dark repair DNA and the high Epstein-Barr virus contamination. Thus, the said high frequency of micronucleated lymphocytes in individuals exposed to minor radiation doses for a long time can be explained by the accumulation of chromosome damages induced by various mutagens (such as agricultural, chemical mutagens, smoking habit etc.) together with the decreased DNA-repair and the high Epstein-Barr virus contamination.

Also, the significantly high frequency of micronucleated T lymphocytes can be associated with a considerable industrial pollution of the environment in the south of the Kuzbass region. We suppose that the radiation contamination of the environment have induced the disbalance of genetic homeostasis and DNA repair systems in the body so that any, even the weakest, chemical mutagen can induce an increase in the frequency of micronucleated cells.

References

1. Ilyinskikh, N.N. , Adam, A.M., Novitskii, V.V., Ilyinskikh, E.N. and Ilyin, S. (1995) Mutagenic consequences of radiating pollution of Siberia. Siberian Med. University Press (Russia), p.254.
2. Ilyinskikh, N.N., Eremich, A.V., Ivanchuk, I.I. and Ilyinskikh, E.N. (1997) Micronucleus test of erythrocytes in the blood of the Altai region residents living near the Semipalatinsk atomic proving ground. *Mutation Res.* (in press).
3. Shevchenko, V.A., Snigiryova, G.P., Suskov, I.I., Akayeva, E.A., Elisova, T.N., Iofa, E.L., Nilova, I.N., Kostina, L.N., Novitskaya, N.N., Sidorova, V.F. and Nazins, E.D. (1995) The cytogenetics effects among the Altai region population exposed to ionizing radiation resulting from Semipalatinsk nuclear tests. *Radiation biology. Radioecology* **35**, 588-596.
4. Ilyinskikh, N.N., Eremich, A.V., Ivanchuk, I.I. and Ilyinskikh, E.N. (1996) Micronucleus test of erythrocytes and lymphocytes in the blood of the people living in the radiation pollution zone as a result of the accident at the Siberian chemical plant on April 6, 1993. *Mutation Res.* **361**, 173-178.

5. Ilyinskikh, N.N., Isaeva, T.M., Ivanchuk, I.I. and Ilyinskikh, E.N. (1996) Breaks of chromosomes in the region of oncogenes and circulation of Epstein-Barr virus among the local population that had been subjected to the effects of radiating deposits. *Journal of BUON* **1**, 101-104.
6. Ilyinskikh, N.N., Novitsky, V.V., Vanchugova, N.N. and Ilyinskikh, I.N. (1992): "Micronucleus test and cytogenetic instability", Tomsk University Press (Russia), p.272.
7. Almassy, Z., Krepinsky, A.B., Bianco, A. and Kotels, G.J. (1987) The present state and perspectives of micronucleus assay in radiation protection: A review. *Appl.Radiat.Isol.* **38**, 241-249.
8. Prosser, J.S., Mognet, L.E., Lloid, D.C. and Edwards, A.A. (1989) Radiation induction of micronuclei in human lymphocytes. *Mutation Res.* **199**, 37-45.
9. Littlefield, J.G., Sayer, A.M. and Frome, E.J. (1989) Comparison of dose-response parameters for radiation-induced acentric fragments and micronuclei observed in cytokinesis-arrested lymphocytes. *Mutagenesis* **4**, 265-270.
10. Epstein, M.A. (1985) "Herpesviruses" London: Chapman and Hall, p 337.
11. Hart, J.W. and Engber-Pedersen, H. (1983) Statistics of the mouse bonnemarrow micronucleus test counting, distribution and evaluation of results. *Mutation Res*: **11**, 195-198.
12. SAS Institute Inc. SAS/STAT ™ User's Guide, Version 6. (1989) Fourth Edition. Cary NC: SAS Institute Inc.
13. Nichols, W.W. (1963) Relationship of viruses, chromosomes and carcinogenesis. *Hereditas* **50**, 53-80.
14. Ilyinskikh, N.N., Bocharov, Ye F. and Ilyinskikh, I.N. (1982) "Infection mutagenesis", Novosibirsk: Nauka (Russia), p,185.
15. Ilyinskikh, N.N. , Ilyinskikh, I.N. and Bocharov, Ye F. (1984) "Cytogenetic homeostasis and immunity", Novosibirsk: Nauka (Russia), p.256.
16. Dillner, J. (1986) An Epstein-Barr virus determined nuclear antigen partly encoded by the transformation - associated BAM WYN region of EBV DNA: preferential expression in lymphoblastoid cell lines. *Proc. Nat. Acad. Sci USA* **83**, 6641-6645.
17. Petti, L. and Kieff, E. (1988) A sixth Epstein-Barr virus nuclear protein (EBNA 38) is expressed in latently infected growth - transformated lymphocytes. *J. Virol.* **62**, 2173-2178.
18. Rowe, D.T., Forrel, J.J. and Milled, G. (1987) Novel nuclear antigens recognized by human sera in lymphocytes latently infected by Epstein-Barr virus. *Virology* **156**, 153-162.

7. USING VARIOUS BIOMARKERS FOR THE ESTIMATION OF THE SITUATION AFTER AN ACCIDENT AT SIBERIAN CHEMICAL PLANT

N.N. ILYINSKIKH[1], A.T. NATARAJAN[2], I.I. SUSKOV[3], S.N. KOLYUBAEVA[4], I.I. DANILENKO[5], L.N. SMIRENNII[6], A. Yu. YURKIN[1], E.N. ILYINSKIKH[1]

[1]Siberian Med. Univ., Tomsk, Russia, [2]Dept. Radiat. Genet. & Chem. Mutag., Leiden Univ., Leiden, Netherlands, [3]Institute Gen. Genet. Moscow, Russia, [4]Institute Radiologic, St Petersburg, Russia, [5]Inst. Epidemilogic, Kiev, Ukraine, [6]Research Centre of Spacecraft Radiation Safety, Moscow, Russia

Abstract

On April 6, 1993 near the town of Tomsk (Russia) there was an accident at the Siberian chemical plant (SCP) as a result of which an area of 200 square kilometres was polluted with radionuclides. Especially dangerous was the fallout of the plutonium-239; there were also traces of caesium-137 and cobalt-60. Scientists from Russia, Ukraine and The Netherlands have participated in the investigations of radiation doses. A cytogenetic method and investigation of tooth enamel by the method of electronic spin resonance (ESR), as well as micronuclei test were used to estimate the radioactive doses received by the population. The ESR signal intensity and cytogenetic aberration frequency in lymphocytes of the tooth donors showed a good correlation. The data obtained testify that 15% of the inhabitants of the Samus settlement received a dose of radiation effect exceeding 100 centrigray (cGy). In 87% of these cases there was reasonable agreement between laboratories. The distinctions concerned the results of the examinations of the fishermen where the method of ESR gave high results (80-210 cGy) and chromosomic method and micronuclei gave low ones (8-52 cGy). A large number of cytogenetically aberrated cells were especially observed in the people born between 1961 and 1969. It was found that during these years, serious failures at the Siberian chemical combine occurred causing radiation pollution of the district. The number of cells with cytogenetic aberrations was considerably less in the people arriving in Samus after 1980. Our experience shows the importance of collecting detailed information on the donors from whom samples were taken for radiation dose reconstruction.

Keywords: Siberian Chemical Plant, radionuclides, micronucleus test, tooth enamel.

D. B. Peakall et al. (eds.),
Biomarkers: A Pragmatic Basis for Remediation of Severe Pollution in Eastern Europe, 101–110.

7.1. Introduction

On April 6, 1993 there was an accident at the Siberian Chemical Plant (SCP). As a result an area of 200 sq.km was polluted by different radionuclides [1,2]). The first accident at SCP was in 1961. There have been about 30 accidents since [2]. The most considerable pollution occurred at the settlement of Samus (located 12km from the radiochemical plant of SCP on the bank of Tom river). Also, it was proved that SCP discarded radioactive industrial wastes into the Tom river, therefore the ways to absorb radionuclides into inhabitants body can include not only soil and air but water and fish. According to a state law of the Russian Federation adopted in 1995, people who have received the dose above 50 millisivert (mSv) are entitled to certain privileges and compensations. The ecological committee of the Tomsk region were faced with the problem of estimating the dose that had been received by inhabitants of settlement of Samus as a result of SCP action. It is clear that we cannot determine doses of irradiation of these people by physical methods, because most of the radionuclides had disintegrated in the environment by this time, and the level of radiation now found in this area is not significantly different from the natural one. The Committee asked independent experts in Russia and abroad to determine these doses by retrospective methods of biodosimetry. The analysis of unstable chromosome aberrations is the best validated and most powerful method [3,4,5,6,7,8]. We can also do a micronucleus test [9,10,11,12,13], and ESR of tooth enamel [6,14,15,16,17]. The goal of this work was a collaborative determination of doses received by inhabitants of Samus (by different methods of biodosimetry), who had been exposed to minor doses of irradiation as a result of environmental pollution by radionuclides from SCP.

7.2. Materials and Methods

As many as 77 people, inhabitants of the settlement of Samus, were examined. In no case had individuals been exposed to radiation professionally or medically. These were workers at a shipbuilding yard, seasonal workers, fishermen, farmers, school teachers, secondary and professional school students in this group. Also a group of infants of one to three years old was also examined.

The determination of effective equivalent doses of irradiation (EED) was done by six independent research groups, through analysis of unstable chromosome aberrations, micronuclei in peripheral blood lymphocytes and ESR-spectrometry of enamel. Samples for investigations were obtained between June 6 and July 6 1993, i.e. 2-3 months after the accident at SCP. In the case of inhabitants of the settlement who underwent extraction of molars for medical reasons, blood samples were simultaneously taken for cytogenetic analysis. Extracted teeth were sent to the Research Centre of Spacecraft Radiation Safety, Moscow. The lymphocytic mass (from the blood of the same people) was derived and cultivated in RPMI-1640 medium (Sigma) with phytohaemoagglutinin (Difco) and serum of calf 20%. In a 48-52 hours standard chromosome preparations (with colchicin [18]) and one for micronucleus test (without colchicin, according to Kolubaeva *et al.* [11]) was prepared. The chromosome preparations were sent to Prof. A.T. Natarajan (Leiden, Netherlands), Dr I.I. Suskov (Moscow), Prof. I.I. Danilenko (Kiev, Ukraine) and Prof. N.N. Ilyinskikh (Tomsk) for analysis. The micronucleus test of blood lymphocytes was performed by Dr S.B. Kolubaeva (St Petersburg) and Prof. N.N. Ilyinskikh (Tomsk).

For determination of doses on the base of chromosome analysis an equation was used: $Y=8.7D+3.88D^2$ (where Y = frequency of dicentric and circular chromosomes by 100 cells, and D = dose [7]). To calculate doses the recommendations of Beninson *et al* [3] was taken into consideration. For micronucleus test an equation was used as follows: $Y=0.012+0.025D$, where Y = the quantity of micronuclei by one cell, D = dose [11]. In the micronucleus tests 1,000 lymphocytes were investigated as nearby 100 of one in chromosome analysis. Statistical analysis of the data was performed.

Retrospective dose reconstruction was carried out according to the branch method recommendation "Individual radiation dose determination using tooth enamel ESP spectrometry" developed in Research Centre of Spacecraft Radiation Safety (Russian Federation).

The ESR assay detects radicals (unpaired electrons) caused by radiation exposure and tooth enamel and bones are suited for this purpose. Because bones are continuously remodelled and are not readily accessible, mainly teeth have been used in ESR studies to date. Enamel, the covering of the tooth surface, consists mostly of hydroxyapatite (a compound of crystalline structure consisting of calcium and phosphate) and is free from any metabolism. Tooth enamel is a unique inorganic body structure. Although laboratory studies about ESR have been published since the 1960s, in the past decade ESR has received renewed attention. In Japan, M. Ikeya now at Osaka University has actively promoted ESR studies [14].

For the enamel separation, a disc-shaped diamond cutter was used with running water. This is entirely a mechanical separation with no chemical treatment. The enamel was then crushed in an agate mortar until the grain size was roughly homogeneous (0.5-1.4mm in diameter). A field modulation of 100 kHz in frequency and of 0.32 mT in width was used at a microwave of 5 mW, with a time constant of 0.1 sec and a field sweep of 10 mT in 16 min. The enamel sample and internally located manganese marker were measured simultaneously.

ESR measurements were carried out using X-band ESP spectrometers BRUKER-ER-420 and Radiopan-SE/X-2544 with PC for execution of ESR spectra accumulation program. An additive irradiation of researched samples was performed on gamma irradiation devices "Start" (Cs-137) or of the same type with dose rate of 30 rad/min. The necessary dose of any value was received by variation of time of radiation and the distance between the gamma-source and a sample. ESR-spectrometers and radiation devices were certificated by Gosstandard of Russian Federation. The results of 77 sample measurements were subjected to systematisation and computer treatment using different methods of mathematical statistic with the help of STATGRAPHICS system (version 2.6). The statistical analysis shows that the distribution of probability density is described by Veibull law:

$$f(x) = 1/\lambda * (x/\lambda)^{1-\alpha} \exp(-(x/\lambda)^{\alpha})$$

the parameters of distribution $\alpha = 1.47$ and $\lambda = 43$, mode 12 cGy, median 38 cGy and mean arithmetic value 39 cGy.

χ^2-criterion was used to test the correlation between experimental results and chosen approximation distribution.

In evaluation of irradiation doses obtained by inhabitants of Samus we have taken into consideration the data [2] according to those in this area the natural dose is 0.1 cGy per year.

7.3. Results and discussion

All scientists taking part in this research concluded that most of the inhabitants of Samus had been subjected to considerable irradiation. The abnormally high quantity of lymphocytes with dicentric and ring chromosomes in blood and results of ESR-spectroscopy of tooth enamel led to this conclusion. The elevation in the level of T-lymphocytes with micronuclei in blood confirms it indirectly too. The indices of EED recorded by all three methods significantly ($p<0.01$) correlate with each other ($r = 0.78$ to 0.82).

According to data in Figure 1 from 77 inhabitants examined, 70 have a clear increase of EED indicators in comparison with the value expected in individuals exposed to natural radiation during a lifetime (0.1 cGy per year). All seasonal workers that were absent from Samus at the time of the accident on April 6 1993 had a level of EED the same as in controls. The chromosome analysis made it possible to estimate the dose in a group of infants born in the years between 1990 and 1992 as 10-20 cGy, but ESR-spectrometry, did not show it. Probably, the damage of stem haemopoetic cells in infants leads to persistence of high levels of cytogenetic aberrated cells, whereas for accumulation of changes in enamel one needed a longer period. The ESR-spectrometry method recorded the largest increase of EEC in fishermen but cytogenetic methods did not show this.

The chromosome and micronucleus tests reveal a level of EED in these groups as high as 14.2 ± 0.9 cGy and 24.9 ± 1.0 cGy respectively. The reconstructed dose determined by ESR-spectrometry of enamel was 136.0 ± 3.8 cGy at the same time. We therefore suggest that inhabitants of Samus received a major part of their dose by consuming contaminated fish from the river Tom. Our previous data demonstrates that fish and molluscs in the Tom river basin have different radionuclides in their body (arising from the disposal of radioactive wastes from SCP [1]). In the case of two individual methods revealed a most material elevation of EED indices to 160-200 and 190-300 cGy respectively. The source of such a high level of EED was not established. The data obtained demonstrate that micronucleus tests correlate with chromosome analysis and ESR-spectrometry of enamel when reconstructing dose but in the older generation group (born 1928-1958) this method gives understated results of EED.

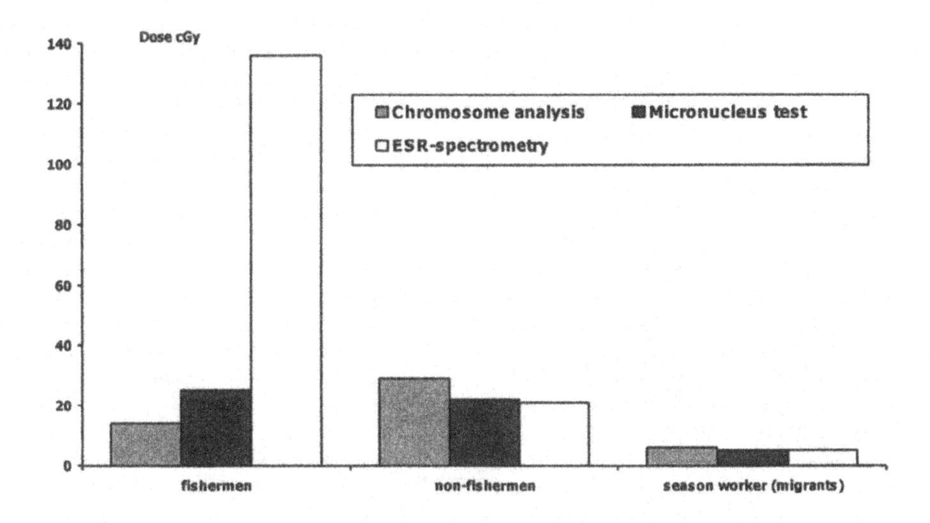

Figure 1. The effective equivalent dose of irradiation in fishermen and other inhabitants of Samus

The chromosome analysis and ESR-spectrometry of enamel testify to the present of a clear gradient - the highest level of EED was recorded in old people. The younger the people examined, the lower the level of EED found. The first radiation accidents at SCP occurred in 1961 and 1963. In connection with this, the accumulation of dose had probably started in 1961. Persons born in 1961 had EED levels practically the same as for individuals born between 1961 and 1963, according to the results of chromosome and ESR-spectrometry methods. Our findings allow us to suggest that a high level of micronucleated lymphocytes can be maintained after a very long time after irradiation when it was in prenatal and early postnatal periods. This appears to be confirmed by data in Figure 2, obtained by the micronucleus test.

The analysis of chromosomal aberration levels shows that dicentric chromosomes often do not have pair fragments. This occurs as a rule when a considerable period of time has passed from irradiation. We suggest that two factors made an important contribution to dose formation in the settlement of Samus - the accidents at SCP from 1961 to 1963 and the consumption of fish by inhabitants. The consequences of the accident in 1993 are less important, but were well registered by chromosome and micronucleus tests in children born between 1990 and 1992. Proceeding from the dose that can be reconstructed by cytogenetic methods in infants it might be possible to conclude that the accident in 1993 caused an irradiation of inhabitants in a dose of the order of 10-20 cGy.-

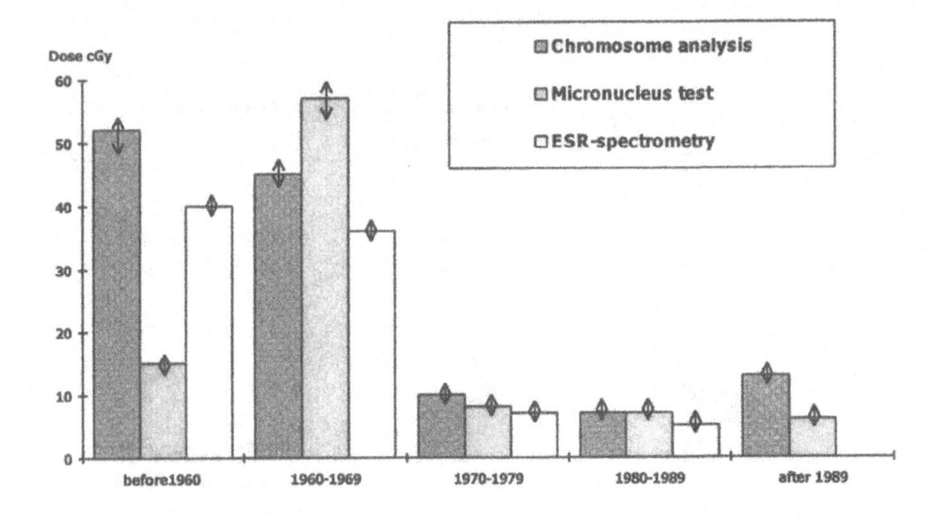

Figure 2. The effective equivalent doses of irradiation of Samus population according to date of birth

The estimation of chromosome aberrations in cultivated lymphocytes of peripheral blood is a universally recognised method of biological dosimetry in cases of even irradiation of body in doses above 0.5 cGy [18]. At the same time, in the case of minor doses or chronic irradiation, different and unsatisfactory data were obtained (in the sense of biological dosimetry) with cytogenetic assays. There is no general agreement about the nature of induction of chromosome aberrations under the influence of doses less than 0.5 gray (Gy). The first point of view arises from a linear concept with no threshold (as a part of general linear - quadratic concept of chromosome aberration under ionising radiation). The second opinion emphasises changes in dose-effect ratios at different ranges of minor doses that give no opportunity to extrapolate the data from higher doses to lower ones [5,6,19]. Nevertheless, the reconstruction of irradiation doses by chromosome analysis was successfully applied to examine a population lived nearby atomic proving ground of Semipalatinsk [8], where the explosions in the air had ceased more than 30 years ago. The high correlation of doses reconstructed was also demonstrated by Nakamura *et al* [15] by simultaneous chromosome and ESR-spectrometry testing of people that had been irradiated after atomic bomb explosion, 50 years ago.

In our opinion, the main cause of discrepancy between the results obtained with the help of model for dose radiation forming of population and direct measurements using tooth enamel ESR-spectrometry is connected with definite limitation of those prerequisites which have been used in model calculations of expected densities of area contamination and population radiation dose. The reconstruction of past events using today's data is not easy. The problem of reconstruction is mathematically described with the help of Bayes theorem. It can be written in such form:

$$P(Hi/A) = \frac{P(Hi)\,P(A/Hi)}{\sum_i P(Hi)\,P(A/Hi)}$$

Here on the left side of equation is the probability of the cause Hi when its sign A is observed. P(A/Hi) is the probability of sign A observation connected with the event Hi; P(Hi) is *a priori* probability of the cause Hi.

The crux of the theorem is that all today's observations P(A/Hi) can only specify the value of the *a priori* probability of the cause P(Hi). Using this theorem one can state that the more multistage process of reconstruction are researched the more *a priori* values of P(Hi/A) that are introduced into the probability of its existence. An increase in the quantity of probable causes Hi considerably decreases both the trustworthiness of P(Hi/A) and the precision of its reproducibility. As to the problem of dose reconstruction by different methods one can say that they have in common *a priori* probability which states that "radiation is the accident at the Siberian chemical plant on April 6 1993" (and there is no other reason). Besides those methods of dose reconstruction (which for example are based on analysis of daughter products of decay or on the comparison of meteorological situations for different times) require an introduction of falls characteristic, an isotope composition of radioactive cloud and decay chains, and involve an introduction of several *a priori* probabilities, that considerably decreases the trustworthiness of these methods in comparison with direct dose measurements.

In view of the applicability of ESR for dating fossil teeth, it seems very likely that tooth enamel can properly accumulate radiation doses imparted at extremely low dose rates. This means that human teeth may be distinctive natural biodosimeters not only for acute exposures such as atomic-bomb radiation but also for repeated small doses or chronic γ-ray exposures such as radiation workers and people residing in contaminated environment. One serious disadvantage of tooth ESR is that the availability of teeth totally depends on chance. In contrast, blood samples are readily available from virtually anybody, and thus cytogenetic examination is superior for dose reconstruction on each individual. The drawback to using chromosome aberration data for dose reconstruction of chronically exposed people is the lack of information on the dose rate factor. In theory, a linear quadratic model predicts that the quadratic component disappears when the dose rate is very reduced. Thus, one may use the fitted linear coefficient of acute dose response curve of chromosome aberration induction obtained *in vitro*. However, it is desirable to collect information on both tooth ESR and chromosome aberration data on the same donors to substantiate the theory. Then, another problem arises: most of the chromosome aberrations remaining in human body many years after the radiation exposure are stable type aberrations. These aberrations are known to require considerable skill for proper detection, and interlaboratory comparison cannot be made without careful standardisation of the technique. Fortunately, recently developed fluorescence-in-situ-hybridisation (FISH) technique is well suited to objective scoring of stable-type aberrations, although it is quite expensive. Last, it is noted that the conventional staining method (i.e. simple Giemsa staining method) can detect as much as 70-80% of stable chromosome aberrations when compared with FISH method [15].

Because the conventional method does not require any expensive reagents and fluorescence microscopes, training of cytogenetic eye is an alternative choice for large scale survey.

We suggest that simultaneous use of different methods of biodosimetry make it possible to determine, with high precision, the doses of irradiation that have been accumulated by man within a long period of irradiation by minor doses, especially when radionuclides are incorporated into the organism.

7.4. Conclusions

Thus, our results in a retrospective reconstruction of equivalent effective doses obtained by inhabitants of Samus (done by six independent research groups through analysis of unstable chromosome aberrations and micronucleus test and ESR-spectrometry of tooth enamel) allow us to conclude that:

1. All findings show that most of the inhabitants of the settlement of Samus received doses of irradiation which exceeded natural levels.

2. The highest doses of irradiation were experienced by fishermen. We suggest that this fact can be explained by consumption of fish contaminated by radionuclides.

3. The data obtained suggest that accidents at SCP within the years 1961 and 1963 must have made a major contribution to the dose received by inhabitants of the settlement of Samus.

4. The accident in 1993 at SCP led to irradiation of inhabitants of the settlement of Samus in a dose 10-20 cGy, that is 100-fold higher than natural dose in that area. (This estimate is based on the dicentric chromosome level and micronuclei in lymphocytes of children born between 1990-1992.)

References

1. Ilyinskikh, N.N. (1995) Radiating ecogenetics of Tomsk area. Tomsk, Siberian Med. Univ., 80pp.
2. Richvanov, L.P. (1994) Environmental state and population health in Siberia Chemical plant zone. State ecological committee of Tomsk region. p.84.
3. Beninson, D., Lloid, D.C. and Natarajan, A.T. (1986) Biological dosimetry: chromosomal aberration analysis for dose assessment. IAEA, Vienna (Technic. Reports Ser. No. 260).
4. Edwards, A.A., Lloid, D.C., Prosser, J.S. (1989) Chromosome aberrations in human lymphocytes - A radiobiological review. In: *Low Dose Radiat.: Biol. Bases Risk Assessment.* London, 423-432.

5. Lloid, D.C., Edwards, A.A. (1983) Chromosome aberrations in human lymphocytes: effect of radiation quality, dose and dose rate. In: *Radiation-induced chromosome damage in man.* New York: Alan R. Liss Inc., 23-49.
6. Lloid, D.C., Edward, A.A., Leonard, A. (1988) Frequencies of chromosomal aberrations induced in human blood lymphocytes by low doses of X-rays. *Int. J. Radiat. Biol.*, **53**(1), 49-55.
7. Piatkin, E.K., Filiushkin, I.V. and Nugia, V. Ju. (1986) The evaluation of evenity of irradiation by man peripheral blood lymphocytes cytogenetic investigation. *Therapevtichesky archiv (Russia)*, **58**, 30-33.
8. Shevchenko, V.A., Snigiryova, G.P., Suskov, I.I., Akayeva, E.A., Elisova, T.N., Iofa, E.L., Nivola, I.N., Kostina, L.N., Novitskaya, N.N., Sidorova, V.F., Nazins, E.D. (1995) The cytogenetic effects among the Altai region population exposed to ionising radiation resulting from Semipalatinsk nuclear tests. *Radiation biology. Radioecology*, **35**(5), 588-596.
9. Almassy, Z., Krepinsky, A.B., Bianco, A. and Kutles, G.J. (1987) The present state and perspectives of micronucleus assay in radiation protection. A review. *Appl. Radiat. Isol.*, **38**(4), 241-249.
10. Fenech, M., Morley, A.A., (1986) Cytokinesis-block micronucleus method in human lymphocytes: effect of *in vivo* ageing and low dose X-irradiation. *Mut. Res.*, **161**, 193-198.
11. Kolubaeva, S.N., Miasnikova, L.V., Raketskaja, V.V., Komar, V.V. (1991) Using of micronucleus test for indication of postradioactive effects in man. Recommendations of the Central Scientific Research Institute of Radiology. St Petersburg, p.14.
12. Littlefield, J.G., Sayer, A.M., Frome, E.J. (1989) Comparison of dise-response parameters for radiation-induced acentric fragments and micronuclei observed in cytokinesis-arrested lymphocytes. *Mutagenesis*, **4**(4), 265-270.
13. Prosser, J.S., Mognet, J.E., Lloid, D.C., Edwards, A.A. (1989) Radiation induction of micronuclei in human lymphocytes. *Mut. Res.* **199**(1), 37-45.
14. Ikeya, M., Miyajima, J., Okajiama, S. (1984) ESR dosimetry for atomic bomb survivors using shell buttons and tooth enamel. *Jap. J. Appl. Rhys.*, **23**(9), 687-699.
15. Nakamura, N., Iwasaki, M., Miyazawa, C., Niwa, K., Akiyama, M., Sawada, S., Awa, A.A. (1996) Radiation dose assessment by electron spin resonance measurement on tooth enamel from atomic-bomb survivors. *Radiation and Risk*, **7**, 253-258.
16. Pass, B., Aldrich, J.E. (1985) Dental enamel as *in vivo* radiation dosimeter. *Med. Phys.*, **12**(3), 305-307.
17. Skvortzov, V.G., Ivannikov, A.I., Eichhoff, U. (1995) Assessment of individual accumulated irradiation doses using EPR-spectroscopy of tooth enamel. *J. Molec. Struc.*, **347**, 321-329.
18. Buckton, K.E., Evans, H.J. (1973) (eds.) *Methods for the analysis of human chromosome aberrations*. WHO, Geneva.

19. Sevankaev, A.V. (1991) The contemporary condition of quantitative evaluation of cytogenetic effects in the range of minor radiation doses. *Radiology,* **31**(4), 600-605.

8. ANIMALS OF THE KATOWICE ADMINISTRATIVE DISTRICT (KAD) POLLUTION IN THE PAST AND PRESENT

KRZYSZTOF DMOWSKI[a], ANDRZEJ KÊDZIORSKI[b], PAULINA KRAMARZ[c]

[a]*Department of Ecology, University of Warsaw, 00-927 Warsaw, Krakowskie Przedmiescie 26/28*

[b]*Department of Human and Animal Physiology, Silesian University 40-007 Katowice, Bankowa 9*

[c]*Institute of Environmental Biology, Jagiellonian University, 30-060 Krakow, Ingardena 6*

Abstract

Very high levels of heavy metals have been found in the Katowice Administrative District (KAD). This problem has been investigated by analyzing vertebrates and invertebrates from the area. These species serve the function of bioindicators. High levels of Pb, Cd and Zn have been found in insects sampled close to zinc smelters and steelworks. Relatively high levels of Cu, Cd and Zn have been reported in spiders from the area and of Cd, Pb and Zn in earthworms. Amongst the vertebrates, analyses of heavy metals has been performed on small mammals, cattle and deer and upon birds. Further evidence was obtained of relatively high concentrations of Pb, Cd and Zn in these species. Also, surprisingly high levels of Tl and Se were reported from certain areas. There is a strong case for the continuation of biomonitoring for metals in this region.

Keywords: Animals, pollution, Katowice, DDT, PCB.

8.1 Introduction

The KAD is a unique region of Poland where numerous emitters of various pollutants have been concentrated within a relatively small area. This gives the opportunity for studies of reaction of individuals, populations and even communities in conditions of extreme pollution. An interesting feature of the KAD is the occurrence of natural and semi-natural patches of ecosystems as well as totally degraded by industrial activity or created artificially thus providing the opportunity of comparative studies of biocoenoses in areas at different stage of man-made transformation.

D. B. Peakall et al. (eds.),
Biomarkers: A Pragmatic Basis for Remediation of Severe Pollution in Eastern Europe, 111–122.

Very severe pollution of this region was noted in the 1960s, and even more in the 1970s. This was due to intensive industrialistion with a philosophy of production at any price and the associated neglect of many indispensable preventive activities. In the same period the studies of contamination of animals began in the KAD which was associated with simultaneous development of advanced analytical methods.

The data quoted below are an attempt to present information on a broad spectrum of groups of animals studied within the KAD. High concentrations of pollutants were found in animal tissues, indicating a very high level of pollution in previous times which still exists to some extent due to a constant but lower emissions and accumulation in soil as well as in sediments

Trace elements, mostly heavy metals, are the main toxicological problem for animals in the KAD. The presence of organic pollutants such as PCBs and pesticides in the KAD environment does not differ basically from other Polish regions. Nyholm *et al.* (1) did not find the presence of HCH in nestlings of great tits (*Parus major*) of the region of the Bukowno zinc smelter, and PCBs levels were similar to those found in samples from control areas in Sweden. Σ DDT was distinctly higher but resulted rather from general presence of DDT derivatives (especially pp'-DDE) in samples from Polish agrocoenoses. Porowska (2) found very low concentrations of HCH residues and DDE but no PCBs in tissues of rodents and shrews from the Miasteczko Slaskie zinc smelter and other Silesian forest regions. Derivatives of DDT, α,β, γ-HCH, dieldrin, endrin, aldrin, heptachlor and mixtures of PCBs were also analysed in liver, muscle and subcutaneous fat of rodents from other KAD smelters and no compounds were found in significantly raised amounts (Lukowski and Mikoszewski, unpublished data).

Altmeyer (3) studied the concentrations of various xenobiotics in eggs of pigeons (*Columba livia domestica*) from German and Polish towns. She found in samples from Katowice very low levels of PCBs, 3 times lower than in samples from Munich and 5-25 times lower than in Saarbrucken. No HCB as well as α, and β-HCH were found in eggs from Katowice.

8.2. Invertebrates - bioindicators of trace element pollution

8.2.1. TERRESTRIAL ECOSYSTEMS

8.2.1.1. *Insects*

The problem of contamination of bees has gained a good deal of attention in the KAD for many years on account of the economic value of these insects as well as their significance in the correct functioning of ecosystems. In the 1970s up to 200-600 μg/g Pb d.w., 6-20 μg/g Cd and 150-600 μg/g Zn were found in bodies of bees living close to some zinc smelters and steelworks of the KAD. Dead bees from the vicinities of certain zinc smelters were also found to have high concentrations of Cu and As (4). At that time many zinc and lead smelters of the KAD were forced to pay financial compensations due to poisoned foragers and the destruction of swarms (managers of the Miasteczko Slaskie zinc smelter; oral information). Further studies of bees and their products (5) also confirmed the accumulation of heavy metals by foragers operating in variously polluted sites. Relatively low concentrations were found in honey (except of Cd in few cases). According to the authors a pollinator can serve as a fairly good bioindicator.

In the 1970s concentrations of Pb, Cd, Zn in various invertebrates from surroundings of zinc smelters were many times higher than in invertebrates inhabiting similar environments of other Polish regions. The species composition was poorer and the number of invertebrates captured was lower (6,7). The diminishing emissions between 1970s and 1980s were reflected in the heavy metal content in terrestrial invertebrates. For example the concentrations of heavy metals in herbivorous orthopterans and coleopterans from the most polluted protection zone of the smelter "Miasteczko Slaskie" decreased between 1975 and 1987: for Pb from ~ 230 µg/g d.w to ~ 55 µg/g, for Cd from ~ 12 µg/g d.w to ~ 2 µg/g, and for Zn from ~ 850 µgg d.w to ~ 270 µg/g (6,8). Probably the most contaminated insects in the region were ants. The concentrations in 1987 in foragers were still extremely high : ~80 µg /g d.w Cd and ~1650 µg /g d.w Zn (8). Similarly high heavy metal concentrations were also reported from other polluted areas by other workers (9,10) and were considered to be derived from the honeydew of aphids. Migula and Glowacka (11) showed that the cumulation depended on the activity of ants outside nest. The highest levels were found in foragers of the KAD returning to nests, while the lowest level were in insects staying inside.

The polluted Silesian regions were also used as experimental plots for examinations of the important ecotoxicological topic of accumulation of heavy metals in trophic chains (12,13,14). Studies of forest-litter invertebrates from an area of the Bukowno zinc smelter and other Polish regions found (15) the highest concentrations of Pb and Cd in *Staphylinidae* (*Coleoptera*) and representatives of other arthropod groups - arachnids and chilopods. The fact that carnivorous carabids accumulated only relatively small amounts of heavy metals despite a high pollution in the Bukowno region thus indicating that some heavy metals are not biomagnified. Similar contents of Zn and Pb in carabids from the KAD as in herbivores were also noted (16).

The studies of Czylok *et al.* (17) provide an example of bioindication studies on insects of the KAD at the community level where it was attempted to classify Upper Silesia and Kraków region for the levels of pollution on the basis of aphids communities on common birch trees (*Betula pendula*). The authors assumed that the fewer aphid species occurring on the trees, the more polluted is the environment. They drew a map showing the location of variously contaminated areas and concluded that it is possible to use such communities as biomarkers of general degradation of the environment, but not for qualitative assessment of the pollution.

8.2.1.2 *Arachnids*

In recent years spiders attracted attention as possible bioindicators of pollution (18). Species-specific and site-specific significant differences in metal concentrations were studied among orb-web *Araneus diadematus*, *A. marmoreus* and *Meta segmentata*, sheet-web *Linyphia triangularis* and wolf-ground *Pardosa sp.* collected in the KAD areas (19). In the spiders analysed there were relatively high levels of zinc (300-500µg/g) and copper (40 - 95µg/g). The highest Cd concentration (~25 µg/g d.w) were found in *A. diadematus* collected in the vicinity of the „Katowice" steelworks . High concentrations of the metals in spiders from these steelworks were also confirmed by Wilczek and Migula (20). Wandering wolf-spiders were more heavily loaded with cadmium than the web-spinning species and Cd body burden was positively correlated with the metal content in dust particles.

Oribatid mites (*Acari, Oribatida*) communities were studied for many years in the KAD (21,22). In degraded areas there was a high percentage of omnivorous phytophages and species with wide range of tolerance to environmental factors.

8.2.1.3 *Earthworms*

Elevated concentrations of heavy metals were reported in tissues of earthworms, mostly due to feeding habits (detrivores) and permanent contact with soils (6,23,24). Migula *et al.* (14) found in earthworms 10-100 µg/g d.w. Cd, 10-40 µg/g d.w Pb and 150-1000 µg/g d.w Zn, depending on the pollution of the site .In earthworms collected in soils along the main roads crossing the KAD average lead concentrations ranged between 4 - 23.5 µg/g (fresh tissue), zinc concentrations ranged between 66.4 - 338.3 µg/g f. w. and the levels both elements positively correlated with the metal content in the soil (25).

8.2.2. INVERTEBRATES OF FRESHWATER ECOSYSTEMS

The KAD region is devoid of natural lakes, but there are reservoirs, sinkhole ponds and sand pits. Studies on rotifers in 1980s revealed that in almost all reservoirs there were communities typical for eutrophic waters but species diversity was relatively low (26). Even in heavily polluted rivers many rotifer species were found (Bielanska-Grajner, unpublished).

Even 15-years studies on freshwater snails did not allow the selection any species that could be a bioindicator of environmental changes. However, during that time a gradual decrease of species diversity in the water bodies was observed. Out of 20 species recorded in late 70s only 14 were found ten years later (27). On the other hand, species that originally were not common at all, increased significantly in number becoming dominant species. One of them is *Potamopyrgus jenkinsi*, an exotic snail which can easily colonise aquatic environments unfavourable for other snails (27).

The installation of many new facilities for flue gas treatment and sewage treatment after 1989 has caused significant amelioration of water quality in the KAD. Nevertheless until the late 1980s the reservoirs in the region were extremely polluted for example those surrounding the Miasteczko Slaskie" zinc smelter (28). In reservoirs situated 5-10 kms from the smelter the concentrations of Pb and Cd in adult and immature insects were only slightly above controls. The larva of *Ephemeroptera* and *Chironomidae* always accumulated the highest levels because of permanent contact with polluted sediments. *Ephemeroptera* from the pond nearest to the smelter (Zyglinek) contained even up to 1080 µg/g d.w Pb and ~60 µg/g Cd. Very high pollution of this site was confirmed in the same year by studies of tadpoles (8).

8.3. Vertebrates - bioindicators of trace elements pollution

8.3.1. DOMESTIC AND GAME ANIMALS

The studies of Zmudzki (29) documented in the 1970s that Pb and Cd levels in tissues from the KAD could cause sub-clinic intoxication. The bad condition of cattle near the Bukowno zinc smelter as well as high concentrations of heavy metals in blood and hairs of cows and dogs from the region were described by Wachalewski *et al.* (30).

Enhanced amounts of harmful elements were found in the milk of cows (e.g. Cr, Pb, Zn) in the surroundings of the "Katowice" steelworks; (31). The calculated uptake of Pb by the animals was 1-20 times higher than the maximum uptake recommended for men. The studies of feed (mostly grasses) carried out in 29 regions of KAD in 1986-87 revealed serious contamination in the areas of: Piekary Slaskie, Bobrowniki, Bêdzin, Siemianowice Slaskie, Tarnowskie Gory, Trzebinia and Bukowno (32). All these areas are affected by emissions from non-ferrous metal industry. The growing of grass (for hay) and vegetables is inadvisable in these areas in the opinion of the authors. Domestic animals of short breeding cycle (pigs, poultry) and raised in a half-open system are recommended.

Studies of deer antlers has enabled assessment of the changes of pollution in Silesian forests in previous decades. Higher Pb and Cd concentrations in roe deer (*Capreolus capreolus*) antlers were shown in the period 1951-1973 in comparison to 1938-1950 (33,34) and analyses of red deer (*Cervus elaphus*) antlers have provided evidence for a constant increase of pollution. Sawicka-Kapusta and Zych (35) found that the levels of these two metals were much lower in 1975-79 than those in 1985-88. When compared with the samples from Great Britain it turned out that antlers of deer from Upper Silesia contained up to 18 times more cadmium and 30 times more lead. Nevertheless cases of poisoning of game animals were more frequent earlier - in the 1960s and 1970s (36).

8.3.2. BIRDS

The quality of the environment of the KAD was also reflected in heavy metal content in internal tissues and feathers of birds living permanently or breeding in the region. There were studies of the house sparrow *Passer domesticus* in the

Rybnik coal region (37), great tits *Parus major* in Jaworzno area (38), moorhens *Gallinula chloropus* (39) and pheasants *Phasianus colchicus* in Siemianowice, Jaworzno and Bytom areas (40), as well as many passerine birds in surroundings of the Miasteczko Slaskie zinc smelter (6,41). The last contained very high levels of heavy metals in bodies, feathers and faeces. Dmowski (41) found that these levels decreased distinctly in breeding birds with distance from the zinc smelter.

Breeding success of great tits in the region of the Bukowno zinc smelter was very low (42,43). The percentage of dead nestlings varied from 40-62, i.e. was 3-4 times more than in unpolluted regions. An experiment with starlings *Sturnus vulgaris* captured in unpolluted site, exposed for 8 days in the zinc smelter area and fed with local food had shown that nomadic or migratory passerine birds in extremely polluted areas may accumulate amounts of Pb and Cd dangerous to birds and their potential consumers (44). Particularly marked enhancement of lead content was noted in livers, brain and kidneys, as well as in the cadmium content in kidneys and livers.

Much significant data on heavy metal pollution of the Upper Silesia region in comparison to other Polish areas were obtained through studies of magpie *Pica pica* tissues, mostly with external tail feathers but also internal tissues - livers and kidneys. This species shows residential behaviour and can accumulate trace elements within a relatively small individual territory. The magpies were caught during the period 1989-1993 in more than 20 Polish regions with various types of industry and urbanisation (45,46). Twenty elements were determined (Pb, Ba, Cd, Zn, Mn, Fe, Hg, Tl, Sb, Sn, Mo, Rb, Se, As, Ge, Cu, Ni, Co, Cr, V) with inductively coupled plasma mass spectrometry (ICP-MS). Among the studied areas were the surroundings of the 3 largest Polish zinc smelters - Szopienice (SZO), Miasteczko Slaskie (MSL) and Bukowno (BUK). In the SZO and MSL areas the magpies were caught twice - in 1989 and 1993. Greatly enhanced concentrations of some elements, especially Pb, Cd, Zn, Sn, Sb, Ge were found in the feather samples from the vicinities of all the studied zinc smelters. These were at least a dozen and sometimes scores of times higher than those in the samples from other industrialised areas, so far considered as seriously polluted by heavy metals in Poland. The levels of lead and cadmium were particularly high (Table 1). Greatly enhanced concentrations of selenium and copper in SZO (due to processing of rolled sheet copper), and thallium in BUK were found.

If the areas surrounding the zinc smelters from the KAD are recognised as the most polluted with heavy metals in Poland, worst case of all is the Szopienice area. The concentrations of many elements were the highest there (Pb, Cd, Zn, Se, Cu, Ge, Sn, Sb), especially in 1989.

TABLE 1. Content of lead and cadmium in tail feathers R_6 of magpies from the KAD towns and smelter areas compared with data from other polluted and unpolluted Polish areas (mean values; µg/g d.w) (45,55)

	Pb	Cd
KAD towns		
Zabrze	66.0	3.05
Katowice	228,0	4.78
Chorzow	169.2	9.53
Bytom	190.4	10.55
KAD smelters		
Miasteczko Slaskie 1993	305.4	7.4
Miasteczko Slaskie 1989	1002.0	44.0
Szopienice 1993	463.4	23.47
Szopienice 1989	1282.8	66.0
Bukowno	1301.9	22.5
other Polish areas		
Glogow - copper plant area	229.3	1.1
Krakow - steelworks area	61.3	1.2
Warszawa - steelworks area	181.9	2.7
Plock petrol refinery area	51.5	1.3
Belchatow - brown coal power plant area	55.1	0.8
Bialowieza - national park area	20.1	0.4

An installation of more effective filters and a distinct decrease of production in Miasteczko Slaskie zinc smelter in the early 1990s caused a notable reduction of elements in the samples from 1993 in comparison to 1989. The concentrations of Pb, Zn, Cd. and Sn in the samples of magpies captured in 1993 were also significantly reduced (Table 1) in comparison to the feathers collected in 1989. There were probably similar reasons for the differences between the samples from Szopienice in 1989 and Szopienice in 1993. Although the emissions have became lower in SZO and MSL between 1989 and 1993, the quantities of heavy metal accumulated in soils and transferred through food chains were at the same level. This was indicated by unchanged or even increasing contents of Pb and Cd in livers and kidneys (Dmowski; unpublished data).

The very high levels of thallium in the feather samples from the SZO area was surprising. Woodland rodents were caught in 1994-95 at various distances from the "Bukowno" smelter in order to eliminate a possibility of incidental contact with rat-poison (containing Tl) in villages as a cause of feather contamination and to assess the possibility of long-distance thallium emissions by smelter chimneys in the direction of the Upper Silesia. An increasing content of Tl in livers and kidneys in sub-populations of rodents allowed a source of thallium to be found [47]. It was 70m-mine dump which contained floatation tailings. The particles with Tl compounds can easily be spread from there all over the neighbourhood as a drying dust. The presence of extremely high thallium amounts was confirmed by analyses of the tailings (Institute of Geology PAS; unpublished data) and the presence in the region of rodents with rear half of the body entirely hairless [47], one of the symptoms of thallium poisoning is progressive baldness. Moreover the analyses of thallium in magpie kidneys and livers have revealed its presence not only in Bukowno but also in Szopienice and its permanent presence there between 1989 and 1993. This is another example of the extremely serious contamination in the vicinity of the Szopienice zinc smelter.

8.3.3. SMALL MAMMALS

Small mammals belong to one of the most studied groups of vertebrates in the KAD. The maximum concentrations of cadmium were found in livers and kidneys of common shrews *Sorex araneus* from the Bukowno area (107.8 and 70.8 mg/g d.w respectively (48) and livers of the same species from the Miasteczko Slaskie area 127.8 on average (49). The highest lead concentrations were noted in bones and kidneys of common shrews, 97.4 and 38.7 respectively (50). Bank voles *Clethrionomys glareolus* were 2-3 times less loaded with heavy metals.

The rodents and insectivorous mammals studied in the other KAD regions had relatively small but still enhanced concentrations of heavy metals in tissues - e.g. bank voles from the Katowice steelworks area (24), bank voles from the Rybnik Coal Region (51), yellow-necked mice *Apodemus flavicollis* and bank voles from the Dulowa Forest (52,53). Dmowski (47) found very dangerous amounts of thallium accumulated bank voles in the region affected by dusts from huge mine dump gathering floatation failings in the Bukowno region - in kidneys 20.6 and in livers 4.8 mg/g d.w.

The rapid rotation of generations of small mammals enables observations to be on the fluctuations of populations and communities with a short period of time. The influence of the environmental quality on the ecological characteristics of the mammals was not clear in moderately polluted regions of the KAD, but in extremely polluted regions the effects were very significant (49). The densities of small mammals in the area around the Miasteczko Slaskie zinc smelter were clearly low (e.g. that of bank vole (3 ind./ha in comparison to (50 ind./ha in unpolluted regions). The age structure of the studied bank vole population was completely reversed in polluted regions with the highest percentage of old individuals and the lowest percentage of juveniles. The high number of reproducing females (about 75%) indicates that small share of juveniles in population has been mostly caused by high mortality rate of suckling and/or weaning individuals due to disturbances of maternal behaviour. Similar changes of ecological characteristics of small mammal populations were observed also in the Bukowno area (47). In such conditions of so-called "sink habitats" (54) these populations are probably in constant danger of extinction and their continued existence around zinc smelters is strongly dependant from immigration of surplus individuals, so called "sink habitats".

8.4. Conclusions

The results presented here provide evidence that despite some decrease of the emissions in the KAD region, constant biomonitoring must be continued there with necessary support from biomarker studies. Particularly the surroundings of the Silesian zinc smelters and steelworks require special treatment and routine multielemental control, taking into account the longevity and severity of this pollution and the proximity to a large human population.

References

1. Nyholm, N., Sawicka-Kapusta, K., Swiergosz, R., and Laczewska, B. (1995) Effects of environmental pollution on breeding populations of birds in southern Poland. *Water, Air and Soil Poll.* **8**, 829-834.
2. Porowska, A. (1995) Contamination of Bank vole (Clethrionomys glareolus) and shrews (*Sorex* sp.) with chlorinated hydrocarbons in two chosen areas of Poland. M.Sc Thesis, 52 pp, Warsaw University.
3. Altmeyer, M. (1993) Biomonitoring mit Stadttaubeneiern zur Erfassung von Chemikalien und deren Wirkungen in Verdichtungsraumen, Dr Thesis, 300 pp, Univ. des Saarlandes.
4. Jêdruszuk, A. (1986) Honey bees and their products as indicators of environmental pollution, *Med. Wet.* **43**, 352-356 (in Polish).
5. Migula, P., Binkowska, K., Kafel, A., Kedziorski, A., and Nakonieczny, M. (1989) Heavy metals contents and adenylate charge in insects from industrialized region as indices of environmental stress. in J. Boha and V. Ruzicka (eds.) Proc. Vth Int. Conf. Bioindicatores deteriorisationis Regionis. Ceské Budějovice, p. 340-349.
6. Dmowski, K. and Karolewski, M. (1979) Cumulation of zinc, cadmium and lead in invertebrates and in some vertebrates according to the degree of an area contamination., *Ekol.Pol.* **27**, 333-349.
7. Chlodny, J., Matuszczyk, I., Styfi-Bartkiewicz, B., and Syrek, D. (1987) Catchability of the epigeal fauna of pine stands as a bioindicator of industrial pollution of forests, *Ekol. Pol.* **35**, 271-290
8. Dmowski, K., Dobrowolski, K. and Karolewski, M. (1991) Extremely high levels of heavy metals in the neighbourhood of the zinc smelting works "Miasteczko Slaskie" - summary of 15 tears studies - Abstracts, papers, posters of Int. East-West Symp. on Contaminated Areas in Eastern Europe; Origin, Monitoring, Sanitation. Gosen/Berlin Nov. 1991.
9. Fangmeier, A., and Steubing, L. (1986) Cadmium and lead in the food web of a forest ecosystem in H.-W. Georgii (ed.) Atmospheric pollutants in forest areas. Kluwer Acad. Publ., Dordrecht
10. Stary, P., and Kubiznakova, J. (1988) Content and transfer of heavy metal air pollutants in populations of Formica spp. wood ants (*Hym.,Formicidae*), *J. Appl. Ent.* **104**, 1-10

11. Migula, P., and Glowacka, E. (1996) Heavy metals as stressing factors in the red wood ants (*Formica polyctena*) from industrially polluted forests, *Fresenius J. Anal. Chem.* **354**, 653-659.
12. Laskowski, R., and Maryanski, M. (1993) Heavy metals in epigeic fauna: trophic-chain and physiological hypotheses, *Bull. Env. Cont. Toxicol.* **59**, 232-240.
13. Luczak, J. (1984) Spiders of industrial areas, *Pol. Ecol. Stud.*, **10**, 157-185.
14. Kramarz, P. (1996) Biomagnification or detoxification - methods of regulation of heavy metal levels in epigeic invertebrates, *Wiad. Ekol.* **42**, 79-89 (in Polish).
15. Niklińska, M., Maryański, M., and Kramarz, P. (1995) Heavy metal concentrations in forest-litter invertebrates of different trophic levels and taxonomic positions - *Acta Biol. Debr. Oecol. Hung.* **5**, 265-277.
16. Migula, P., Dolezych, B., Kielan, Z., Laszczyca, P., and Howaniec, M. (1990) Responses to stressors in animals inhabiting industrially polluted areas (in Polish) in S. Godzik (ed.) Zagroaenia i stan rodowiska przyrodniczego rejonu Slasko-krakowskiego. Wydawnictwo SGGW-AR, Warszawa, pp: 108-129.
17. Czylok, A., Herczek, A., Klimaszewski, S.M., Wojciechowski, W. (1990) Associations of aphids in birch (*Betula pendula*) as a marker of environmental quality in areas of industrial influence of the Silesia and Krakow agglomerations in S. Godzik (ed.) *Hazards and state of the environment of the Cracow-Silesia Region.* 130-135. CPBP 04.10.SGGW AR Warszawa nr 62 (in Polish).
18. Marczyk, G., Migula, P., and Trzcionka, E. (1993) Physiological responses of spiders to environmental pollution in the Silesian Region (Southern Poland), *Sci. Total Environ. Suppl.*, **2**, 1315-1322.
19. Maryanski , M., Niklinska, M., Kramarz, P., Laskowski, R., and Nuorteva, P. (1995) Zinc and copper concentrations in litter invertebrates of different trophic levels and taxonomic positions, *Arch. Ochr. Srodowiska* **3-4**, 207-212.
20. Wilczek, G., and Migula, P. (1996) Metal body burdens and detoxifying enzymes in spiders from industrially polluted areas, *Fresenius J. Anal. Chem.* **354**, 643-647.
21. Skubala, P. (1995) Moss mites (*Acarina: Oribatida*) on industrial dumps of different ages, *Pedobiologia* **39**, 170-184.
22. Skubala, P. (1997) The structure of oribatid mite communities (*Acari: Oribatida*) on mine dumps and reclaimed land, *Zool. Bietr. N.F.* **38**, 1-22.
23. Pytasz, M., Wielgus-Serafinska, E., and Kawka-Serwecinska, E. (1980) Histochemical localization and cumulation of zinc and lead in selected invertebrates from the ironwork "Katowice" region, *Acta Biologica (Katowice)* **8**, 50-56.
24. Jethon, M. (1983) Physiological functions of animals in conditions of changed environment. Content of zinc and lead in tissues of selected animal species from the "Katowice" steelworks region Report 1983 of the Inst. Environ. Engineering, grant 07.01.07.00 (in Polish).

25. Falkus, B. and Pytasz, M. (1980) The cumulation of zinc and lead in bodies of earthworms from the roadside areas in Upper Silesian Region, *Acta Biologica (Katowice)* **8**, 57-72.
26. Bielańska-Grajner, I. (1987) The comparison of rotifer communities (*Rotatoria*) in various types of reservoirs within Upper Silesia, *Przeglad Zoologiczny* **31**, 37-47 (in Polish).
27. Serafinski, W., Strzelec, M.,and Czekaj, D. (1993) Changes in the freshwater snail fauna in an over-industrialized area in Poland, *Walkerana* 7, 11-13.
28. Weigle, A. (1990) Contamination of chosen waterbodies with heavy metals (P and Cd) in a zone affected by the zinc smelting work "Miasteczko Saskie" (in Polish), M.Sc. thesis, Biol. Fac., Warsaw Univ.
29. Zmudzki, J. (1978) Content of cadmium, zinc, copper and iron in tissues of farm animals with special regard to agricultural and industrial regions, Ph.D. thesis, Inst. Vet., Pulawy.
30. Wachalewski, T., Fertig, and Kawecki, J. (1978) Environmental hazard in regions of increased emission of zinc and lead compounds, *Arch. Ochr. Srod.* **2**, 115-134.
31. Kolczak, T. (1989) Contamination of plant feed and animal products with heavy metals in the region of the protective zone of the Huta Katowice steelworks, *Arch. Och. Srod.* **3-4**, 91-112 (in Polish).
32. Kucharski, R., Marchwinska, E., Piesak, Z., Nikodemska, E., and Wita a, B. (1988) Contamination of forage plants with lead and cadmium in chosen regions of Katowice voivodeship, *Med. Wet.* **45**, 162-166 (in Polish).
33. Sawicka-Kapusta, K. (1978) Estimation of the contents of heavy metals in antlers of roe deer from Silesian woods, *Arch. Ochr. Srod.* **1**, 107-121 (in Polish).
34. Sawicka-Kapusta, K. (1979) Roe deer antlers as bioindicators of environmental pollution in southern Poland, *Environ. Pollut.* **19**, 283-289.
35. Sawicka-Kapusta, K., and Zych, J. (1991) Heavy metals concentrations in red deer antlers from Southern Poland, Proc. 8th Int. Conf. on Heavy Metals in the Environment, Edinburgh **2**, 298-301.
36. Kozyra, J. (1970) Foreseen changes in now-a-day structure of forest districts: Typescript report from the Zinc Smelter "Miasteczko Slaskie" (in Polish) Brynica, Swierklaniec, Zyglinek .
37. Pinowska, B., Krasnicki, K., and Pinowski, J. (1981) Estimation of the degree of contamination of granivorous birds with heavy metals in agricultural and industrial landscape, *Ecol. Pol.* **29**, 137-149.
38. Sawicka-Kapusta, K., Kozlowski, J., and Sokolowska, T. (1986) Heavy metals in tits from polluted forests in southern Poland, *Environ. Pollut.* **42**, 297-310.
39. Betleja, J., Cempulik, P., Kwapuliński, J., Lebedeva, N., Loska, K., and Wiechula, D. (1993) Ecotoxicological characteristics of the winter habitat of the moorhen (*Gallinula chloropus*), *Pollutants in Environments* **3**, 142-147
40. Swiergosz, R. (1993) Heavy metal distribution in the tissues of pheasants *Phasianus colchicus* - Proc. IU Game Biologists. , Halifax, Canada **2**, 14-19.

41. Dmowski, K. (1985) Dynamics of Passerine birds in a lake littoral zone and their role in transport of heavy metals between ecosystems, PhD Thesis, Warsaw Univ. (in Polish).
42. Sawicka-Kapusta, K. (1995) Uptake and concentrations of lead and cadmium and their effects on birds and mammals, *Arch. Ochr. Srod.* 39-52.
43. Sawicka-Kapusta, K., Laczewska, B., and Kowalska, A. (1995) Heavy metal concentrations in great tit nestlings from polluted forest in southern Poland, Proc. 10 Int. Conf. Heavy Metals in the Environment, Hamburg, Germany **1**, 244-247.
44. Dmowski, K. (1993) Lead and cadmium contamination of passerine birds (Starlings) during their migration through a zinc smelter area, *Acta Orn.* **28**, 1-9.
45. Dmowski, K. (1995) Feathers of Magpie (*Pica pica*) as a bioindicator material for an assessment of heavy metal pollution. Report for the Commitee of Scientific Research (KBN) (in Polish).
46. Dmowski, K. (1997) Biomonitoring with the use of Magpie *Pica pica* feathers: heavy metal pollution in the vicinity of zinc smelters and national parks in Poland, *Acta Oorn.* **32**, 15-23.
47. Dmowski, K., Kozakiewicz, A., and Kozakiewicz, M. (in press) Small mammal populations and community in conditions of extremely high thallium contamination in the environment, *Ecotox. Env. Safety* **40.**
48. Sawicka-Kapusta K., Zakrzewska M., Kowalska A., and Stachowicz, A. (1995) Heavy metals in small mammals from different polluted forests in southern Poland, Abst. 2 Eur. Congress of Mammalogy, Southampton, England
49. Dmowski, K., Kozakiewicz, A., and Kozakiewicz, M. (1995) Ecological effects of heavy metal pollution (Pb, Cd., Zn) on small mammal populations and communities, *Bull. Pol. Acad. Sci.* **41**, 1-10.
50. Sawicka-Kapusta, K., and Stachowicz, A. (1993) Heavy metals in small mammals from contaminated forest in southern Poland, Abst. SETAC World Congress - Lisbon, Portugal.
51. Wlostowski, T. (1987) Heavy metals in the liver of Clethrionomys glareolus (Schreber,1780) and Apodemus agrarius (Pallas, 1771) from forests contaminated with coal-industry fumes, *Ekol. Pol.* **35**, 115-129.
52. Sawicka-Kapusta, K., Gorecki, A., and Lange, R. (1987) Heavy metals in rodents from polluted forests in southern Poland, *Ekol. Pol.* **35**, 345-354.
53. Sawicka-Kapusta, Swiergosz, R., and Zakrzewska, M. (1990) Bank voles as monitors of environmental contamination by heavy metals. A remote wilderness area in Poland imperilled, *Environ. Pollut.* **67**, 315-324.
54. Pulliam, H. (1988) Sources, sinks, and population regulation, *Am. Nat.* **132**, 652-661.
55. Dmowski, K., and Golimowski, J. (1993) Feathers of the Magpie (*Pica pica*) as a bioindicator material for heavy metal pollution assessment, *Sci. Total Env.* **139/140**, 251-258.

9. THE USE OF BIOMARKERS IN HAZARD ASSESSMENT

D.B. PEAKALL
17 St Mary's Road
Wimbledon, London SW19 7BZ, UK

Abstract

Definitions of biomarkers have varied considerably and have been frequently so broad as to include any biological change caused by chemical agents. Recently the concept that this term should be restricted to organisational levels of the intact organism and below has gained support. On this basis biomarkers can be defined as "any biological response to an environmental chemical at the individual level or below demonstrating a departure from normal status". Changes at organisational levels above that of the individual are referred to as bioindicators. Biomarkers have the advantage over chemical analysis in that they can demonstrate whether or not an organism is meaningfully exposed. Analytical chemistry is now so sophisticated that many pollutants, especially the organochlorines, can be detected in almost all samples. But unless the biological significance of chemical residues is known, this information is of very limited value. With biomarkers, it is possible to determine if the physiology of the organism is significantly different from normal. If the parameter measured is outside the normal physiology range for that species, then the organism can be considered to be meaningfully exposed, and equally importantly, if the physiology is not significantly different, then the organism can be considered not to be meaningfully exposed even though the chemical(s) can be detected. The ability to determine whether or not an organism is meaningfully exposed is important in the decision-making process about the necessity or success of remedial action.

Keywords. Biomarkers, Indicator Species, Hazard Assessment.

9.1. Introduction

Definitions of biomarkers have varied considerably and have been frequently so broad as to include any biological change. Recently the concept that this term should be restricted to organisational levels of the intact organism and below has gained support [1,2]. On this basis biomarkers can be defined as "any biological response to an environmental chemical at the individual level or below

D. B. Peakall et al. (eds.),
***Biomarkers: A Pragmatic Basis for Remediation of Severe Pollution in Eastern Europe*, 123–133.**

demonstrating a departure from normal status". Changes at organisational level above that of the individual are referred to ecological indicators or bioindicators.

Beside the various definitions of biomarkers there have been various attempts at the classification of biomarkers. The recently launched journal Biomarkers subdivided biomarkers into three areas with the following definitions:

> Biomarkers of exposure: covering detection and measurement of internal exposure or dose with respect to chemicals.
>
> Biomarkers of effect: including measurement of endogenous substances or parameters indicative of biological change in response to chemicals, such as alterations in enzymes in tissues or body fluids.
>
> Biomarkers of susceptibility or sensitivity: including genetic factors and changes in receptors which alter the susceptibility of an organism to exposure to a chemical substance.

To me the sub-divisions of biomarkers that have been proposed do not seem helpful. The definition of biomarkers of exposure proposed above is too broad since it would cover residue level data without any biological changes being measured, although this does not seem to have been the intention of the editors. Since all biomarkers involve biological change all biomarkers are biomarkers of exposure. Further, susceptibility is not an effect and it suggested that the term "biomarker of susceptibility or sensitivity" be dropped. Here the terms "biomarker" and "ecological indicator" are used without further sub-division.

9.2. Use of Biomarkers

Biomarkers have the advantage over chemical analysis that they can demonstrate whether or not an organism is meaningfully exposed. Analytical chemistry is now so sophisticated that many pollutants, especially the organochlorines, can be detected in almost all samples. But, unless the physiological significance is known, this information is of very limited value. Further with heavy metals these are naturally occurring and the main question to be tackled is whether or not they are in sufficient excess to cause adverse effects.

9.2.1 CRITERIA FOR MEANINGFUL EXPOSURE

With biomarkers, it is possible to determine if the biochemistry or physiology of the organism is significantly different from normal. If the parameter measured is outside the normal physiological range for that species, then the organism can be considered to be meaningfully exposed, and equally importantly, if the physiology is not significantly different, then the organism can be considered not to be meaningfully exposed even though the chemical(s) can be detected. The ability to determine whether or not an organism is meaningfully exposed is central to the decision-making process about the necessity or success of remedial action. This question is the main focus of this meeting.

The following criteria need to be met before this concept can be used. These are:

(a) Data must be available on what is the normal range of values for each biomarker under a variety of environmental conditions. While it is important to select the best indicator species for a specific problem, it is obviously impossible to have background data on all species and there is merit in concentration on a limited number of indicator species. A centralised data base to collect and verify this baseline information would be valuable.

(b) There needs to be good biomarkers to indicate the health of the major functions of the organism and to be able to assess the impact of the major classes of chemicals of concern. While we have not reached this point yet, the rate of progress towards the goal is encouraging.

Biomarkers are well established in demonstrating effects at the individual organism level, particularly when a suite of selected responses is measured. Biomarkers also can be used to predict future trends and to obtain insights into causal relationship between stress and effects that may occur at higher levels of ecological organisation.

9.2.2 SPECIFICITY OF BIOMARKERS

The specificity of biomarkers to chemicals varies greatly. Both specific and non-specific biomarkers have their place in environmental risk assessment. A non-specific biomarker can tell one that a pollutant is present in a meaningful concentration but does not tell one which chemical is present. Based on this information detailed chemical investigation can be justified. In contrast a specific biomarker tells one which chemical is present, but gives no information on the presence of other chemicals.

One of the best studied non-specific biomarkers are the heme-containing enzymes known as the mixed function oxidases (MFOs). This large family of enzymes are a major component of the defences of organisms against toxic chemicals in their environment. Originally evolved, perhaps as long as two billion years ago [3], to handle naturally occurring toxic compounds, they now play an important role in the detoxification of a wide range of man-made chemicals. The usefulness of MFOs as a biological monitor has been clearly demonstrated for both polynuclear aromatic hydrocarbon and organochlorine pollution in a wide range of organisms. Since the response is caused by a wide variety of chemicals it means that the system is capable of showing exposure sufficient to cause a biological response to many xenobiotics. Conversely it tells little about the causative agent(s), but can be used to delimit an area for which it is worth the time and expense of more detailed investigations. One successful use of the MFO system has been development of dioxin-equivalents using induction of the cytochrome P_{450} 1A1 to explaining the ecotoxicology of polychlorinated biphenyls, polychlorinated dibenzofurans and polychlorinated dibenzodioxins (4). It has been found that the co-planar congeners are responsible for most of the activity.

In contract the inhibition of the enzyme of the heme biosynthetic pathway d-aminolevulinic acid dehydratase (ALAD) is a sensitive and specific indicator of lead exposure (5). A close relationship between the concentration of lead in the blood and the degree of inhibition of ALAD exists, especially when this is expressed as an activity ratio (6).

9.2.3 CRITERIA FOR THE USE OF BIOMARKERS

Before biomarkers can be usefully employed they must meet certain criteria, modified from (7):

> Biological Specificity. It is important to know which classes of organisms can be used for which biomarker. For example the inhibition of AChE can be applied throughout the animal kingdom whereas the induction of vitellogenin is confined to those vertebrates that lay eggs.
>
> Clarity of Interpretation. How clear-cut is the end-point as an indicator of exposure to anthropogenic stress? Can the responses that are measured by clearly distinguished from those due to natural stresses? It is valuable to know the mechanism of response to the chemical in assessing this point.
>
> Time of responses. The response time of different biomarkers can vary widely from nearly instantaneous to years. DNA adducts are formed rapidly but the subsequent damage to the genetic material make take years to become apparent.

Permanence of Response. Similarly it is important to know how long the response lasts. If it is transient it can readily be missed. The inhibition of AChE, especially in blood, is transient (24-48 hours) and thus it is necessary to know when the exposure occurred to assess the importance of the degree of inhibition. In contrast, the inhibition of ALAD persists for months.

Reliability. This can be considered under two headings: (a) environmental influences that modulate the organism's response to a chemical and (b) inherent variation in the biological response to a given exposure. It is important to know the extent of all variation in order to have a reliable biomarker.

Methodological Considerations. Important considerations here are precision (analytical reproducibility of the method) and cost and ease of the assay. Although many reliable assays have been developed there is a need for standardization, along the lines used in analytical chemistry, so that results from different laboratories are comparable.

Relative Sensitivity. It is important that the biomarker be sensitive compared to other endpoints, such as mortality or reproductive impairment and it is important to know the relative sensitivity compared to other biomarkers.

Validation in the Field. For a biomarker to be useful in environmental assessment it must be validated in the field. Organisms in the field are subjected to a much wider range of variables than occurs in the laboratory.

Linkage to Higher Level Effects. A biomarker is more useful if there is clear linkage to effects at higher levels of organisation. Studies on invertebrates have been particularly fruitful as population changes occur more rapidly than in higher species. This concept is considered in more detail elsewhere in this volume [8].

9.3. Selection of Indicator Species

One of the main difficulties of ecotoxicology tests is the selection of indicator species, or in other words which species should be studied. Estimates of the total number of species in the world varies widely ranging from a few million to as high as eighty million [9]. Obviously the number of species that can be studied is extremely small compared to the total and thus great care should be taken in their selection. While one wants to study the most relevant species for a specific problem one also needs a great deal of background information for each species that is investigated.

Two of the concepts for selection of species that have been put forward are "keystone species" and the "most sensitive species". The concept of "keystone" species, that is species which, if removed, cause the collapse of may other species, was put forward by Paine [10] and has attracted considerable interest in ecological circles. Despite some successes, especially on tidal pool communities, the concept has not become widely established and consequently has not been taken up by ecotoxicologists in hazard assessment exercises [11]. The concept of the most sensitive species was put forward by ecotoxicologists on the grounds that if this species was protected then the remaining species would also be protected. Regrettably it has not been possible to find such a species. In a broad survey, [12] it was found that *Daphnia* were, on the whole, more sensitive than fish, which in turn were more sensitive than algae. However, there are numerous exceptions to this generalisation.

There are many instances where a particular group of organisms has been shown to be extremely sensitive in a way that could not have been predicted from standard tests. One example is the sensitivity of molluscs to tributyltin. This compound, used in anti-fouling pains on boats, was found to cause reproductive failure in the dog whelk (*Nucella lapillus*) at concentrations as low as a few nanograms per litre [13] whereas effects on other classes of aquatic organisms only occur at concentrations several orders of magnitude higher.

Moore [14] put forward the following criteria for choice of indicator species:

(a) The species should be widely distributed, relatively abundant and easy to collect.

(b) If monitoring is to be based on the organs of single animals each animal should be large enough for adequate samples to be taken.

(c) Ideally it should be possible to ascertain the age of the animal by inspection. To indicate contamination occurring just before collection, young or short-lived animals should be used. Where one needs information on accumulative build-up, long-lived animals are required.

(d) Species with residue levels between the limit of detection and that causing toxicological effects should be selected.

(e) If measurement of local changes are required, indicator species must be sedentary in their habits. When contamination of a large area is to be studied, then species with extensive ranges are more appropriate. In either case, the range of the indicator species must be known.

Although these criteria were generated for the measurement of chemical residues they fit well into criteria for the selection of indicator species for biomarkers studies. These considerations are important in relation to the bioaccumulation potential of the chemicals of concern. If the bioaccumulation factor, often measured in the laboratory as the octanol:water coefficient, is low then it is necessary to study a species directly affected. For example, in the case of water pollution from nonylphenols, which are not strongly bioaccumulated, it would be necessary to study fish as exposure of fish-eating birds such as herons would be low. On the other hand, organochlorine compounds such as PCBs bioaccumulate in the food chain so top predators such as the heron would be highly exposed and therefore are good indicator species.

There has been increasing interest in using invertebrates in ecological risk assessment and the use to which this important component, which accounts for 95% of all animal species, is considered in more detail elsewhere in this volume [8]. Plants have not been used as widely; the current status of the field is reviewed elsewhere in this volume [15].

9.4. Non-destructive Biomarkers

Concerns over the numbers of animals used in experiments has become a political issue. There is the paradox of growing public demand for greater safety regarding the effects of chemicals on human health and the environment and at the same time a demand for fewer or even no animal experiments. As far as field studies are concerned there is increasing difficulty in obtaining permits for collection, especially of birds and mammals. Such concerns are often illogical since the numbers involved are usually very small compared to natural mortality. Nevertheless there is increasing pressure to use either non-destructive tests or tests on 'lower' organisms since our ethical considerations are largely phylogenetic, ie our outlook is anthropomorphic.

Non-destructive testing has invoked considerable interest in recent years [16,17]. The approach has the advantages that serial measurements can be made on the same individual and that measurements can be made on species, such as whales, where conservation interests are paramount. The disadvantage of non-destructive measurements of biomarkers is that it often excludes the key organ: for example the inhibition of acetylcholinesterase is best measured in the brain, which is the target organ, further the inhibition of this enzyme in the blood is much more transient.

There has been considerable interest in the possibility of replacing tests on vertebrates with bacteria tests, such as the Microtox test, for both cost and ethical reasons. One of the most widely used bacterial test is the Microtox test.[18].

9.5. Single Species *v.* Multi-species Testing

The use of single-species toxicity tests by regulatory agencies have been repeatedly criticised on ecological grounds. Simply stated, ecosystems are greater than the sum of their parts. Single species tests ignore the interactions between species such as competition and predation, as well as ecosystem functions such as nutrient flow or carbon cycling.

Although the ecological objections to single species testing are obvious enough, nevertheless the extent to which a system consisting of a few species of bacteria, algae and invertebrates can be considered to mimic natural systems is open to debate and many larger and more complex systems have been used. The basic difficulty is that as the size and complexity of the system increases so does the cost while the ability to replicate and control the variables decrease. Nevertheless a vast amount of work has been done on microcosms and mesocosms.

9.5.1. MICROCOSMS AND MESOCOSMS

The original distinction between microcosms and mesocosms has become blurred.

Small scale models or portions of natural systems housed in artificial containers in the laboratory are clearly defined as microcosms. Odum [19] defined mesocosms as bounded and partially enclosed outdoor experimental units that stimulate the natural environment. But the difference between laboratory and outside experiments is not universally accepted. Although both terms continue to be used it is probably best to treat them as a continuum. An extensive listing of both microcosms and mesocosms, treating them as a continuum has been given by Kennedy and co-workers [20].

Model streams, groups of artificial ponds and enclosures that isolate a portion of a larger body of water have all been used. A recent review on the use of stream microcosms has been presented by Pontasch [21]. He concluded that although they have been useful in studying many specific pollution problems the progress towards standardised for regulatory purposes has been limited.

Artificial ponds have been widely used to study the effects of chemicals on aquatic systems and these sometimes include predatory fish such as the rainbow trout (*Oncorrhynchus mykiss*). Pond mesocosm studies have been used in support of pesticide registration in the US under the Federal Insecticide, Fungicide and Rodenticide Act. Both limnocorrals (enclosures in open waters) and littoral corrals (enclosure systems on the border of lakes) have been widely used to study the fate and effect of a variety of pollutants [22,23]. However, despite the valid objections to single species testing they still remain the basis for most regulations.

9.5.2. USE OF CAGED ORGANISMS

A useful approach to making single-species testing more realistic is the use of caged animals; it has been widely used for fish but has also been used with other orders. A recent example of its application has been the use of caged rainbow trout in sewage effluent to look for the presence of estrogenic chemicals by measuring the induction of vitellogenin formation in make fish [24]. The protein vitellogenin is a major component of fish eggs and is normally found only in female fish, thus its presence in males is evidence for the presence of estrogenic chemicals in the environment. The technique is particularly useful if the chemicals of concern have not been identified (as was the case in the study cited above) or where complex mixtures of chemicals are involved. However, the exposure may not be realistic as the animal's movements are restricted and other parameters such as diet and behaviour differ from their free-living counterpart.

9.5.3. CORRELATION WITH POPULATION CHANGES

A few biomarkers have been conclusively correlated with population changes; for example avian eggshell thinning, with many species of raptorial and fish-eating birds showing marked population declines over wide areas [25]. Although ecologists are largely interested in protection at the population or community level, environmental regulations are often directed at the level of the individual. Maximum protection would be provided if the physiology of all organisms living in any area are not significantly altered by the presence of the chemical pollutants. This would be the equivalent of human health considerations where concentrations of pollutants that cause physiological alterations are not tolerated.

9.6. Use of Biomarkers in Hazard Assessment

In the case of the environment, the fundamental question is how much change we are prepared to tolerate. Once one moves away from the absolute of allowing no alterations the situation becomes less clear-cut and one is driven by political or societal concerns rather than scientific considerations. The problem can be discussed in terms of space, time and species. Conservation concerns come into play for the protection of habitats and species of particular concern.

Obviously the larger the area affected, the greater our concern. While we might be prepared to tolerate some damage a few meters from the end of the outlet pipe we are unlikely to tolerate damage throughout a river system. This brings us to the critical question of how large an area are we prepared to tolerate pollution? The concept of an 'exclusion zone' has been proposed as a means of answering this problem [26]. An 'exclusion zone' is an area where it is decided that pollution will be allowed in amounts that will alter the physiology of some species living there. However, it is difficult to see how an impartial scientific basis for determining the size of the exclusion zone could be established, but it is equally difficult to see how pollution control can proceed without it. If one takes the example of an industrial port, we cannot expect the entire area of the port to maintain the same populations of the same animals under the same physiological conditions as a pristine area. Nor, on the other hand, is it acceptable for the port to pollute the entire estuary and nearby lakes or ocean.

To come up with a solution, one has to have a method of defining both the end points to be measured and the size of the exclusion zone. The combination of these two factors has been suggested to achieve a practical end point [26]. The end point suggested is 'that the physiological functions of organisms, outside the exclusion zone, should be within normal limits'.

The major advantages of the approach are:

(a) It is, at least initially, independent of the pollutants involved and thus avoids the problem of mixtures and unknown substances. Only if the studies outside the exclusion area reveal abnormalities will detailed investigations be needed.

(b) It does not require, again at least initially, proof that the effect seen is deleterious. If the area over which the effect is seen is large, it may be argued that such proof is necessary, but it is not a basic precondition.

(c) Philosophically it is a defensible position. Although pollutants are present, the functioning of the animals living in the area is normal.

There are, naturally, limitations to this approach. The most important being:

(a) It implies that we have a good enough battery of tests to be confident that the physiology is indeed normal and that this battery of tests covers all the major classes of pollutants.

(b) It implies that we can separate effects caused by chemicals from those occurring due to natural causes.

(c) A difficult question to answer is which and how many species should be tested. However, the problem of inter-species differences exists with all other approaches.

(d) It does not tackle the question 'Is harm caused by the abnormal physiological state?'

The concept of using physiological normality outside an exclusion zone is a mixture of science and politics, but that is an inevitable part of the environmental assessment process. Will this concept hold up under the conditions of severe pollution that have occurred in eastern Europe? It may well be that physiological normality is an unrealistic goal under these circumstances. The answer to this question is to be found in the reports of the working groups.

References

1. Van Gestel, C.A.M. and van Brummelen, T.C. (1996) Incorporation of the biomarker concept in ecotoxicology calls for a definition of terms, *Ecotoxicol.* **5**, 217-255.
2. Peakall, D.B. and Walker, C.H. (1996) Comment. *Ecotoxicol.* **5**, 227.
3. Nebert, D.W., Nelson, D.R. and Feyereisen, R. (1989) Evolution of the cytochrome P_{450} genes. *Xenobiotica* **19**, 1149-1160.
4. Metcalfe, C.D. and Haffner, G.D. (1995) The ecotoxicology of coplanar polychlorinated biphenyls. *Environ. Rev.* **3**, 171-190.
5. Scheuhammer, A.M. (1987) Erythrocyte d-aminolevulinic acid dehydratase in birds. I. The effects of lead and other metals *in vitro, Toxicol.* **45**, 155-163.
6. Scheuhammer, A.M. (1987) Erythrocyte d-aminolevulinic acid dehydratase in birds. II. The effects of lead and other metals *in vivo, Toxicol.* **45**, 165-175.
7. Huggett, R.J., Kimerle, R.A., Mehrle, P.M. Jr. and Bergman, H.L. (1992) (eds.), *Biomarkers. Biochemical, Physiological, and Histological Markers of Anthropogenic Stress.* Lewis Publ. Boca Roca, 347pp.
8. Lagadic, L. (1998) Biomarkers in invertebrates. (This volume)
9. Stork, N. and Gaston, K. (1990) Counting species one by one. *New Scientist*, 11 August, p.43-47.
10. Paine, R.T. (1966) Food web complexity and species diversity, *Amer. Nat.* **100**, 65-75.
11. Mills, L.S., Soule, M.E. and Doak, D.F. (1993) The keystone-species concept in ecology and conservation, ***Bioscience*** **43**, 219-223.
12. Kenaga, E.E. (1981) Comparative toxicity of 131,596 chemicals to plant seeds, *Ecotoxicol. Environ. Safety* **5**, 469-475.
13. Gibbs, P.E. and Bryan, G.W. (1986) Reproductive failure in populations of the dog-whelk, *Nucilla lapillus*, caused by imposex induced by tributyltin from antifouling paints, *J. Mar. Biol. Ass. UK*, **66**, 767-777.

14. Moore, N.W. (1966) A pesticide monitoring system with special reference to the selection of indicator species, *J. Appl. Ecol.* **3**, 261-269.
15. Ernst, W.H.O. (1998) Biomarkers in plants. (This volume.)
16. Fossi, M.C. and Leonzio, C. (eds.) (1994) Nondestructive Biomarkers in Vertebrates. Lewis, Boca Raton, pp.345.
17. Walker, C., Kaiser, K., Klein, W., Lagadic, L., Peakall, D., Sheffield, S., Soldan, T. and Yasuno, M. (1998) Alternative Testing Methodologies for Ecotoxicology, *Environ. Health Perspectives*, **106** (Suppl. 2), 441-451.
18. Calow, P. (ed.) (1993) Handbook of Ecotoxicology, Vol. 1. Blackwell, London. Pp.478.
19. Odum, E.P. (1984) The mesocosm, *BioScience* **34**, 558-562.
20. Kennedy, J.H., Johnson, Z.B., Wise, P.D. and Johnson, P.C. (1995) Model aquatic ecosystems in ecotoxicological research: consideration of design, implementation and analysis, in D.J. Hoffman, B.A. Rattner, G.A. Burton, Jr. and J. Cairns, Jr. (eds.) *Handbook of Ecotoxicology,* Lewis Publ., Boca Raton, 1995, pp.117-162.
21. Pontasch, K.W. (1995) The use of stream microcosms in multispecies testing, in J. Cairns, Jr. and B.R. Niederlehner (eds.) *Ecological Toxicity Testing*, Lewis Publ., Boca Raton, pp.169-191.
22. Servos, M.R., Muir, D.C.G. and Webster, G.R.B. (1992) Bioavailability of polychlorinated dibenzo-p-dioxins in lake enclosures, *Can. J. Fish. Aquat. Sci.* **49**, 735.742.
23. Brazner, J.C., Heinis, L.J. and Jensen, D.A. (1989) A littoral enclosure for replicated field experiments, *Environ. Toxicol. Chem.* **8**, 1209-1215.
24. Purdom, C.E., Hardiman, P.A., Bye, V.J., Eno, N.C., Tyler, C.R. and Sumpter, J.P. (1994) Estrogenic effects of effluents from sewage treatment works, *Chemistry and Ecology* **8**, 275-285.
25. Peakall, D.B. (1993) DDE-induced eggshell thinning: an environmental detective story, *Environ. Rev.* **1**, 13-20.
26. Peakall, D.B. (1992) Animal Biomarkers as Pollution Indicators. Chapman and Hall, London, 291pp.

10. BIOMARKERS IN PLANTS

WILFRIED H.O. ERNST
Department of Ecology and Ecotoxicology
Faculty of Biology, Vrije Universiteit,
De Boelelaan 1087, 1081 HV Amsterdam, The Netherlands

Abstract

Plants can react rapidly to adverse environmental conditions ranging from invisible biochemical processes up to the development of visible symptoms of injury, independent of the primary target of the stress the best recognized in leaves in the field. Biochemical biomarkers may be very specific for one or a group of compounds or a general indication of an environmental stress. Biophysical biomarkers are based on changes in chlorophyll content and fluorescence, visual biomarkers on changes in leaf pigments. Exposure time, exposure concentration and the stability of the biomarker are the most important aspects of the quantitative biomarkers.

Keywords: *Betula pendula* *Brassica napus* cadmium chlorophyll fluorescence chlorosis copper dwarfism fluor /fluorocitrate heavy metals infrared necrosis needles peroxidases phytochelatins selenoproteins zinc

10.1. Introduction

The term "biomarker" is often defined in a broad as well as in a narrow scope. After evaluation of the publications, based on a biomarker workshop of the European Science Foundation [1-5], van Gestel and van Brummelen [6] restricted the application of the term "biomarker" to "any biological response to an environmental chemical at the below-individual level, measured inside an organism or in its products (urine, faeces, hairs, feathers, etc), indicating a departure from the normal status, that cannot be detected from the intact organism", but disregarding the individual level [7]. This narrow scope is nearly identical to that presented for plants by Ernst and Peterson [2] in the above-mentioned workshop with the following extensions: in plants the detection of an injury may be possible from outside the intact organism and the adverse environmental factors include not only chemicals, but also gamma and

D. B. Peakall et al. (eds.),
Biomarkers: A Pragmatic Basis for Remediation of Severe Pollution in Eastern Europe, 135–151.

ultraviolet radiation. Melanomes in man are excellent biomarkers of an excess of UV-B radiation. All of the above-mentioned concepts [1-6] consider mostly qualitative biomarkers, i.e. a timely recognition of the impact of an excess of a chemical in the environment on an organism. To decide upon the degree of a departure from a normal situation demands a quantitative approach. The quantitative biomarker concept has to consider three important components which determine the concentration of a biomarker in a specific tissue of an organism: (1) a dose-response relationship between the concentration of the biomarker and that of the environmental compound, (2) a time-response relationship between the duration of the exposure and the synthesis of the biomarker and (3) the stability of the biomarker over time.

Adverse environmental conditions can cause the appearance of visible symptoms of injury in plants relatively rapidly and thus broaden the application of the biomarker concept to the intact individual. Zoologists [7] came up with the statement "plant biomarkers are not as well advanced as animal biomarkers"; this seems an inevitable consequence of the very much larger resources put into medical, i.e. vertebrate toxicology than into ecotoxicology. Nevertheless, the specificity of visible symptoms of plant injury may be as low as eggshell thinning in birds caused not only by DDT and DDE [8], but also by decalcification of the soil due to acid precipitation [9]. The visible biomarker is a non-destructive biomarker [1]; in plants it was formerly called "symptomatology" and has been applied for longterm to responses of plants to both an environmental deficiency and surplus of a chemical, mostly in agricultural and horticultural systems [10,11], but relatively recently also in forestry [12], and sometimes in natural ecosystems [13].

In the context of this workshop I will restrict the biomarker concepts to the excess of an environmental compound and will elaborate on qualitative and quantitative biomarkers.

10.2. Biochemical Biomarkers

Plants are able to explore all abiotic environmental compartments. Therefore adverse conditions or changes in these compartment will affect a plant's metabolism; long-lasting exposure, however, will result in the selection and evolution of adapted genotypes [14]. An example is the change in the intensity of the UV-B radiation due to the breakdown of the stratospheric ozone layer [15]. Upon exposure to UV-B plants will enhance the synthesis of flavonoids; accumulation of flavonoids in the epidermal layer act as a UV-B screen [16]. But an increase of the concentration of this biomarker only occurs in those plants which have not experienced a long-term exposure to UV-B radiation during their at least recent evolutionary history [17]. This general biological memory has to be kept in mind in all application of biomarker concepts.

In contrast to animals, plants are mostly able to keep organic contaminants away from the living cell by adsorption to surfaces of roots and shoots, except those which were primarily designed to destroy plant growth such as herbicides. Therefore metabolic processes in plants are not very suitable to detect a surplus of organic contaminants such as PAHs, dioxines etc. [18] or estrogenic compounds [19]. If a negative response is detectable, then plants react in a species-specific manner; the reaction may last only for several hours as shown for the impact of nitrobenzene on photosynthesis [20].

All inorganic components of the environment, however, can be taken up by plants dependent on their bioavailablility. Therefore biomarkers in plants may be expected to indicate the principle chemical elements causing the injury.

As reviewed ealier [2], there are several biochemical biomarkers which are specific for one or a group of chemical elements. The advantage of a biochemical biomarker is the indication of the bioavailability of a chemical element in the environment and its biologically effective concentration in the plant cell after compartmentation and speciation. At a high specificity a biochemical biomarker will be more informative than the analysis of chemical elements in plants by advanced instruments [21].

10.2.1. SPECIFIC BIOCHEMICAL BIOMARKERS

Specific biomarkers, i.e. a unique biochemical pathway or endproduct, allow the identification of a chemical element without further chemical element analysis. The multiple interactions between chemical elements and a plant's metabolism are the reason for the scarcity of such biomarkers in plants.

A very specific biomarker is the synthesis of fluorocitrate with exposure of plants to fluorides, present in the soil or in the atmosphere.

10.2.1.1. Fluorocitrate as Specific Biochemical Biomarker

Plants can incorporate fluoride by synthesizing fluoroacetyl-CoA and convert this compound via the tricarboxylic acid (TCA) cycle to fluorocitrate; this compound inactivates the enzyme aconitase so that the TCA is blocked and fluorocitrate accumulates [22]. It is a very reliable biomarker for fluor poisoning, but a dose-response relationship is not yet established. Even its qualitative application in field situations is only documented by one published report [23]. For the quantification of the impact of fluoride emission on plants, exposure of sensitive plants and cultivars (active biomonitoring) is still preferred, obviously due to the specific visible symptoms of leaf necrosis [24] combined with or even without fluor analysis of the leaves [25, 26].

10.2.1.2. Seleno-proteins as Specific Biochemical Biomarker

Another example of a specific quantitative biomarker is the synthesis of seleno-proteins with exposure of plants to selenium. Due to the chemical similarity between selenate and sulphate, the ion uptake mechanism of plant roots cannot differentiate between both elements. Whereas Se-resistant plants only synthesize selenopeptides and non-protein amino acids [27], Se-sensitive plants incorporate them into proteins [28]. Due to the low stability of the Se-Se bridges compared to the S-S bridges, the metabolism is deregulated.

10.2.2. BIOCHEMICAL BIOMARKER FOR GROUPS OF ELEMENTS: PHYTOCHELATINS (PCs)

Several environmental compounds can cause a similar metabolic reaction. A well known example is the synthesis of phytochelatins (gamma-Glu-Cys)nGly [29] in most higher plants upon exposure of the cytoplasm to a surplus of free metal ions; in certain grasses the glycine at the terminal position is substituted by serine [30], in Fabaceae by beta-alanine [31]. Due to the progress and emphasis on PCs as biomarkers they are very suitable to demonstrate the advantages and the pitfalls of a group-specific biomarker.

PCs are components of a special pathway of the sulfur metabolism in plants. The synthesis of phytochelatins (PCs) is catalyzed from glutathione (GSH) by the specific, but constitutive enzyme gamma-glutamyl-cysteine dipeptidyl transpeptidase (Fig. 1), named phytochelatin synthase [32]. At the same exposure metal concentration there is a decreasing order: Hg > Cd > As > Te > Ag > Cu > Ni > Sb > Au > Sn > Se > Bi > Pb > W > Zn. The chain length of PCs increases and the GSH concentration decreases with increasing exposure time and/or exposure concentration [33].

10.2.2.1. Synthesis of Phytochelatins under Controlled Conditions

Heavy metals which are essential in the metabolism of all plants, i.e. Cu and Zn, cause a different response pattern from the non-essential heavy metals, Hg, Cd and all other above mentioned elements. Therefore Cd is the most important metal in PC research. The first target of a soil-borne metal is the root which shows also the primary effect. Cd does not only stimulate the activity of the PC synthase, it also increases already the activity of the first enzyme in the bioactivation of sulfate, i.e. the ATP sulfurylase, and the activity of the gamma glutamyl-cysteine synthase [40]. With exposure to moderate external Cd levels there is an inverse linear relation between root length and PC concentration [34, 35]. In Cd-exposure experiments under controlled conditions the synthesis of PCs is a process which is dependent on exposure time and exposure concentration during the initial period of exposure [34], thus PCs have potential as qualitative biochemical biomarkers.

Some restrictions, however, have to be obeyed. PC production in vitro ceases after the free heavy metal ions are completely chelated by PCs so that PC accumulation indicates the limit of the detoxification process [36]. There is strong evidence that phytochelatins are accumulated in the vacuole [37, 38, 39], but it is uncertain how long the Cd-phytochelatin complex remains stable. Three days after terminating the exposure of roots to Cd the concentration of PCs declined in roots, indicating either a breakdown or a translocation to the shoot [40]. In terrestrial environments a sudden decrease of the external Cd concentration will not occur due to the binding of Cd to organic matter of the soil. In aquatic environments, especially in rivers and streams, however, metal exposure events of short duration are typical for discharge accidents: they will stimulate the synthesis of PCs in aquatic organisms [41], but a rapid breakdown may affect the reliability of PCs as biomarkers. Above a certain exposure concentration, however, affected plants are not able to increase the PC synthesis, thus modifying the dose-response relationship; at high Cd exposure the synthesis of PCs is even negatively related to the Cd concentration [42]. The presence of free metal ions in the cytosol will depend on the rate of the detoxification and of the metal compartmentation which is determined by species-specific metal-demands in the case of essential mineral nutrients, the degree of metal-sensitivity of plant species [43] and within a species of the genotypes [52, 44].

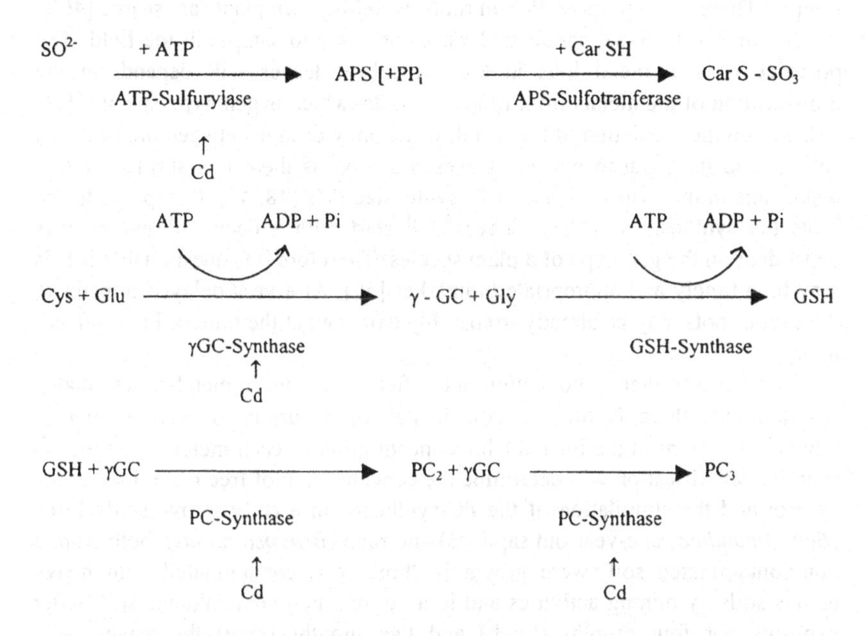

Figure 1. The pathway from sulfate to phytochelatins (PC2, PC3). arrows indicate the site of cadmium (Cd) interaction. ATP = adenosine 5'-triphosphate, APS = adenosine 5'-phosphosulfate, ADP = adenosine 5'-diphosphate, Car-SH = carrier thiol, Cys = cysteine, Glu = glutamic acid, GC = gamma glutamyl-cysteine, Gly = glycine, GSH = glutathione.

10.2.2.2. Phytochelatins as Biomarkers in the Field

Most of the phytochelatin research has been restricted to laboratory experiments with a preference for root analysis. A biomarker is only operational if it can be detected in material appropriately collected in the field. Roots sampled in the real world confront the scientist with the following problems [45]:
(1) Due to the intimate association of roots with soil particles it is very difficult to separate the root from very fine soil and humus particles, thus hampering a proper determination of the root mass. (2) Mycorrhizal fungi as well as free-living micro-organisms of the rhizosphere contribute to the root mass so that the quantitative analysis of a biomarker is impaired. (3) In many species the roots are either very fine or extend to great depth system and will be difficult to sample. Therefore reports on PCs in roots of field-grown plants are scarce [46].

In contrast to roots, shoots and leaves are easy to sample in the field. The presence of free metal ions in the cytosol of leaves will depend on the translocation of the metals from roots to shoots, which is genotype-specific [34, 47], and on the speciation of the metal, which may change between the first root contact and its entrance into the leaves. As soon as there is a surplus of free metal ions in the cytosol, leaves will synthesize PCs [48, 49]. Compared to the roots the synthesis of PCs in leaves will start with a delay of several days dependent on the genotype of a plant species. Therefore it is questionable if PCs may be a timely and appropriate biomarker [50]. At a very delayed transfer to the shoot, roots may be already irreparably damaged at the time of PC synthesis in leaves.

There is a further complication in the field situation. In metal-contaminated environments there is often a coincidence of a surplus of several metals, however different in the bioavailable concentration of each metal. The species-specific detoxification will determine the concentration of free metal ions in the cytosol and the stimulation of the PC-synthesis. In a comparative study birch (*Betula pendula*, one-year old saplings) and rape (*Brassica napus*), both from a non-contaminated soil, were grown in three soils contaminated with heavy metals soils by mining activities and in a normal, non-contaminated soil. After exposure for four months (birch) and two months (rape) the leaves were analysed for heavy metals and phytochelatins (Table 1). Compared to those from non-contaminated soil the leaves of both plants from contaminated soil contained phytochelatins, the most in rape leaves. Thus PCs indicated the presence of free metal ions in the cytosol, i.e. PCs as qualitative biomarkers in a species-specific manner. Whereas birches could obviously regulate the influx of metals and the concentration of free metal ions in the cytosol, a high zinc content in the soil deregulated the detoxification in rape which resulted in an enormous increase in the PC concentration. The low PC concentration of rape on the copper soils is related to an exhaustion of the glutathion pool due to the high copper exposure [cf. 42].

There was however, no quantitative relation between concentrations of phytochelatins and heavy metals in the leaves. Metal analysis of whole leaves cannot discriminate between the compartmentation and speciation of metals.

Therefore a lack of reasonable metal/PC correlation may be explained by species-specific differences in metal compartmentation, i.e. precipitation in the cell wall and accumulation in the vacuole [51], thus making it difficult or impossible to use PCs as quantitative biomarkers across species of an ecosystem. The discussion of the role of biomarkers in biological impact assessment [52] has to consider the cellular compartmentation and the longevity of a biomarker, at least for plants and for the plant/herbivore interaction. In the case of phytochelatins the rapid complexing of free metals is not only important for the plants, but also for the evaluation of a metal's toxicity in the food chain, because cadmium bound to phytochelatins is less toxic than the ionic form of the metal [53].

10.2.3. GENERAL BIOCHEMICAL BIOMARKERS

Different environmental stressors can have the same impact on a plant's metabolism. Well known is the stimulation of the activity of peroxidases and the expression of particular isoenzymes of peroxidases after exposure to a complex of air pollutants [54, 55], acid rain [56], specific air pollutants such as ammonia [57] and ozone [58], and heavy metals [56, 59].

TABLE 1. Concentration of heavy metals (micromol per g dry mass) and phytochelatins (PC, nanomol per g dry mass) in leaves of birch (Betula pendula Roth) and rape (Brassica napus L.) grown in four soils: A (Marsberg, rich in Cu); B (Innerste, rich in Cd, Pb, and Zn); C (Welfesholz, rich in Cu and Zn); D (dune soil, poor in Cd and Pb, normal in Cu and Zn). n.d. = not determined. Unpubl. data from W.H.O. Ernst.

Soil	Plant species	Metal concentration				Phytochelatin concentration
		Zn	Cu	Cd	Pb	
A	*Betula pendula*	8.5	0.18	0.03	0.19	17.5
	Brassica napus	5.5	1.52	0.04	n.d.	<2.0
B	*Betula pendula*	21.7	0.12	0.27	0.49	7.7
	Brassica napus	14.7	0.21	0.09	n.d.	660
C	*Betula pendula*	15.5	0.09	0.08	0.12	9.6
	Brassica napus	12.6	0.87	0.07	n.d.	1490
D	*Betula pendula*	8.8	0.10	0.00	0.03	<2.0
	Brassica napus	9.7	0.29	0.00	n.d.	<2.0

The adverse effects of environmental stressors on the synthesis of amino acid and on the protein metabolism are numerous. Physiological desiccation causes an increase in the level of free proline, regardless of whether the stressor is salt [60], drought [61, 62], high and low temperature [63], or a heavy metal [64 - 67]. In the case of proline it has been shown that Cu-tolerant plants which grow in environments with episodic droughts have a high constitutive proline level which can be further increased after exposure to Cd, but not after that to Cu and Zn [66]. In addition to increased levels of polyamines the synthesis of stress proteins such as the heat shock proteins (HSP 70 family) is a metabolic response not only to a heat shock [67] or wounding [68] but also to a surplus of heavy metals [69]. All of these biochemical responses have in common that they indicate a deregulation of the metabolism; thus they are qualitative biochemical bio-markers. When there is an excess of heavy metals there are various first-order actions in a plant cell (Table 2) which may start at the plasma membrane by disturbing membrane integrity and blocking aquaporins, then stimulate in the cytosol the synthesis of specific metabolites such as the phytochelatins and end up in the nucleus by delaying cell division. A quantification of the response in relation to the intensity of the stressor is very difficult.

TABLE 2. First-order action of a surplus of various metals in plant cells at an effect concentration of 25 %. The absolute external metal concentrations can vary by a factor of 100 between metal-sensitive and metal-tolerant plants.

Metal(s)	First-order action	Evidence	Plant species
Cd, Cu (Ag, Hg)	stimulation of PC synthesis	PC content	more than 30 plant species, genotypes
Cu, Ag, Hg	disturbance of plasma membrane integrity	K+-efflux, malonaldehyde content	*Agrostis capillaris* *Mimulus guttatus* *Silene vulgaris*
Cu, Hg	blocking aquaporins	indirect by injection of plant mRNA in Xenopus oocytes	*Arabidopsis thaliana*
		volume flow of water	*Lycopersicon esculentum*
Zn	delay of cell division	number, length of mitotic cycle	*Festuca rubra*
Zn, Ni, Cd	vacuole compartmentation	accumulation in vacuoles	*Silene vulgaris* *Thlaspi coerulescens*

10.3. Biophysical biomarkers

Excess of a chemical present in the soil will first affect the roots and other organisms in the rhizosphere, but due to the intimate metabolic relation between root and shoot it will with a certain delay finally also affect the shoot. Leaves make the largest contribution to the biomass, the harvest of solar energy (photosynthesis) and the water transfer of a plant from the soil to the atmosphere (transpiration) and are freely accessable to physical instruments.

10.3.1. CHLOROPHYLL CONCENTRATION IN THE INFRARED SPECTRUM

Within the leaf the photosynthetic system of the chloroplasts is very responsive to chemicals, which has been exploited as a target of modern weed biocides [70]. Two properties of a leaf can be detected by the use of biophysical markers. Leaves reflect the incoming radiation in the infrared spectrum (0.7 - 0.9 micrometer) four to five times more than in the visible spectrum (0.4 - 0.7 micrometer). The amount of chlorophyll per unit leaf area will alter the reflectance of radiation by the leaf surface. The healthier a tree is, the more intense red appears a tree crown in false colour photographs. The first application of false colour as a biophysical biomarker was the analysis of trees in the city of Amsterdam, The Netherlands, exposed to a surplus of manganese [71]. A change from relatively moist city gas (32.5% methane) to dry natural gas (81.6 % methane) caused leakage of pipelines; methane oxidizing bacteria consumed so much oxygen that they created an anoxic environment resulting finally in reduction of manganese and iron and their enhanced availability to plants. The final result of this leakage was manganese toxicity, indicated by the weak red colour of crowns of affected trees compared to the dark red colour of healthy trees, identified by chemical analysis. It is an example of a qualitative biomarker without the elaboration of a dose- or time-dependent relation of the Mn concentration in plants and of the intensity of the red colour.

Infrared and multispectral analysis by remote sensing is another appropriate tool to localize adverse environmental conditions in trees, e.g. forest decline in Europe [72]. Using infrared images the adverse effect of the herbicide diuron, experimentally applied to two pine species, could be detected 16 days prior to the first visible symptom [73]. Timely signalling of an adverse environmental situation is thus the advantage of this qualitative physical biomarker, but it can not identify the compound and its concentration.

10.3.2. CHLOROPHYLL FLUORESCENCE

As soon as a leaf is exposed to stress, the photosynthetic quantum conversion to chemical energy declines, and heat emission and chlorophyll fluorescence increase considerably [74]. The latter change in chloroplast response has been used by the chlorophyll fluorescence technique to analyse invisible injury of leaves, especially water and temperature stress [75,76], but recently also for aluminium [77] and copper stress [78]. The development of a high resolution fluorescence imaging system [79] may even improve the application of chlorophyll fluorescence as a powerful tool for detecting the occurrence of stress responses in plants *in vivo* and *in situ*. A chemical analysis, however, is necessary to identify the stress factor. Information on the time difference between invisible damage and visible injury is not yet available.

10.4. Visual biomarkers

Adverse environmental conditions can cause a lot of changes in a plant's metabolism which can be expressed by changes of colour in various plant parts or by morphological traits; both can qualitatively be registered from outside the plants.

10.4.1. VISUAL PHYSIOLOGICAL BIOMARKERS

As soon as an environmental stress affects the structure of the chloroplast and/or the synthesis of chloroplast pigments the green colour of the leaf will change to either pale green and yellow (lack of chlorophyll synthesis, chlorosis) or to brown (breakdown of chloroplast pigments, necrosis). Chlorosis may be the result of a deficiency of boron, iron, magnesium, nitrogen, sulphur and zinc [80], but may also indicate a surplus of the heavy metals copper [13, 81, 82], lead [83], manganese [84], nickel [85] and zinc [13], and injury caused by viruses, herbicides, and flooding. Chlorosis is only a qualitative, and not a very timely biomarker because it takes some time to desintegrate the structure of the chloroplast; by that time the roots are already severely damaged.

Necrosis is the result of die-off of leaf tissue either indicating nutrient deficiency [80] or a surplus of elements such as fluorides, sulphur dioxide [86], heavy metals [13] and salt spray [87]. The appearance of a necrosis is mostly more delayed than that of chlorosis, except in the case of the impact of fluoride [88], where the specific form of the necrosis is a qualitative biomarker applied in the assessment of the degree of economic damage in horticulture in the vicinity of F-emitting industries.

An enhanced synthesis of anthocyanidins can occur as a result of nitrogen, phosphorus and zinc deficiency [11], a surplus of arsenic causing P-deficiency [89], copper, zinc [13] and salt, or high UV-A radiation [90]. The appearance of visual anthocyanidins in a plant's leaf may occur as early as two days after germination at high zinc exposure, but as late as some months after exposure to salt so that this visual biomarker is not very reliable as early warning system.

10.4.2. VISUAL MORPHOLOGICAL BIOMARKERS

A severe impact on the metabolism of a plant will finally demand a reorganization of sink/source relationships. Dwarfism of a whole plant or of plant parts is the most obvious response to nutrient deficiency [80] or to a surplus of chemical elements in the soil [13] (Fig. 2), air pollutants [86,91], beta and gamma radiation [13,92]. Visual morphological biomarkers are very useful in identifying stressed environments, i.e. environmental quality, but without information on the kind and the quantity of the stressor. The importance of a visual biomarker, i.e. the malformation of leaves of trees, has been demonstrated by monitoring the impact of the radionuclides released during the Chernobyl accident.

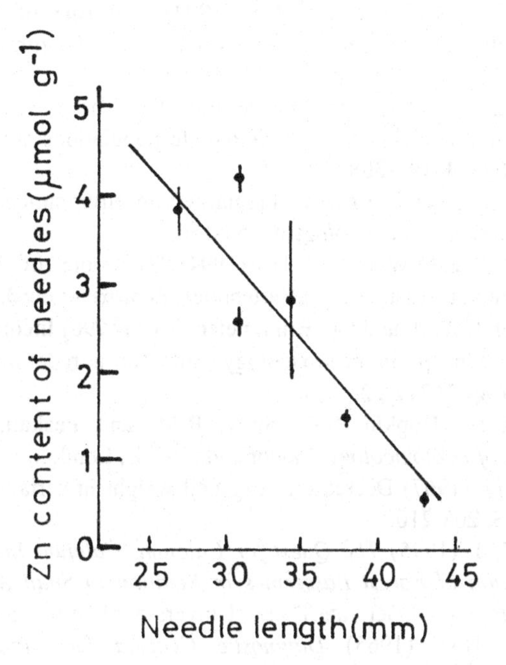

Figure 2. The quantitative relationship between the length of needles of *Pinus sylvestris* (x) and the Zn concentration in the needles (y) in the vicinity of a Zn smelter in the Netherlands. The relation can be described by the function $y = 9.7 - 0.2 x$, $r = 0.90$. Unpublished data from W.H.O. Ernst

10.5. Conclusion

Various biomarkers are elaborated in laboratory experiments under controlled conditions which is a necessary start for a sound understanding of the physiological basis of metabolic changes at exposure to adverse conditions. The question still remains as to whether the biomarker will be useful in the field where there is a coincidence of many environmental factors with positive and negative effects on plant growth. In every case measurable amounts of a biomarker in the field will indicate that there is a stress, but only together with other biological and chemical analyses will it be possible to judge upon the degree of the stress and to incorporate plant biomarkers in environmental assessment.

References

1. Depledge, M.H. and Fossi, M.C. (1994) The role of biomarkers in environmental assessment. (2) Invertebrates, *Ecotoxicology* **3**, 161-172.
2. Ernst, W.H.O. and Peterson, P.J. (1994) The role of biomarkers in environmental assessment. (4) Terrestrial plants, *Ecotoxicology* **3**, 180-192.
3. Lagadic, L., Caquet, T. and Ramade, F. (1994) The role of biomarkers in environmental assessment. (5) Invertebrate populations and communities, *Ecotoxicology* **3**, 193-208.
4. Peakall, D. (1994) The role of biomarkers in environmental assessment. (1) Introduction, *Ecotoxicology* **3**, 157-160.
5. Peakall, D. and Walker, C.H. (1994) The role of biomarkers in environmental assessment. (3) Vertebrates, *Ecotoxicology* **3**, 173- 179.
6. Van Gestel, C.A.M. and van Brummelen, T.C. (1996) Incorporation of the biomarker concept in ecotoxicology calls for a redefinition of terms, *Ecotoxicology* **5**, 217-225.
7. Walker, C.H., Hopkin, S.P., Sibly, R.M. and Peakall, D.B. (1996) *Principles of Ecotoxicology*, Taylor and Francis, London.
8. Ratcliffe, D. (1967) Decrease in eggshell weight in certain birds of prey, *Nature* **215**, 208-210.
9. Graveland, J. (1995) *The Quest for Calcium. Calcium Limitation in the Reproduction of Forest Passerines in Relation to Snail Abundance and Soil Acidification.* Doctorate Thesis, University of Groningen.
10. Chapman, H.D. (1965) *Diagnostic Criteria for Plants and Soils*, University of California, Riverside.
11. Bergmann, W. (1983) *Ernährungsstörungen bei Kulturpflanzen*, Gustav Fischer Verlag, Stuttgart.
12. Hartmann, G., Nienhaus, F. and Butin, H. (1988) *Farbatlas der Waldschäden. Diagnose von Baumkrankheiten*, Verlag Eugen Ulmer, Stuttgart.
13. Ernst, W. (1974) *Schwermetallvegetation der Erde*, Gustav Fischer Verlag, Stuttgart.
14. Ernst, W.H.O. (1993) Population dynamics, evolution and environment: adaptation to environmental stress, in L. Fowden, T. Mansfield and J. Stoddart (eds.) *Plant Adaptation to Environmental Stress*, Chapman & Hall, London, pp. 19-44.
15. Rozema, J., Van de Staaij, J., Björn, L.O. and Caldwell, M. (1997) UV-B as an environmental factor in plant life: stress and regulation, *Trends Ecol. Evol.* **12**, 22-28.
16. Braun, J. and Tevini, M. (1993) Regulation of UV-protective pigment synthesis in the epidermal layer of rye seedlings (Secale cereale L. cv. kustro), *Photochem. Photobiol.* **57**, 318-323.
17. Ernst, W.H.O., Van De Staaij, J.W.M. and Nelissen, H.J.M. (1997) Reaction of savanna plants form Botswana on UV-B radiation, *Plant*

Ecology **128**, 162-170.
18. Verkleij, J.A.C. and Ernst, W.H.O. (1991) Dangerous environmental compounds and their effect on higher plants (in Dutch), in G. P. Hekstra and F.J.M. Van Linden (eds.) *Flora en Fauna Chemisch onder Druk*, Pudoc, Wageningen, pp. 81-102.
19. Shore, L.S., Kapulnik, Y., Gurevich, M., Wininger, S., Badamy, H. and Shemesh, M. (1995) Induction of phytoestrogen production in Medicago sativa leaves by irrigation with sewage water, *Environ. Exp. Bot.* **35**, 363-369.
20. McFarlane, C., Pfleeger, T. and Fletcher, J. (1990). Effect, uptake and disposition of nitrobenzene in several terrestrial plants, *Environ. Tox. Chem.* **9**, 513-520.
21. Markert, B. (1996) *Instrumental Element and Multi-Element Analysis of Plant Samples*, J. Wiley and Sons, Chichester.
22. Meyer, J.J.M., Grobbelaar, N., Vleggaar, R. and Louw, A.J. (1992) Fluoroacetyl-coenzyme A hydrolase-like activity in Dichapetalum cymosum, *J. Plant Physiol.* **139**, 369-372.
23. Twigg, L.E. and King, D.R. (1991): The impact of fluoroacetate-bearing vegetation on native Australian fauna: a review, *Oikos* **61**, 412-430.
24. Klumpp, G., Klumpp, A., Domingos, M. and Guderian, R. (1995) Hemerocallis as bioindicator of fluoride pollution in tropical countries, *Environ. Monit. Assessment* **35**, 27-42.
25. Arnesen, A.K.M. (1997) Availability of fluoride to plants grown in contaminated soil, *Plant Soil* **191**, 13-25.
26. Klumpp, A., Domingos, M. and Klumpp, G. (1997) Assessment of the vegetation risk by fluoride emissions from fertiliser industries at Cubatao, Brazil, *Sci. Total Environ.* **192**, 219-228.
27. Peterson, P.J. and Butler, G.W. (1967) Significance of selenocyctathionine in an Australian selenium-accumulating plant Neptunia amplexicaulis. *Nature* **219**, 599-600.
28. Burnell, J.N. (1981) Selenium metabolism in Neptunia amplexicaulis, *Plant Physiol.* **67**, 316-324.
29. Grill, E., Winnacker, E.L. and Zenk, M.H. (1985) Phytochelatins: the principal heavy-metal complexing peptides of higher plants, *Science* **230**, 674-676.
30. Klapheck, S., Fliegner, W. and Zimmer, I. (1994) Hydroxymethyl-phytochelatins ((gamma-glutamylcysteine)n-serine) are metal-induced peptides in Poaceae, *Plant Physiol.* **104**, 1325-1332.
31. Grill, W., Gekeler, W.K., Winnacker, E.L. and Zenk, M.H. (1986) Homophytochelatins are the heavy metal-binding peptides of homo-glutathione containing Fabaceae, *FEBS* **205**, 47-50.
32. Grill, E., Loeffler, E.L., Winnacker, E.L. and Zenk, M.H. (1989): Heavy metal-binding peptides of plants are synthesized from glutathione by a specific gamma-glutamylcysteine dipeptidyl transpeptidase (phytochelatin synthase), *Proc. Natl. Acad. Sci. USA* **86**, 6838-6842.

33. Tukendorf, A. and Rauser, W.E. (1990) Changes in glutathione and phytochelatin in roots of maize seedlings exposed to cadmium, *Plant Sci.* **70**, 155-166.
34. De Knecht, J.A. (1994) *Cadmium Tolerance and Phytochelatin Production in Silene vulgaris*, Doctorate Thesis, Vrije Universiteit Amsterdam.
35. Ernst, W.H.O. (1997) Effects of heavy metals in plants at the cellular and organismic level, in: G. Schüürmann and B. Markert (eds.) *Ecotoxicology*, J. Wiley, London pp. 587-620.
36. Schat, H. and Kalf, M.A. (1992) Are phytochelatins involved in differential metal tolerance or do they merely reflect metal-imposed strain?, *Plant Physiol.* **99**, 1475-1480.
37. Vögeli-Lange, R. and Wagner, G.J. (1990) Subcellular localization of cadmium and cadmium-binding peptides in tobacco leaves, *Plant Physiol.* **92**, 1086-1093.
38. Salt, D.E. and Wagner, G.J. (1993) Cadmium transport across tonoplast of vesicles from oat roots, *J. Biol. Chem.* **268**, 12297-12302.
39. Lichtenberger, O. and Neumann, D. (1997) Analytical electron microscopy as a powerful tool in plant cell biology: examples using electron energy loss spectroscopy and X-ray microanalysis, *Eur. J. Cell Biol.* **73**, 378-386.
40. De Knecht, J.A., Van Baren, N. , Ten Bookum, W.M., Wong Fong Sang, H.W., Koevoets, P.L.M., Schat, H. and Verkleij, J.A.C. (1995) Synthesis and degradation of phytochelatins in cadmium-sensitive and cadmium-tolerant Silene vulgaris, *Plant Sci.* **106**, 9-18.
41. Gekeler, W., Grill, E., Winnacker, E.L. and Żenk, M.H. (1988) Algae sequester heavy metals via synthesis of phytochelatin complexes, *Arch. Microbiol.* **150**, 197-202.
42. De Vos, C.H.R., Vonk, M.J., Vooijs, R. and Schat, H. (1992) Glutathion depletion due to copper-induced phytochelatin synthesis causes oxidative stress in Silene cücubalus. *Plant Physiol.* **98**, 853-858.
43. Inouhe, M., Ninomiya, S., Tohoyama, H., Joho, M. and Murayama, T. (1994) Different characteristics of roots in the cadmium-tolerance and Cd-binding complex formation betweén mono- and dicotyledonous plants, *J. Plant Res.* **107**, 201-207.
44. De Knecht, J.A., Van Dillen, M., Koevoets, P.J.M., Schat, H., Verkleij, J.A.C. and Ernst, W.H.O. (1994) Phytochelatins in cadmium-sensitive and cadmium-tolerant Silene vulgaris, *Plant. Physiol.* **104**, 255-161.
45. Ernst, W.H.O. (1995) Sampling of plant material for chemical analysis, *Sci. Total Environ.* **176**, 15-24.
46. Grill, E., Winnacker, E.L. and Zenk, M.H. (1988) Occurrence of heavy metal binding phytochelatins in plants growing in a mining refuse area, *Experientia* **44**, 539-540.
47. Lolkema, P.C., Donker, M.H., Schouten, A.J. and Ernst, W.H.O. (1984) The possible role of metallothioneins in copper tolerance of Silene cucubalus, *Planta* **162**, 174-179.

48. Vögeli-Lange, R and Wagner, G.J. (1996) Relationship between cadmium, glutathione and cadmium-binding peptides (phytochelatins) in leaves of intact tobacco seedlings, *Plant Sci.* **114**, 11-18.
49. Gallego, S.M., Benavides, M.P. and Tomaro, M.L. (1996) Oxidative damage caused by cadmium chloride in sunflower (Helianthus annuus L.) plants, *Phyton* **58**, 41-52.
50. McLaughlin, S.B. and Wang, Y.C. (1996) Foliar phytochelatin levels as indicators of forest decline, *Trend Plant Sci.* **1**, 249-250.
51. Ernst, W.H.O., Verkleij, J.A.C. and Schat, H. (1992) Metal tolerance in plants, *Acta Bot. Neerl.* **41**, 229-248.
52. Depledge, M.H., Aagaard, A. and Gyorkos, P. (1995) Assessment of trace metal toxicity using molecular, physiological and behavioural biomarkers, *Mar. Pollut. Bull. 31, Spec.Iss.* **1**, 19-27.
53. Fujita, Y., El-Belbasi, H.I., Min, K.S., Onosaka, S., Okada, Y., Matsumoto, Y., Mutoh, N. and Tanaka, K. (1993) Fate of cadmium bound to phytochelatin in rats. *Res. Commun. Chem. Pathol. Pharm.* **82**, 357-365.
54. Keller, Th. (1974) The use of peroxidase activity for monitoring and mapping air pollution areas, *Eur. J. Forest Pathol.* **4**, 11-19.
55. Ranieri, A., Schenone, G., Lencioni, L., and Soldatini, G.F. (1994) Detoxificant enzymes in pumpkin grown in polluted ambient air, *J. Environ. Qual.* **23**, 360-364.
56. Koricheva, J., Roy, S., Vranjic, J.A., Haukioja, E., Hughes, P.R. and Hanninen, O. (1997) Antioxidant responses to simulated acid rain and heavy metal deposition in birch seedlings, *Environ. Pollut.* **95**, 249-258.
57. Perez Soba, M., Van der Eerden, L.J.M., Stulen, I. and Kuiper, P.J.C. (1994) Gaseous ammonia counteracts the response of Scots pine needles to elevated atmospheric carbon dioxide, *New Phytol.* **128**, 307-313.
58. Rao, M.V., Paliyath, G. and Omrod, D.P. (1996) Ultraviolet-B- and ozone-induced biochemical changes in antioxidant enzymes of Arabidopsis thaliana, *Plant Physiol.* **110**, 125-136.
59. Weckx, J.E.J. and Clijsters, H.M.M. (1997) Zn phytotoxicity induces oxidative stress in primary leaves of Phaseolus vulgaris, *Plant Physiol. Biochem.* **35**, 405-410.
60. Stewart, G.R. and Lee, J.A. (1974) The role of proline accumulation in halophytes, *Planta* **120**,279-289.
61. Aspinall, D. and Paleg, L.G. (1981) Proline accumulation: physiological aspects, in L.G. Paleg and D. Aspinall (eds.) *The Physiology and Biochemistry of Drought resistance in Plants*, Acadmic Press, Sydney, pp. 206-241.
62. Köhl, K. (1996) Population-specific traits and their implication for the evolution of a drought-adapted ecotype in Armeria maritima, *Bot. Acta* **109**, 206-215.
63. Öztürk, M. and Szaniawski, R.K. (1981) Root temperature stress and proline content in leaves and roots of two ecologically different plant species, *Z. Pflanzenphysiol.* **102**, 375-377.

64. Farago, M.E. (1981) Metal tolerant plants, Coord. Chem. Rev. **36**, 155-162.
65. Bassi, R. and Sharma, S.S. (1993) Changes in proline content accompanying the uptake of zinc and copper by Lemna minor, *Ann. Bot.* **72**, 151-154.
66. Costa, G. and Morel, J.L. (1994) Water relations, gas exchange and amino acid content in Cd-treated lettuce, *Plant Physiol. Biochem.* **32**, 561-570.
67. Schat, H., Sharma, S.S. and Vooijs, R. (1997) Heavy metal-induced accumulation of free proline in a metal-tolerant and a nontolerant ecotype of Silene vulgaris, *Physiol. Plant.* **101**, 477-482.
68. Heikkila, J.J., Papp, J.E.T., Schultz, G.A. and Bewley, D. (1984) Induction of heat shock protein messenger RNA in maize mesocotyls by water stress, abscisic acid and wounding, *Plant Physiol.* **76**, 270-274.
69. Neumann, D., Lichtenberger, O., Guenther, D., Tschiersch, K. and Nover, L. (1994) Heat-shock proteins induce heavy-metal tolerance in higher plants, *Planta* **194**, 360-367.
70. Trebst, A. (1991) The molecular basis of resistance of photosystem II inhibitors, in J.C. Caseley, G.W. Cussans and R.K. Atkin (eds.) *Herbicide Resistance in Weeds and Crops*, Butterworth-Heinemann, Oxford, pp. 145-164.
71. Hoeks, J. (1972) Effect of leaking natural gas on soil and vegetation in urban areas. *Agric. Res. Rep.* Wageningen, 778.
72. Lambert, H.J., Ardo, J., Rock, B.N. and Vogelmann, J.E. (1995) Spectral characterization and regression-based classification of forest damage in Norway spruce stands in the Czech Republic using Landsat Thematic Mapper data. *Internat. J. Remote Sensing* **16**, 1261-1287.
73. Carter, G.A., Cibula, W.G. and Miller, R.L. (1996) Narrow-band reflectance imagery compared with thermal imagery for early detection of plant stress, *J. Plant Physiol.* **148**, 515-522.
74. Lichtenthaler, H.K. (1996) Vegetation stress: an introduction to the stress concept in plants, *J. Plant Physiol.* **148**, 4-14.
75. Govindjee, Downton, W.J.S., Fork, D.C. and Armond, P.A. (1981) Chlorophyll a fluorescence transient as an indicator of water potential of leaves. *Plant Sci. Lett.* **20**, 191-194.
76. Havaux, M. and Lannoye, R. (1984) Effects of chilling temperature on prompt and delayed chlorophyll fluorescence in maize and barley leaves, *Photosynthetica* **18**, 117-127.
77. Moustakas, M., Ouzounidou, G. and Lannoye, R. (1994) Rapid screening for aluminium tolerance of cereals by the use of chlorophyll fluorescence test, *Plant Breed.* **111**, 343-346.
78. Ouzounidou, G. (1993) Changes in variable chlorophyll fluorescence as a result of Cu-treatment: Dose-response relationship in Silene and Thlaspi, *Photosynthetica* **29**, 455-462.
79. Lichtenthaler, H.K., Lang, M., Sowinska, M., Heisel, F. and Miehé, J.A. (1996) Detection of vegetation stress via a new high resolution

fluorescence imaging system, *J. Plant Physiol.* **148**, 599 - 612.
80. Marschner, H. (1995) *Mineral Nutrition of Higher Plants*, Academic Press, London.
81. Ouzounidou, G. (1994) Root growth and pigment composition in relationship to element uptake in Silene compacta plants treated with copper, *J. Plant Nutr.* **17**, 933-943.
82. Mocquot, B., Vangronsveld, J., Clijsters, H. and Mench, M. (1996) Copper toxicity in young maize (Zea mays L.) plants: effects on growth mineral and chlorophyll contents, and enzyme activities, *Plant Soil* **182**, 287-300.
83. Moustakas, M., Lanaras, T., Symeonidis, L. and Karataglis, S. (1994) Growth and some photosynthetic characteristics of field grown Avena sativa under copper and lead stress, *Photosynthetica* **30**, 389-396.
84. Horst, W.J. and Marschner, H. (1978) Einfluss von Silizium auf den Bindungszustand von Mangan im Blattgewebe von Bohnen (Phaseolus vulgaris), *Z. Pflanzenernähr. Bodenkd.* **141**, 487-497.
85. Crooke, W.M. (1955) Further aspects of the relationship between nickel toxicity and iron supply, *Ann. appl. Biol.* **43**, 465-476.
86. Dässler, H.G. (1976) *Einfluss von Luftverunreinigungen auf die Vegetation*, G.Fischer Verlag, Jena.
87. Ernst, W.H.O. and Feldermann, D. (1975) Auswirkungen der Wintersalzstreuung auf den Mineralstoffhaushalt von Linden, *Z. Pflanzenernähr. Bodenkde* **1975**, 629-640.
88. Hitchcock, A.E., Zimmerman, P.W. and Coe, R.R. (1962) Results of ten years work (1951 - 1960) on the effect of fluoride on Gladiolus, *Contr. Boyce Thompson Inst.* **21**, 303-344.
89. Otte, M.L. and Ernst, W.H.O. (1994) Arsenic in vegetation of wetlands, in J.O. Nriagu (ed.) *Arsenic in the Environment, Part 1: Cycling and Characterization*, J. Wiley & Sons, New York, pp. 365-379.
90. Larcher, W. (1993) *Ökophysiologie der Pflanzen*, 5. Ed., Ulmer Verlag, Stuttgart.
91. Leith, I.D., Sheppard, L.J. and Cape, J.N. (1995) Effects of acid mist on needles from mature Sitka spruce grafts. 2. Influence of developmental stage, age and needle morphology on visible damage, *Environ. Pollut.* **90**, 363-370.
92. Savchenko, V.K. (1995) *The Ecology of the Chernobyl Catastrophe*, Parthenon Publishing Group, New York.

11. BIOMARKERS IN INVERTEBRATES

Evaluating the effects of chemicals on populations and communities from biochemical and physiological changes in individuals

LAURENT LAGADIC
Institut National de la Recherche Agronomique (INRA)
Station Commune de Recherche en Ichtyophysiologie,
Biodiversité et Environnement (SCRIBE)
Unité d'Ecotoxicologie Aquatique
Campus de Beaulieu - F-35042 Rennes cedex - France

Abstract

Invertebrates offer a number of advantages for *in situ* measurement using animal biomarkers. Their ability to colonize many different compartments of terrestrial and aquatic ecosystems mainly result from both the development of various strategies for resource exploitation and a high biotic potential as expressed by fast growth and high reproduction rates for most species. Due to their position at different levels in food webs, invertebrates play functional key-roles in ecosystems and when affected by chemicals, they can be responsible for dramatic changes in community structure and function. Therefore, they display a great potential for evaluating the ecological impact of pollutants. Adaptability is probably a key-feature of invertebrates in heavily deteriorated environments. Invertebrates can thus be found in highly polluted habitats where vertebrates are virtually absent. In this case, they frequently represent the only way to evaluate the effects of pollutants on animal biological systems. Biomarkers measured in invertebrates allow diagnosis of long-term effects of chemicals. The development of resistance to toxic chemicals, physiological adaptation involving changes in metabolic energy allocation, occurrence of morphological or anatomical abnormalities, are situations in which changes induced by long-term exposure of the individuals have consequences on population structure and dynamics. Biomarkers measured at the individual level may thus be linked with population changes from which further impact on community structure and function may be expected. Such an approach can be used to classify polluted areas according to the level of ecological damage in order to identify priority cases for remediation. The same biomarkers can then be used as early signs of biological restoration of damaged ecosystems. For example, enzymatic biomarkers can indicate decreased levels of resistance within exposed populations as a result of decreased selection pressure on individuals. Similarly, changes in energy allocation, as assessed through individual evaluation of scope for growth and scope for reproduction, can occur as a

D. B. Peakall et al. (eds.),
Biomarkers: A Pragmatic Basis for Remediation of Severe Pollution in Eastern Europe, **153–175.**

response to better environmental conditions and may result in changes in population dynamics. Measurement of biomarkers in invertebrates certainly provide valuable information on both the ecological impact of long-term chemical contamination (diagnostic approach) and the conditions of biological restoration of damaged ecosystems (predictive approach). Both approaches however require a pertinent choice of invertebrate species and the assessment of linkable responses at different levels of biological organisation.

Keywords: invertebrate population , resistance, endocrine disruptors, remediation.

11.1. Introduction

The general framework of the monitoring of ecosystem quality can be defined on the basis of the following four questions [1]:

How bad is the situation ?
Is it getting better or worse ?
Why do we have these problems ?
What can we do to improve the situation ?

The first question refers to the assessment of ecosystem quality, whereas the second implies a dynamic approach which tries to establish how the ecosystem quality is changing. The third question refers to the origin of environmental changes, and may result in a more fundamental approach based on experimental works. The last question has at least two major components: a scientific component which may result in proposals for remediation, and a socio-economic component which involves efforts from public authorities and industries. This last question will not be further developed in the present paper. Instead, emphasis will be placed on how invertebrates can help to answering the first three questions.

Invertebrates have long been used in biological methods of assessment of environment quality. Beside their widespread use as bioindicators [2-8], invertebrate species have received particular attention as sources of biomarkers [9,10]. The present paper discusses reasons for doing research on biomarkers in invertebrates, leading to the development and regular use of some of them for the diagnosis of environmental quality. Those reasons mainly refer to ecological and biological characteristics of invertebrates that render them useful (A) in many different compartments of terrestrial and aquatic ecosystems, and (B) for the extrapolation of biochemical/physiological disturbances in individuals to changes at the population level, with putative or actual consequences on community structure and functions. In this respect, invertebrate biomarkers may support a predictive approach which may help to answer the additional question: *how is the situation likely to evolve once remediation has been undertaken ?*

11.2. The Biomarker Approach

11.2.1. SIMPLIFIED PRESENTATION

Broadly speaking, biomarkers are biochemical or physiological parameters that can be measured in individuals to indicate exposure to environmental chemicals and, for some of them, to detect toxic effects [11-15]. As such, biomarkers describe individual health, and therefore can be considered as diagnostic tools. When individual changes revealed by biomarkers can be connected to actual or potential changes at the population level, those biomarkers can be considered as predictive tools [9,16]. Both approaches are important components of the assessment of the extent of ecological damage, and of the evolution of ecosystem quality, especially when remediation has been undertaken (Fig. 1).

A number of biomarkers have been identified in invertebrates [9,12], and some of them, such as cytochrome P450, metallothioneins, esterases and DNA damage, which have the advantages of specificity, sensitivity and applicability are at an advanced stage of usage or development as diagnostic tools [17]. However, the demonstration of mechanistic links between biomarkers and population changes is not straightforward [18,19]. Although invertebrates appear as suitable animal models for such a purpose, only a few studies, that will be further developed, have assigned a predictive role to biomarkers.

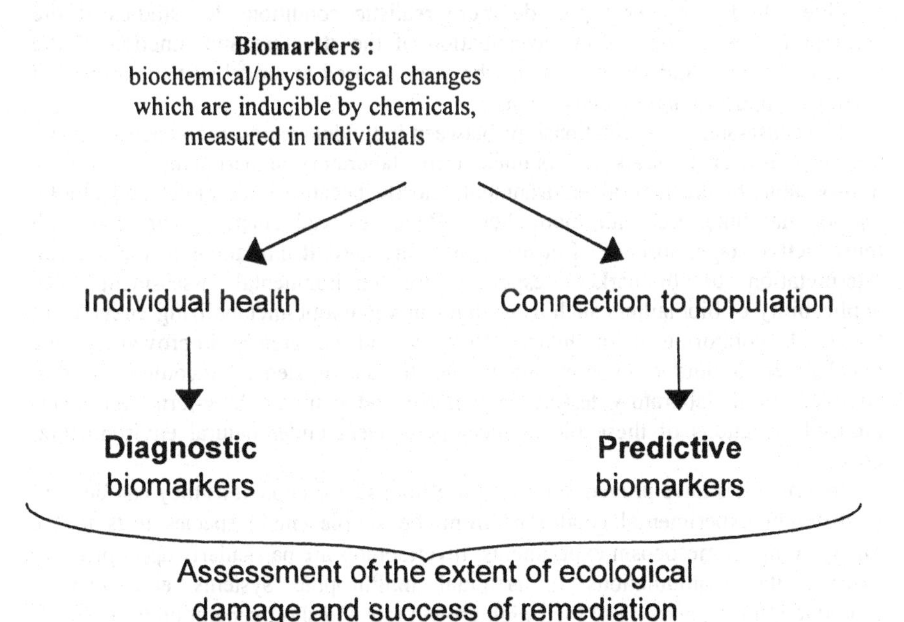

Figure 1. Biomarkers as diagnostic and predictive tools for environmental quality assessment

11.2.2. SIGNIFICANCE OF BIOMARKERS IN LABORATORY AND FIELD CONDITIONS

Biomarkers can be measured in different experimental contexts. The significance of biomarkers very much depends on the situation in which they are measured. Figure 2 illustrates this in the case of the evaluation of chemical impact in the aquatic environment.

Laboratory conditions usually provide high replicability and reproducibility, but rather low ecological realism, even though indoor systems can be fairly complex as in the case of microcosms. Outdoor systems are poorly replicable and reproducible, but provide more realistic conditions of exposure to chemicals. In any of these situations, biochemical or physiological measurements can provide useful information to establish diagnosis in individuals.

Single species tests and experiments on a limited number of species usually allow investigations on the mode of action of chemicals, and may provide basic information on the origin of effects. They may help to answer the question '*why do we have these problems?*' However, such experiments are usually restricted to one chemical whereas environmental problems frequently arise from mixtures of contaminants. Moreover, tests performed on laboratory strains often under-estimate inter-individual variability of the response to chemical exposure.

As the systems become more complex, investigations on effects at the population level and on effects upon interactions between selected species are made possible. Outdoor systems provide more realistic conditions for studies at the community level. They allow investigation of the structure and function of the whole ecosystem, and for this reason, they can be used as model systems to predict chemical impact in natural ecosystems.

The cause-and-effect relationships between individual exposure to chemicals and the response of biomarkers obtained from laboratory experiments are hardly transposable to the natural environment, mostly because both biotic and abiotic factors may interfere with biomarkers. Thus, seasonal changes, variation with reproductive stage, absence of dose-response are current limitations to the use and interpretation of biomarker responses for environmental assessment. The applicability of biomarkers such as changes in xenobiotic-metabolizing enzymes to the field monitoring of pollutant effects would be greatly improved by the development of outdoor experimental reconstituted ecosystems that could be used in complement to laboratory testing to validate and calibrate long-term, pollutant-induced variations of these biochemical parameters under natural environmental conditions.

Invertebrates have proved to be very useful in such a context as they can be used in different experimental conditions from the simple single species tests to the highly complex mesocosm experiments. Invertebrates are particularly appropriate to experimental manipulations in artificial multitrophic systems and also to reproduction tests because of their high growth rate and short generation time.

11.3. The use of invertebrates in environmental quality assessment

The main advantages and disadvantages of using invertebrates in the assessment of the effects of environmental chemicals have recently been reviewed [20]. Only the most important features are summarized below.

11.3.1. INVERTEBRATES IN LABORATORY TESTING

Because of their small size and high reproductive rate, large numbers of invertebrates can be maintained for many generations in fairly limited space. This, and the use of simplified, controlled diet, reduce the cost of maintenance and improve subsequent toxicological and biochemical investigations.

Toxicity testing of chemicals in invertebrates is improved by many replicates, by the homogeneity of individual size, age and development stage, and by the rapidity of response, even at the population level since, for some species, individuals can reproduce within a few weeks. The use of invertebrates in toxicological experiments raises considerably less public concern than does the use of vertebrates; however, the results of invertebrate toxicity tests are not always directly transferable to higher animals and to man.

Most of the biomarkers which have been identified in vertebrates can potentially be measured in invertebrates, the only exceptions being biomolecules, structures or processes which have no equivalents in the latter (e.g. thyroid hormones, endoskeletal structures, kidney or lung functions, etc). However, many essential biological molecules and processes are highly conserved throughout the Animal Kingdom so that methodological and functional similarities allow common interpretations of the molecular basis of toxicant action in individuals. The expression and significance of molecular/biochemical changes at the individual level may, however, differ when observed in invertebrates or in vertebrates. The size of individuals remains a limit to biochemical investigations in invertebrates. In spite of highly sensitive molecular techniques, it is still often necessary to pool several individuals for one given analysis, making it impossible to estimate inter-individual variability [21].

11.3.2. INVERTEBRATES FOR FIELD ASSESSMENT OF THE EFFECTS OF CHEMICALS

Invertebrates are also most useful for field investigations of chemical impact [2-8,17]. Field data obtained in invertebrates usually have high ecological relevance because the large number of species and diversity of their functional roles allow investigations of critical compartments or processes of ecosystems, the first important critical step that can be thoroughly investigated in invertebrates being reproduction.

In spite of :

- small individual size, which can complicate sampling even when it is counterbalanced by individual abundance,
- taxonomy, which is quite difficult and often unstable,

- crossbreeding between individuals from different subspecies and strains, but also from species described as taxonomically distinct,
- seasonal abundance which may reduce sampling to particular time periods,

invertebrates fulfil most of the characteristics of sentinel species, as defined by Lower and Kendall [22]:

- wide geographic and ecological distribution,
- restricted home range of individuals,
- presence of a particular species at a critical location
- availability of easily measured and/or significant biological endpoint(s) in a species,
- ability of a species to be used as a probe of a particular aspect of the environment (niche, habitat, trophic position, etc).

In addition, due to their positions in food webs, invertebrates are important in biomagnification processes. For species that have identificable generations, changes in population structure can reflect effects of pollutants. Invertebrates tend to be more 'tolerant' towards pollutants and other stressors than vertebrates, due to lower energy requirements, passive protection mechanisms (e.g. quiescence, dormancy), the existence of less vulnerable stages (e.g. pupae or cysts), and the ability to modify metabolism (e.g. shift to anaerobic metabolism). Furthermore, invertebrates display a high, genetically-based adaptability to environmental changes.

11.4. Long-term changes in invertebrate populations

Typical case-studies of long-term population changes in invertebrates include :

- abnormalities and morphological deformities described in chironomids [23-27] and molluscs [28,29] exposed to metals, hydrocarbons, or pesticides;
- changes in energy allocation to growth and reproduction in numerous species for wide range of pollutants [30];
- damage to the genetic material in molluscs [31-38], polychaetes [39], earthworms [40-42] or echinoderms [43] exposed to hydrocarbons, pesticides or metals;
- resistance to pesticides identified in many insects, mites, and nematodes (see 11.4.1);
- change in sex-ratio and reproductive performances demonstrated in several species of marine molluscs exposed to organotin compounds (see 11.5.2).

Some of these changes induced by chemicals can be directly assessed in field populations.

11.4.1. DIRECT ASSESSMENT OF LONG-TERM POPULATION CHANGES : INSECT RESISTANCE TO PESTICIDES

In susceptible organisms, pesticides exert their toxic effect by acting on specific targets, thus inducing metabolic and physiological disorders. Like many other organisms, invertebrates are capable of genetic adaptation conferring resistance to most toxicants when under long-term selection pressure [44-46]. In resistant organisms, various mechanisms prevent the adverse effects from occurring. The most important resistance mechanisms are changes in target site sensitivity, and enhanced detoxication which increases the clearance of the toxicant.

According to the NAS/NRC classification of biological markers [47], resistance represents a marker of susceptibility to the effects of toxicant exposure. When used in populations of invertebrate pests, it may indicate that the threshold of efficacy of pesticides is reached, and that increases of the spray rates and/or number of applications would only contribute to an excessive contamination of the environment. From the ecotoxicological point of view, resistance to pesticides is the only population-level effect that can be directly correlated with genetically determined *biochemical changes* in individuals. Biochemical techniques offer the possibility of detecting the initial stages of resistance in a population and the mechanism(s) of resistance involved.

Resistance mechanisms have been characterised in a number of species [48-50]. Insensitive acetylcholinesterase, modified sodium channel proteins, and changes in GABA receptor amino-acids are common resistance mechanisms at the target site level. Overexpression of esterases and synthesis of isoforms of cytochrome P450 and glutathione *S*-transferases that specifically degrade pesticides are classically recognised as the basis of enhanced detoxication. Also, the genetic basis of each mechanism has been characterised, so that the frequency of resistant genotypes within a population can be determined. Variations in the frequency of resistant genotypes may indicate how the pressure induced by pesticides on the population varies. This may be used to establish a sort of 'toxicological history' of the population [50].

Resistance to pesticides cannot be fully understood without the perspectives of population genetics and ecology ; however, progress in this area has been slow as compared to that on biochemistry of resistance [50-52]. Indeed, it is difficult to study the genetics of resistance in field populations. While some populations of major pests develop resistance to most pesticides to which they are repeatedly exposed, many other populations of target or non-target species remain susceptible, even though exposed to similar spray rates and numbers of applications [51]. Variation in population structure probably contributes to these differences. Population structure depends on behaviour, sex ratio, distribution of resources, population size, the degree of deviation from random mating, and spatial dispersion that determines the flow of genes from one location to another [52].

The development of resistance depends on the genetic variability already present in a population, or arising during the period of the selection. It results from the evolution of mutations that establish resistance alleles in natural populations of a species. [52,53]. Induction of detoxication enzymes may play an important role in this process, by preserving susceptible individuals in which a resistance mechanism is present but unexpressed (e.g. a new mutation or a single copy of a recessive allele conferring resistance) [54,55]. Widespread applications of pesticides propagate resistance alleles through preferential survivals, and they become dispersed throughout the population [46]. There are at least three types of genetic changes that might cause resistance: (A) gene amplification or duplication, (B) changes in particular DNA base pair that modify protein structure, and (C) changes in the number of target sites or the amount of enzymes, possibly through changes in genes that regulate the amount of proteins produced. These mechanisms have been extensively reviewed elsewhere [48,50,52,53].

The various mechanisms conferring resistance to pesticides can be present alone or in combination, resulting in multiple-resistant populations or strains that can resist the majority of pesticides. Polygenic resistance to a single pesticide results from sequential addition of genes selected by repeated exposures to that pesticide or other toxicants [52]. In field selection for chlorpyrifos resistance in *Culex pipiens*, for example, resistance was initially monofactorial and associated strictly with a highly active detoxifying esterase, but there was later evolution which led to the inclusion of at least two more factors, an altered AChE and a microsomal monooxygenase [56,57].

The reversion of resistance classically occurs when the selection pressure induced by pesticide exposure decreases [55]. In south-european populations of *Culex pipiens*, the frequency of genes conferring resistance to insecticides is much lower in untreated areas as compared to those where pesticides are intensively used. Furthermore, in winter, when no insecticides are applied, the frequency of resistant AChE genes decreases from 25% to 9% in females of the hibernating populations suggesting that mortality of the resistant individuals is approximatively 1% per day during this period [58]. Counter-selection of resistant genotypes in natural populations is of primary importance to assess the extent of ecological restoration when concentrations of environmental contaminants are decreasing.

The relationship between resistance to pesticides and ecological fitness has important implications at the population level [59-61]. It is usually considered that high levels of resistance to specific toxicants are associated with reduced ability to tolerate the effects of other pollutants or of natural stressors. Decreased fertility and reduced viability have also been observed in natural populations in which the genetic variability was reduced by stress exposure [62]. However, the reproductive capacity of resistant individuals may not be affected, except when the resistance mechanism is based upon esterase amplification which results in enhanced energy expenditure [50].

Community level effects related to resistance in insect populations remain poorly documented. It has been shown that resistant strains of the aphid *Myzus persicae* may be more exposed to predators than the susceptible ones as a result of decreased sensitivity to alert pheromone [63]. Recent investigations have demonstrated that the development of resistance in *Drosophila melanogaster* may affect its relationship with the parasitoid *Leptopilina boulardi* through changes in the host cellular immune system [64].

Genetic studies in non-target invertebrate species, especially key species in terrestrial ecosystems, should be promoted so that the biochemical and molecular genetic methods can be used to investigate the regulation of resistance genes at the population level and to identify consequences of the development of resistance on communities.

11.4.2. DIRECT ASSESSMENT OF LONG-TERM POPULATION CHANGES : ANOXIC SURVIVAL IN BIVALVE MOLLUSCS

An interesting approach has been developed in marine bivalves exposed to various pollutants, metals, PAHs or PCBs. This approach known as the 'Stress Approach' [65] or 'Survival in Air' [66] is based on the ability of the molluscs to survive in air by shifting to anaerobic metabolism. The anoxic survival time has been shown to be affected by the exposure to xenobiotics, namely cadmium and PCBs [66,67], and when combined to biochemical measurements, it may allow to determine the frequency of affected individuals within a population.

Individual responses to aerial exposure and increased temperature have been studied in *Mytilus edulis* collected from polluted and non-polluted areas using stress indicators ranging from the molecular to the organismal level [68]. Heat-shock protein synthesis, adenylate energy charge, and anoxic survival time were significantly affected in individuals from polluted areas when they were subjected to controlled stress in the laboratory, whereas AEC and the condition index (i.e. the ratio of soft tissue dry weight to total weight) remained unchanged in animals collected in the same polluted sites. This demonstrates that the direct measurement of biomarkers in field populations may not account for the actual health status of the individuals as affected by the exposure to contaminants. In spite of the possible influence of physico-chemical conditions [69] and acclimation of the organisms [70], the 'stress approach' appears as a promising method to detect individual changes induced by pollutants as it improves the reliability of biomarkers.

So far, anoxic survival has been mainly studied in marine bivalves [71-74]. It is likely to be applicable to freshwater bivalves and other invertebrates which are able to tolerate hypoxic or anoxic conditions [75]. However, it is still necessary to look for relationships between the response of such biomarkers and changes in populations. These relationships may not be straightforward because of compensatory mechanisms which regulate population dynamics in natural systems.

In the previous examples, the assessment of chemical impact on invertebrate populations is based on markers measured at the individual level. The frequency of affected individuals provides an evaluation of the impact on the whole population. Extrapolation of the effects of chemicals from the individual to the population can also be made on a physiological basis. Impairments in critical physiological functions may also affect interspecific relationships with possible consequences on community structure and functions.

11.5. Extrapolation from the individual to the population and community

11.5.1. GENERAL OVERVIEW

The action of environmental pollutants on the individual may result in various internal changes (Fig. 3). Once a chemical penetrates the organism, it binds to various molecules within different tissues. Those molecular interactions result in metabolic changes which may affect different functions at the cellular and tissue levels with consequences at the individual level (e.g. disease development, reduced growth, altered behaviour, impaired reproduction). In this respect, impact on the nervous system usually results in serious disorders in individuals and also in changes in inter-individual relationships.

Most metabolic processes and individual functions are mediated by hormones and require energy. Hormonal control and energetic balance are therefore of primary importance as they both determine eventually the reproductive capacity of the individual, and reproduction is the key-process that links the individual to the population. Successful reproduction is essential for the continuation of the species. The reproductive success of individuals determines population characteristics in terms of structure/composition, density/size, and stability/evolution.

Community level effects depend on both population functions and interactions between populations. Population functions result from both the previous demoecological parameters (structure/composition, density/size, stability/evolution) and individual health, and may induce changes in physicochemical conditions. Interactions between populations are considered as an essential component of community stability and ecosystem function. If the external perception of environmental factors in individuals is affected by neurotoxic pollutants, changes in interspecific relationships may occur, with potential consequences on communities.

Biological criteria related to behaviour, hormone metabolism, and bioenergetics have been identified as biomarkers in a number of invertebrate species [76].

Behaviour is the individual outcome of a sequence of neurophysiological events involving sensory neurons, motor neurons, muscular contraction and release of chemical messages. Thus, neurotoxic pesticides will be the main contaminants to elicit behavioural changes although some observations suggest that metals could also significantly affect invertebrate behaviour. There can be alterations in various activities such as avoidance, foraging, feeding or locomotion [77-82]. Behavioural responses can be adaptive, representing an attempt to mitigate the effects of environmental disturbances, or maladaptive, in which a significant deviation from the behavioural norm occurs [83]. Adaptative behavioural changes such as avoidance behaviour are often the initial response of mobile organisms to environmental perturbations and, therefore, are most difficult to assess in the field. Maladaptive behaviours, e.g. changes in foraging, feeding, or reproductive behaviours, can be particularly insidious as the potential for long-term individual or species survival is reduced.

Invertebrate reproduction, as expressed through fecundity and fertility, can be impaired as a result of both direct impact of toxicants on the endocrine system, and indirect effect on energy allocation. Besides alterations in courtship and mating behaviour, changes in reproduction can be caused by changes in maturation of gametes, fertilization process and age of sexual maturity, or by morphological abnormalities in adults. Reduction of fertility results from the effects of contaminants on egg viability, larval development, or hatching of young. Hormonal and energetic biomarkers can be used to detect changes in fecundity and fertility, and can also be related to individual growth [9,30,76,84]. Such biomarkers are therefore considered as potential predictive indicators of changes in populations, with possible consequences for communities. However, only a few studies have so far demonstrated causal links between hormonal or energetic alterations in individuals and changes in populations and communities. This requires a good knowledge of the succession of pollutant effects across levels of biological organisation, as illustrated by the following examples.

11.5.2. EFFECTS OF ENDOCRINE DISRUPTORS IN INVERTEBRATES : CASE-STUDY OF TRIBUTYLTIN (TBT) IN COASTAL ECOSYSTEMS

The effect of pollutants on the endocrine system have received considerable attention over the past five years [85]. Endocrine-disrupting effects of environmental chemicals have long been demonstrated in invertebrates. A number of plant allelochemicals are known to affect growth and reproduction of insects by acting as hormone mimics [86]. Synthetic hormone mimics have also been developed as insecticides against insect pests (e.g. diflubenzuron, teflubenzuron, tebufenozide) [87-89]. More recently, endocrine-disrupting effects of environmental chemicals have been identified in freshwater crustaceans [90-93] and marine molluscs [94,95]. In these latter, tributyltin (TBT) was identified as the compound acting on the reproductive system of individuals, and from a twenty year long field survey, the progression of effects from the molecular level to communities has been established, as summarized below.

TBT is an organotin compound used in antifouling paints applied onto boats. Long-term studies on marine gastropod populations have demonstrated that TBT induces a phenomenon named 'imposex' which results in the development of male characteristics in females. This change was first identified in *Nucella lapillus* [96] and *Nassarius obsoletus* (= *Ilyanassa obsoleta*) [97] and has been reported in more than 40 species of marine mollusc exposed to TBT [98]. The dogwhelk *Nucella lapillus* and the oyster drill *Ocinebrina aciculata* are highly susceptible to TBT and are being largely used for investigating its effects at sub-cellular, individual and population levels (reviews in 94,95,99). In females, TBT may inhibit cytochrome P450-dependent aromatase which converts testosterone to 17β-oestradiol. Inhibition of sulphur conjugation and excretion of testosterone and its metabolites is another putative pathway of TBT mode of action. In any of those situations, the level of male steroids increases, leading to anatomical changes in the reproductive system. Following the development of the penis and vas deferens, the affected females become sterile and eventually die. As a result of reduced reproductive output and individual death, populations decline. Extinctions of populations of *N. lapillus* and *O. aciculata* have even been reported [99,100].

In spite of the role of *N. lapillus* in shore communities as a necrophagous species, as has been established by experimental manipulations of field populations, no community level effects were clearly demonstrated as a consequence of dogwhelk population decline. This may result from the lack of long-term observations of dogwhelks in the context of the whole community on such shores, and also from compensatory mechanisms which are able to maintain the communities functional (e.g. replacement by other necrophagous species such as decapod crustaceans), except in the case of catastrophic events such as oil spills [94].

Since the use of TBT in antifouling paints has been restricted in 1982 in France and in 1987 throughout Europe, North America, Australia and Japan (101-103), the degree of imposex in some dogwhelk populations has decreased (104), but the process of recovery of all affected populations and communities is likely to be slow as bioaccumulation still occurs in some organisms that now recolonize the contaminated areas [94,102,103].

This case-study demonstrates that the effects of reproductive impairement on marine mollusc populations and consequences on communities can conceivably be predicted from individual measurements of biomarkers which represent potent early-warning indicators of pollution by endocrine-disrupting chemicals. The opportunity to detect effects relevant to endocrine disruption from acute and chronic reproduction tests in aquatic and terrestrial invertebrates has recently been considered in ecological test guidelines for industrial chemicals and pesticides [105].

11.5.3. EFFECTS OF POLLUTANTS ON INVERTEBRATE BIOENERGETICS : CHANGES IN THE PHYSIOLOGICAL ENERGETICS OF FRESHWATER CRUSTACEANS

The demonstration of mechanistic linkages between effects at different levels of biological organisation has also been achieved using the freshwater amphipod *Gammarus pulex* in various experimental contexts [106]. Physiological energetics assessed through measurements of scope for growth have successfully been used to predict the concentrations of pollutants that impair growth and reproduction as well as the magnitude of this impairment. Reduced scope for growth which results from lower individual feeding rate have been shown to be correlated with decreased reproductive success, through reduced offspring size and brood viability [107,108]. Energy re-allocation between maintenance, growth and reproduction has also been reported for *Asellus aquaticus* [109] and *Cambarus robustus* [110] exposed to acid effluents containing heavy metals.

Although reduction of metabolic rates seems to be a common consequence of stress resistance [59], recent observations on freshwater crustaceans transferred to ion-poor water indicate that this is not a constant one [111].

Stress-induced reductions in *G. pulex* feeding rate have been shown to be correlated with reductions in the rate of incorporation of leaf organic material into freshwater food webs [108]. Field trials further indicated that between-site differences in *G. pulex* feeding rate were correlated with differences in community structure, but this correlation did not result from causal relationships between *G. pulex* energetics and community structure [106]. Similarly Jones *et al.* [112] have indicated that changes in the feeding behaviour of *Daphnia catawba* induced by sublethal concentrations of toxicants may have more rapid and pronounced effects at the community (i.e. algal populations) than on the cladoceran population itself.

Energetic biomarkers certainly have a real potential for predicting effects of pollutants on invertebrate population structure and dynamics [30,76]. On the basis of experimental and field investigations, a conceptual framework has been elaborated, which supports modelling approaches [113,114]. However, compensatory population mechanisms and natural variations of the parameters related to individual bioenergetics may act as confusing factors so that the regular use of these latter as predictive biomarkers still requires further investigations.

These two examples show that mechanistic links can be established between biomarkers measured in individuals and changes at the population level, the key-process that links the individual to the population being reproduction. In invertebrates, biomarkers related to reproduction through hormone metabolism and energetic balance appear as highly relevant to predict effects at the population level. Prediction of further consequences on communities and ecosystems relies on a sound knowledge of the functional role of the population and of the interactions between populations.

11.6. Assessment of the ecological impact of chemicals through the use of biomarkers in invertebrates : concluding remarks

A large amount of data show that environmental contaminants may exert deleterious effects on organisms and ecological processes. However, even if responses of organisms or processes to pollutants can been demonstrated in controlled conditions (e.g. laboratory, microcosms or mesocosms), this does not necessarily imply that the same responses (or any response) will be observed in the field. Intrinsic unpredictability of the natural environment generally impedes the identification of precise causal relationships between effects of toxicants on individuals and subsequent consequences on populations and communites. The demonstration of such relationships implies that the component of biological responses due to background fluctuations (e.g. natural variations) and to confounding variables (e.g. natural factors) is accurately known [115].

Physiological compensation in individuals and/or natural heterogeneity within populations, that usually ensure survival of species within a community [116], frequently make it difficult to extrapolate from biochemical and physiological changes to population-level effects of exposure [117,118]. Compensatory mechanisms have themselves a cost which can, in turn, interfere with the performance of individuals (Darwinian fitness). Even adaptation or genotype selection may have a cost in terms of population performances (e.g. decrease of tolerance to other stresses in tolerant populations with reduced genotype variability) [119-121].

Therefore, when studies are performed in contaminated sites, it should be remembered that the populations which are currently present are those which survived exposure to pollutants and that they will not necessarily exhibit the same characteristics as laboratory reference species (e.g. physiological adaptation, selection for tolerant genotypes) [122-124].

Surprisingly, population biology is rarely considered in ecotoxicological studies, even though modelling approaches have been proposed in population dynamics, which take into account the effects of pollutants on various biological responses [125-128].

Population-level effects are sometimes measurable only after several generations when affected endpoints are long-term parameters such as juvenile recruitment or age to first reproduction [129,130]. The age-structure of natural populations may also results in age-specific stress that will not necessarily affect the whole population survival and dynamics [131]. On the other hand, field observations suggest that populations are frequently interconnected, especially in large-scale ecosystems (e.g. sea shore) or spatially complex landscapes so that the real object of interest to ecotoxicologists should consist in metapopulations rather than in isolated populations [132].

Therefore, even when laboratory experiments or model outputs predict either population decrease or survival, various natural fluctuation processes (e.g. immigration, food availability, etc) may considerably modify actual demographic parameters, especially in unpredictable or short-lived systems (e.g. temporal ponds and rivers, coastal ecosystems) or when pollution is spatially restricted and that juvenile organisms may colonise from surrounding non-contaminated areas [115,133]. Conversely, populations living in uncontaminated environments may be affected by the recruitment of juveniles from areas which can be polluted. In the case of such a metapopulation approach, the impact of human activities on habitat fragmentation and/or conservation is of primary importance.

The existence of natural confounding factors and compensatory mechanisms at both individual and population level explains why the observation of a response in a contaminated ecosystem or even of a correlation between biological response(s) and contaminant levels is generally not sufficient to prove that toxicants are the true cause of biological adverse effects [115].

Figure 4 provides a general framework for the assessment of variations of the ecological impact of chemicals using invertebrate biomarkers. In such a framework, the advantages of using invertebrates are evident at three levels, and perhaps more.

First, the large number of invertebrate species allows a pertinent choice of the species which can be used for biomarker measurements. In addition to the characteristics of sentinel species which make them ecologically relevant, invertebrates allow investigations at different levels of biological organisation.

The second advantage of using invertebrates arises from the success there has been in combining diagnostic and predictive biomarkers. This has depended on the possibility of aggregation of data obtained at the levels of the individual, population and community, requiring the use of different experimental contexts.

The third advantage of invertebrates is in the context of remediation. Their fast growth and short generation time allow a rapid assessment of the extent of ecological restoration for which biomarkers can be used as primary indicators.

References

1. Wilson, R.C.H., Harding, L.E. and Hirvonen, H. (1995) Marine ecosystem monitoring network design. *Ecosystem Health* **1**, 222-227.
2. Metcalfe, J.L. (1989) Biological water quality assessement of running waters based on macroinvertebrate communities: history and present status in Europe. *Environ. Pollut.* **60**, 101-139.
3. Hopkin, S.P. (1993) In situ biological monitoring of pollution in terrestrial and aquatic ecosystems, in P. Calow (ed.), *Handbook of Ecotoxicology*, Vol 1, Blackwell Scientific Publications, Oxford, pp. 397-427.
4. Phillips, D.J.H. and Rainbow, P.S. (1993) *Biomonitoring of Trace Aquatic Contaminants*. Elsevier Science Publishers, Barking.
5. Wilson, J.G. (1994) The role of bioindicators in estuarine management. *Estuaries* **17**, 94-101.

6. Resh, V.H., Norris, R.H. and Barbour, M.T. (1995) Design and implementation of rapid assessment approaches for water resource monitoring using benthic macroinvertebrates. *Aust. J. Ecol.* **20**, 108-121.
7. Resh, V.H., Myers, M.J. and Hannaford, M.J. (1996) Macroinvertebrates as biotic indicators of environmental quality, in F.R. Hauer and G.A. Lamberti (eds), *Methods in Stream Ecology*, Academic Press, San Diego, pp. 647-667.
8. Verdonschot, P.F.M. (1995) Typology of macrofaunal assemblages: a tool for the management of running waters in The Netherlands. *Hydrobiologia* **297**, 99-122.
9. Lagadic, L., Caquet, Th. and Ramade, F. (1994) The role of biomarkers in environmental assessment (5). Invertebrate populations and communities. *Ecotoxicology* **3**, 193-208.
10. Lagadic, L., Caquet, Th., Amiard, J.-C. and Ramade, F. (eds.) (1997) *Biomarqueurs en Ecotoxicologie : Aspects Fondamentaux*. Collection d'Ecologie, Masson Editeur, Paris.
11. McCarthy, J.F. and Shugart, L.R., 1990. Biomarkers of Environmental Contamination. Lewis Publishers, Boca Raton.
12. Huggett, R.J., Kimerle, R.A., Mehrle, P.M. Jr. and Bergman, H.L. (eds) (1992) *Biomarkers. Biochemical, Physiological, and Histological Markers of Anthropogenic Stress*. Lewis Publishers, Boca Raton.
13. Walker, C.H. (1995) The role of biomarkers in ecotoxicology. *Toxicol. Ecotoxicol. News* **2**, 67.
14. Peakall, D.B. and Walker, C.H. (1996) Comment on van Gestel and van Brummelen. *Ecotoxicology* **5**, 227.
15. Lagadic, L., Caquet, Th. and Amiard, J.-C. (1997) Biomarqueurs en écotoxicologie : principes et définitions, in L. Lagadic, Th. Caquet, J.-C. Amiard, and F. Ramade (eds.), *Biomarqueurs en Ecotoxicologie : Aspects Fondamentaux*. Collection d'Ecologie, Masson Editeur, Paris, pp. 1-9.
16. Depledge, M.H. and Fossi, M.C. (1994) The role of biomarkers in environmental assessment (2). Invertebrates. *Ecotoxicology* **3**, 161-172.
17. Lagadic, L., Caquet, Th., Amiard, J.-C. and Ramade, F. (eds.) (1998) *Utilisation de Biomarqueurs pour la Surveillance de la Qualité de l'Environnement*. Tec & Doc Lavoisier, Paris.
18. Clarke, K.R. and Warwick, R.M. (1994) Similarity-based testing for community pattern: the two-way layout with no replication. *Mar. Biol.* **118**, 167-176.
19. Attrill, M.J. and Depledge, M.H. (1997) Community and population indicators of ecosystem health: targeting links between levels of biological organisation. *Aquat. Toxicol.* **38**, 193-197.
20. Lagadic, L. and Caquet, Th. (1998) Invertebrates in testing of environmental chemicals : are they alternatives? *Environ. Health Perspect.* **106**, suppl. 2 593-612.
21. Stegeman, J.J., Brouwer, M., Di Giulio, R.T., Förlin, L., Fowler, B.A., Sanders, B.M. and Van Veld, P.A. (1992) Molecular responses to environmental contamination: enzyme and protein systems as indicators of chemical exposure and effect, in R.J. Huggett, R.A. Kimerle, P.M. Mehrle Jr and H.L. Bergman (eds), *Biomarkers. Biochemical, Physiological, and Histological Markers of Anthropogenic Stress*, Lewis Publishers, Boca Raton, pp. 235-335.

22. Lower, W.R. and Kendall, R.J. (1990) Sentinel species and sentinel bioassay, in J.F. McCarthy and L.R. Shugart (eds), *Biomarkers of Environmental Contamination*, Lewis Publishers, Boca Raton, pp. 309-331.
23. Warwick, W.F. (1988) Morphological deformities in Chironomidae (Diptera) larvae as biological indicators of toxic stress, in Evans, M.S. (ed), *Toxic Contaminants and Ecosystem Health: A Great Lakes Focus*, John Wiley, New York, pp. 281-320.
24. Dickman, M., Brindle, I. and Benson, M. (1992) Evidence of teratogens in sediments of the Niagara river watershed as reflected by chironomid (Diptera: Chironomidae) deformities. *J. Great Lakes Res.* **18**, 467-480.
25. Janssens de Bisthoven, L.G., Timmermans, K.R. and Ollevier, F. (1992) The concentration of cadmium, lead, copper and zinc in *Chironomus* gr. *thummi* larvae (Diptera, Chironomidae) with deformed *versus* normal menta. *Hydrobiologia* **239**, 141-149.
26. Van Urk, G., Kerkum, F.C.M. and Smit, H. (1992) Life cycle patterns, density, and frequency of deformities in *Chironomus* larvae (Diptera: Chironomidae) over a contaminated sediment gradient. *Can. J Fish. Aq. Sci.* **49**, 2291-2299.
27. Bird, G.A., Rosentreter, M.J. and Schwartz, W.J. (1995) Deformities in the menta of chironomid larvae from the Experimental Lakes Area, Ontario. *Can. J. Fish. Aquat. Sci.* **52**, 2290-2295.
28. Yevich, P.P. and Barszcz, C.A. (1977) Neoplasia in soft-shell clams (*Mya arenaria*) collected from oil-impacted sites. *Ann. N.Y. Acad. Sci.* **298**, 409-426.
29. Gardner, G.R., Yevich, P.P., Harhbarger, J.C. and Malcolm, A.R. (1991) Carcinogenicity of Black Rock Harbor sediment to the eastern oyster and trophic transfer of Black Rock Harbor carcinogens from the blue mussel to the winter flounder. *Environ. Health Perspect.* **90**, 53-66.
30. Le Gal, Y., Lagadic, L., Le Bras, S. and Caquet, Th. (1997) Charge énergétique en adénylates (CEA) et autres biomarqueurs associés au métabolisme énergétique, in L. Lagadic, Th. Caquet, J.-C. Amiard, and F. Ramade (eds.), *Biomarqueurs en Ecotoxicologie : Aspects Fondamentaux*. Collection d'Ecologie, Masson Editeur, Paris, pp. 241-285.
31. Herbert, A. and Zahn, R.K. (1990) Monitoring DNA damage in *Mytilus galloprovincialis* and other aquatic animals. II. Pollution effects on DNA denaturation characteristics. *Germ. J. Appl. Zool.* **76**, 143-167.
32. Vukmirovic, M., Bihari, N., Zahn, R.K., Müller, W.E.G. and Batel, R. (1994) DNA damage in marine mussel *Mytilus galloprovincialis* as a biomarker of environmental contamination. *Mar. Ecol. Progr. Ser.* **109**, 165-171.
33. Black, M.C., Ferrell, J.R., Horning, R.C. and Martin, L.K., Jr. (1996) DNA strand breakage in freshwater mussels (*Anondonta grandis*) exposed to lead in the laboratory and field. *Environ. Toxicol. Chem.* **15**, 802-808.
34. Burgeot, Th., Woll, S. and Galgani, F. (1996) Evaluation of the micronucleus test on *Mytilus galloprovincialis* for monitoring applications along French coasts. *Mar. Pollut. Bull.* **32**, 39-46.
35. Nacci, D.E., Cayula, S. and Jackim, E. (1996) Detection of DNA damage in individual cells from marine organisms using the single cell gel assay. *Aquat. Toxicol.* **35**, 197-210.

36. Mersch, J., Beauvais, M.-N. and Nagel, P. (1996) Induction of micronuclei in haemocytes and gill cells of zebra mussels, *Dreissena polymorpha*, exposed to clastogens. *Mutat. Res.* **371**, 47-55.
37. Pasantes, J.J., Martinez-Exposito, M.J., Torreiro, A. and Mendez, J. (1996) The sister chromatid exchange test as an indicator of marine pollution: some factors affecting SCE frequencies in *Mytilus galloprovincialis*. *Mar. Ecol. Progr. Ser.* **143**, 113-119.
38. Venier, P.; Maron, S. and Canova, S. (1997) Detection of micronuclei in gill cells and haemocytes of mussels exposed to benzo(a)pyrene. *Mutat. Res.* **390**, 33-44.
39. Jha, A.N., Hutchinson, T.H., Mackay, J.M., Elliott, B.M. and Dixon, D.R. (1996) Development of an *in vivo* genotoxicity assay using the marine worm *Platynereis dumerelii* (Polychaeta: Nereidae). *Mutat. Res.* **359**, 141-150.
40. Verschaeve, L. and Gilles, J. (1995) Single cell electrophoresis assay in the earthworm for the detection of genotoxic compounds in soils. *Bull. Environ. Contam. Toxicol.* **54**, 112-119.
41. Šalagovic, J., Gilles, J., Verschaeve, L. and Kalina, I. (1996) The Comet Assay for the detection of genotoxic damage in the earthworms: a promising tool for assessing the biological hazards of polluted sites. *Fol. Biol.* **42**, 17-21.
42. Walsh, P., El Aldouni, C., Nadeau, D., Fournier, M., Coderre, D. and Poirier, G.G. (1997) DNA adducts in earthworms exposed to a contaminated soil. *Soil Biol. Biochem.* **29**, 721-724.
43. Everaarts, J.M. (1997) DNA integrity in *Asterias rubens*: a biomarker reflecting the pollution of the North Sea. *J. Sea Res.* **37**, 123-129.
44. Terriere, L.C. (1983) Enzyme induction, gene amplification and insect resistance to insecticides, in Georghiou, G.P. and Saito, T. (eds), *Pest Resistance to Pesticides*, Plenum Press, New York, pp. 265-298.
45. Sawicki, R.M. and Denholm, I. (1984) Adaptation of insects to insecticides, in Evered, D. and Collins, G.M. (eds), *Origins and Development of Adaptation*, Ciba Foundation Symposium Series No. 102, Pitman Books, London, pp. 152-166.
46. Metcalf, R.L. (1989) Insect resistance to insecticides. *Pestic. Sci.* **26**, 333-358.
47. NAS/NRC (National Academy of Sciences/National Research Council) (1989) *Biologic Markers in Reproductive Toxicology*. National Academy Press, Washington, DC.
48. Brown, T.M. (1990) Biochemical and genetic mechanisms of insecticide resistance, in Green, M.B., LeBaron, H.M. and Moberg, W.K. (eds), *Managing Resistance to Agrochemicals. From Fundamental Research to Practical Strategies*, ACS Symposium Series No. 421. American Chemical Society, Washington, DC, pp. 61-76.
49. Soderlund, D.M. and Bloomquist, J.R. (1990) Molecular mechanisms of insecticide resistance, in Roush, R.T. and Tabashnik, B.E. (eds), *Pesticide Resistance in Arthropods*, Chapman and Hall, New York and London, pp. 58-96.
50. Amichot, M., Bergé, J.-B., Cuany, A., Pasteur, N., Pauron, D. and Raymond, M. (1998) Bases moléculaires, génétiques et populationnelles de la résistance des insectes aux insecticides in L. Lagadic, Th. Caquet, J.-C. Amiard, and F.

Ramade (eds.), *Utilisation de Biomarqueurs pour la Surveillance de la Qualité de l'Environnement*, Tec & Doc Lavoisier, Paris, pp. 241-263.
51. Roush, R.T. and McKenzie, J.A. (1987) Ecological genetics of insecticide and acaricide resistance. *Ann. Rev. Entomol.* **32**, 361-380.
52. Roush, R.T and Daly, J.C. (1990) The role of population genetics in resistance research and management, in Roush, R.T. and Tabashnik, B.E. (eds), *Pesticide Resistance in Arthropods*, Chapman and Hall, New York and London, pp. 97-152.
53. Oppenoorth, F.J. (1985) Biochemistry and genetics of insecticide resistance, in Kerkut, G.A. and Gilbert, L.I. (eds), *Comprehensive Insect Physiology, Biochemistry and Pharmacology*, Vol. 12, Pergamon Press, Oxford, pp. 713-773.
54. Brattsten, L.B. (1988) Potential role of plant allelochemicals in the development of insecticide resistance, in Barbosa, P. and Letourneau, D.K. (eds), *Novel Aspects of Insect-Plant Interactions*, John Wiley, New York pp. 313-348.
55. Poirié, M. and Pasteur, N. (1991) La résistance des insectes aux insecticides. *La Recherche* **22**, 874-882.
56. Raymond, M., Fournier, D., Bride, J.-M., Cuany, A., Bergé, J., Magnin, M. and Pasteur, N. (1986) Identification of resistance mechanisms in *Culex pipiens* (Diptera: Culicidae) from Southern France: Insensitive acetylcholinesterase and detoxifying oxidases. *J. Econ. Entomol.* **79**, 1452-1458.
57. Raymond, M., Pasteur, N. and Georghiou, G.P. (1987) Inheritance of chlorpyrifos resistance in *Culex pipiens* L. (Diptera: Culicidae) and estimation of the number of genes involved. *Heredity* **58**, 351-356.
58. Chevillon, C., Pasteur, N., Marquine, M., Heyse, N. and Raymond, M. (1995) Population structure and dynamics of selected genes in the mosquito *Culex pipiens*. *Evolution* **49**, 997-1007.
59. Hoffmann, A.A. and Parsons, P.A. (1991) *Evolutionary Genetics and Environmental Stress*. Oxford University Press, Oxford.
60. Mueller, L.D., Guo, P. and Ayala, F.J. (1991) Density-dependent natural selection and trade-offs in life history traits. *Science* **253**, 433-435.
61. ffrench-Constant, R.H., Steichen, J.C. and Ode P.J. (1993) Cyclodiene insecticide resistance in *Drosophila melanogaster* (Meigen) is associated with a temperature-sensitive phenotype. *Pestic. Biochem. Physiol.* **46**, 73-77.
62. Rapport, D.J. (1989) Symptoms of pathology in the Gulf of Bothnia (Baltic Sea): Ecosystem response to stress from human activity. *Biol. J. Linn. Soc.* **37**, 33-49.
63. Dawson, G.W., Griffiths, D.C., Pickett, J.A. and Woodcock, C.M. (1983) Decreased response to alarm pheromone by insecticide-resistant aphids. *Naturwissenschaften* **70**, 254-255.
64. Delpuech, J.-M., Frey, F. and Carton, Y. (1996) Action of insecticides on the cellular immune reaction of *Drosophila melanogaster* against the parasitoid *Leptopilina boulardi*. *Environ. Toxicol. Chem.* **15**, 2267-2271.
65. Veldhuizen-Tsoërkan, M.B., Holwerda, D.A. and Zandee, D.I. (1991) Anoxic survival time and metabolic parameters as stress indices in sea mussels exposed to cadmium or polychlorinated biphenyls. *Arch. Environ. Contam. Toxicol.* **20**, 259-265.

66. Eertman, R.H.M., Wagenvoort, A.J., Hummel, H. and Smaal, A.C. (1993) "Survival in air" of the blue mussel *Mytilus edulis* L. as a sensitive response to pollution-induced environmental stress. *J. Exp. Mar. Biol. Ecol.* **170**, 179-195.
67. Veldhuizen-Tsoërkan, M.B., Holwerda, D.A., van der Mast, C.A. and Zandee, D.I. (1991) Synthesis of stress proteins under normal and heat shock conditions in gill tissue of sea mussels after chronic exposure to cadmium. *Comp. Biochem. Physiol.* **100C**, 699-706.
68. Veldhuizen-Tsoërkan, M.B., Holwerda, D.A., de Bont A.M.T., Smaal, A.C. and Zandee, D.I. (1991) A field study on stress indices in the sea mussel, *Mytilus edulis*: Application of the "stress approach" in biomonitoring. *Arch. Environ. Contam. Toxicol.* **21**, 497-504.
69. Eertman, R.H.M. and de Zwaan, A. (1994) Survival of the fittest : Resistance of mussels to aerial exposure, in K.J.M. Kramer (Ed.), *Biomonitoring of Coastal Waters and Estuaries*, CRC Press, Boca Raton, FL, pp. 269-284.
70. Demers, A. and Guderley, H. (1994) Acclimatization to intertidal conditions modifies the physiological response to prolonged air exposure in *Mytilus edulis*. *Mar. Biol.* **118**, 115-122.
71. Oeschger, R. and Storey, K.B. (1993) Impact of anoxia and hydrogen sulphide on the metabolism of *Arctica islandica* L. (Bivalvia). *J. Exp. Mar. Biol. Ecol.* **170**, 213-226.
72. Oeschger, R. and Pedersen, T.F. (1994) Influence of anoxia and hydrogen sulphide on the energy metabolism of *Scrobicularia plana* (da Costa) (Bivalvia). *J. Exp. Mar. Biol. Ecol.* **184**, 255-268.
73. de Zwaan, A., Cortesi, P. and Cattani O. (1995) Resistance of bivalve to anoxia as a response to pollution-induced environmental stress. *Sci. Total Environ.* **171**, 121-125.
74. Devi, U.V (1996) Changes in oxygen consumption and biochemical composition of the marine fouling dreissinid bivalve *Mytilopsis sallei* (Recluz) exposed to mercury. *Ecotox. Environ. Saf.* **33**, 168-174.
75. Paterson, B.D. and Thorne, M.J. (1995) Measurements of oxygen uptake, heart and gill bailer rates of the callianassid burrowing shrimp *Trypae australiensis* Dana and its response to low oxygen tensions. *J. Exp. Mar. Biol. Ecol.* **194**, 39-52.
76. Caquet, Th. and Lagadic, L. (1998) Conséquences des altérations individuelles précoces sur la dynamique des populations et la structuration des communautés et des écosystèmes, in L. Lagadic, Th. Caquet, J.-C. Amiard, and F. Ramade (eds.), *Utilisation de Biomarqueurs pour la Surveillance de la Qualité de l'Environnement*, Tec & Doc Lavoisier, Paris, pp. 265-298.
77. Haynes, K.F. (1988) Sublethal effects of neurotoxic insecticides on insect behaviour. *Ann. Rev. Entomol.* **33**, 149-168.
78. Baatrup, E. and Bayley, M. (1993) Effects of the pyrethroid insecticide cypermethrin on the locomotor activity of the wolf spider *Pardosa amentata*: quantitative analysis employing computer-automated video tracking. *Ecotoxicol. Environ. Saf.* **26**, 138-152.
79. Bayley, M., Baatrup, E., Heimbach, U. and Bjerregaard, P. (1995) Elevated copper levels during larval development cause altered locomotor behavior in the

adult carabid beetle *Pterostichus cupreus* L. (Coleoptera: Carabidae). *Ecotoxicol. Environ. Saf.* **32**, 166-170.

80. Sørensen, F.F., Weeks, J.M. and Baatrup, E. (1996) Altered locomotory behavior in woodlouse (*Oniscus asellus* L.) collected at a polluted site. *Environ. Toxicol. Chem.* **16**, 685-690.
81. Cohn, J. and MacPhail, R.C. (1996) Ethological and experimental approaches to behavior analysis: implications for ecotoxicology. *Environ. Health Persp.* **104** (Suppl. 2), 299-305.
82. Bayley, M., Baatrup, E. and Bjerregaard, P. (1997) Woodlouse locomotor behavior in the assessment of clean and contaminated field sites. *Environ. Toxicol. Chem.* **16**, 2309-2314.
83. Malins, D.C. and Ostrander, G.K. (1991) Perspectives in aquatic toxicology. *Ann. Rev. Pharmacol. Toxicol.* **31**, 371-399.
84. Mayer, F.L., Versteeg, D.J., McKee, M.J., Folmar, L.C., Graney, R.L., McCume, D.C. and Rattner, B.A. (1992) Physiological and nonspecific biomarkers, in Huggett, R.J., Kimerle, R.A., Mehrle Jr., P.M. and Bergman, H.L. (eds), *Biomarkers. Biochemical, Physiological, and Histological Markers of Anthropogenic Stress*, SETAC Special Publication Series, Lewis Publishers, Boca Raton, FL, pp. 5-85.
85. Dickerson, R.L. and Kendall, R.J. (eds) (1998) Endocrine disruptors. *Environ. Toxicol. Chem.*, Annual Review Issue, **17**.
86. Grunwald, C., 1980. Steroids, in E.A. Bell and B.V. Charlwood (eds), *Secondary Plant Products, Encyclopedia of Plant Physiology, Vol. 8* (A. Pirson and M.H. Zimmermann, eds), Springer Verlag, pp. 221-256.
87. Wing, K.D., Slawecki, R.A. and Carlson, G.R. (1988) RH5849, a nonsteroidal ecdysone agonist: effects on larval Lepidoptera. *Science* **241**, 470-472.
88. Sundaram, K.M.S., Holmes, S.B., Kreutzweiser, D.P., Sundaram, A. and Kingsbury, P.D. (1991) Environmental persistence and impact of diflubenzuron in a forest aquatic environment following aerial application. *Arch. Environ. Contam. Toxicol.* **20**, 313-324.
89. Kreutzweiser, D.P. and Thomas, D.R. (1995) Effects of a new molt-inducing insecticide, tebufenozide, on zooplankton communities in lake enclosures. *Ecotoxicology* **4**, 307-328.
90. Baldwin, W.S., Milam, D.L. and LeBlanc, G.A. (1995) Physiological and biochemical perturbation in *Daphnia magna* following exposure to the model environmental estrogen diethylstilbestrol. *Environ. Toxicol. Chem.* **14**, 945-952.
91. Shurin, J.B. and Dodson, S.I. (1997) Sublethal toxic effects of cyanobacteria and nonylphenol on environmental sex determination and development in *Daphnia. Environ. Toxicol. Chem.* **16**, 1269-1276.
92. Baldwin, W.S., Graham, S.E., Shea, D. and LeBlanc, G.A. (1997) Metabolic androgenization of female *Daphnia magna* by the xenoextrogen 4-nonylphenol. *Environ. Toxicol. Chem.* **16**, 1905-1911.
93. Zou, E. and Fingerman, M. (1997) Synthetic estrogenic agents do not interfere with sex differentiation but do inhibit molting of the cladoceran *Daphnia magna. Bull. Environ. Contam. Toxicol.* **58**, 596-602.
94. Hawkins, S.J., Proud, S.V., Spence, S.K. and Southward, A.J. (1994) From the individual to the community and beyond: water quality, stress indicators and key

species in coastal systems, in D.W. Sutcliffe (ed), *Water Quality and Stress Indicators in Marine and Freshwater Ecosystems: Linking Levels of Organisation (Individuals, Populations, Communities*, Freshwater Biological Association, Ambleside, pp. 35-62.

95. Matthiessen, P. and Gibbs, P.E. (1998) Critical appraisal of the evidence for tributyltin-mediated endocrine disruption in mollusks. *Environ. Toxicol. Chem.* **17**, 37-43.
96. Blaber, S.J.M. (1970) The occurrence of a penis-like outgrowth behind the right tentacle in spent females of *Nucella lapillus* (L.). *Proc. Malacol. Soc. Lond.* **39**, 377-378.
97. Smith, B.S. (1971) Sexuality in the American mud snail, *Nassarius obsoletus* Say. *Proc. Malac. Soc. Lond.* **39**, 377-378.
98. Ellis, D.V. and Pattisina, A. (1990) Widespread neogastropod Imposex: a biological indicator of global TBT contamination ? *Mar. Pollut. Bull.* **21**, 248-253.
99. Oehlmann, J., Fioroni, P., Stroben, E. and Markert, B. (1996) Tributyltin (TBT) effects on *Ocinebrina aciculata* (Gastropoda: Muricidae): Imposex development, sterilization, sex change and population decline. *Sci. Total Environ.* **188**, 205-223.
100. Bryan, G.W., Gibbs, P.E., Hummerstone, L.G. and Burt G.R. (1986) The decline of the gastropod *Nucella lapillus* around south-west England: Evidence for the effect of tributyltin from antifouling paints. *J. Mar. Biol. Assoc. UK* **66**, 767-777.
101. Matthiessen, P., Waldock, R., Thai, J.E., Waite, M.E. and Scrope-Howe, S. (1995) Changes in periwinkle (Littorina littorea) populations following the ban of TBT-based antifoulings on small boats in the United Kingdom. *Ecotoxicol. Environ. Saf.* **30**, 180-194.
102. Langston, W.J. (1996) Recent developments in TBT ecotoxicology. *Toxicol. Ecotoxicol. News* **3**, 179-187.
103. Ruiz, J.M., Bachelet, G., Caumette, P. and Donard O.F.X. (1996) Three decades of tributyltin in the coastal environment with emphasis on Arcachon bay, France. *Environ. Pollut.* **93**, 195-203.
104. Evans, S.M., Hutton, A., Kendall, M.A. and Samosir, A.M. (1991) Recovery in populations of dogwhelks *Nucella lapillus* (L.) suffering from Imposex. *Mar. Pollut. Bull.* **22**, 331-333.
105. Kavlock, R.J., Daston, G.P., DeRosa, C., Fenner-Crisp, P., Earl Gray, L., Kaattari, S., Lucier, G., Luster, M., Mac, M.J., Maczka, C., Miller, R., Moore, J., Rolland, R., Scott, G., Sheehan, D.J., Sinks, T. and Tilson, H.A. (1996) Research needs for the risk assessment of health and environmental effects of endocrine disruptors: a report of the U.S. EPA-sponsored workshop. *Environ. Health Perspect.* **104**, suppl 4, 715-740.
106. Maltby, L. (1994) Stress, shredders and streams: using *Gammarus* energetics to assess water quality, in D.W. Sutcliffe (ed), *Water Quality and Stress Indicators in Marine and Freshwater Ecosystems: Linking Levels of Organisation (Individuals, Populations, Communities*, Freshwater Biological Association, Ambleside, pp. 98-110.

107. Maltby, L. and Naylor,C. (1990) Preliminary observations on the ecological relevance of the *Gammarus* "Scope for Growth" assay: effect of zinc on reproduction. *Funct. Ecol.* **4**, 393-397.
108. Tattersfield, L.J. (1993) Direct and indirect effects of copper on the energy budget of a stream detritivore. Proceedings of the first SETAC World Congress, 28-31 March 1993, Lisbon, 88.
109. Maltby, L. (1991) Pollution as a probe of life-history adaptation in *Asellus aquaticus* (Isopoda). *Oikos* **61**, 11-18.
110. Daveikis, V.F. and Alikhan, M.A. (1996) Comparative body measurements, fecundity, oxygen uptake, and ammonia excretion in *Cambarus robustus* (Astacidae, Crustacea) from an acidic and a neutral site in northeastern Ontario, Canada. *Can. J. Zool.* **74**, 1196-1203.
111. Glazier, D.S. and Sparks, B.L. (1997) Energetics of amphipods in ion-poor waters: stress resistance is not invariably linked to low metabolic rates. *Funct. Ecol.*, **11**, 126-128.
112. Jones, M., Folt, C. and Guarda, S. (1991) Characterizing individual, population and community effects of sublethal levels of aquatic toxicants: an experimental case study using *Daphnia. Freshwater Biol.* **26**, 35-44.
113. Sibly, R.M. and Calow, P. (1989) A life-cycle theory of response to stress. *Biol. J. Linn. Soc.* **37**, 101-116.
114. Sibly, R.M. (1994) From organism to population: the role of life-history theory, in D.W. Sutcliffe (ed), *Water Quality and Stress Indicators in Marine and Freshwater Ecosystems: Linking Levels of Organisation (Individuals, Populations, Communities*, Freshwater Biological Association, Ambleside, pp. 63-74.
115. Luoma, S.N. (1996) The developing framework of marine ecotoxicology: pollutants as a variable in marine ecosystems ? *J. Exp. Mar. Biol. Ecol.* **200**, 29-55.
116. Marshall, J.D. and Mellinger, D.L. (1980) Dynamics of cadmium-stressed plankton communities. *Can. J. Fish. Aqua. Sci.* **37**, 403-412.
117. Kluytmans, J.H., Brands, F. and Zandee, D.I. (1988) Interactions of cadmium with the reproductive cycle of *Mytilus edulis* L. *Mar. Environ. Res.* **24**, 189-192.
118. Depledge, M.H., Agard, A. and Gyorkos, P. (1995) Assessment of trace metal toxicity using molecular, physiological and behavioral biomarkers. *Mar. Pollut. Bull.* **31**, 19-27.
119. Guttman S.I. (1994) Population genetic structure and ecotoxicology. *Environ. Health Perspect.* **102** (suppl.12), 97-100.
120. Forbes, V.E. (1996) Chemical stress and genetic variability in invertebrate populations. *Toxicol. Ecotoxicol. News* **3**, 136-141.
121. Hebert, P.D.N. and Luiker, M.M. (1996) Genetic effects of contaminant exposure - Towards an assessment of impacts on animal populations. *Sci. Total Environ.* **191**, 23-58.
122. Bryan, G.W. and Hummerstone, L.G. (1973) Adaptation of the polychaete *Nereis diversicolor* to estuarine sediments containing high concentrations of heavy metals: general observations and adaptations to copper. *J. Mar. Biol. Assoc. U.K.* **51**, 845-863.

123. Luoma, S.N., Cain, D.J., Ho, K. and Hutchinson, A. (1983) Variable tolerance to copper in two species from San Francisco Bay. *Mar. Environ. Res.* **10**, 209-222.
124. Klerks, P.L. and Levinton, J.S. (1989). Effects of heavy metals in a polluted aquatic ecosystem, in S.A. Levin ***et al.*** (eds), *Ecotoxicology, Problems and Approaches*, Springer-Verlag, New York, pp. 41-67.
125. Kooijman, S.A.L.M. and Metz, J.A.J. (1984) On the dynamics of chemically stressed populations: the deduction of population consequences from effects on individuals. *Ecotoxicol. Environ. Saf.* **8**, 254-274.
126. Suer, J.R. and Pendleton, G.W. (1995) Population modeling and ist role in toxicological studies, in D.J. Hoffman, B.A. Rattner, G.A. Burton Jr. and J. Cairns Jr. (eds), *Handbook of Ecotoxicology*, Lewis Publishers, Boca Raton, pp. 681-702.
127. Baveco, J.M. and De Roos, A.M. (1996) Assessing the impact of pesticides on lumbricid populations: an individual-based modelling approach. *J. Appl. Ecol.* **33**, 1451-1468.
128. Klok, C. and De Roos, A.M. (1996) Population level consequences of toxicological influences on individual growth and reproduction of *Lumbricus rubellus* (Lumbricidae, Oligochaeta). *Ecotoxicol. Environ. Saf.* **33**, 118-127.
129. Sundelin, B. (1983) Effects of cadmium on *Pontoporeia affinis* (Crustacea: Amphipoda) in laboratory soft-bottom microcosms. *Mar. Biol.* **74**, 203-212.
130. Scott, J.K. and Redmond, M.S. (1989) The effects of a contaminated dredged material on laboratory populations of the tubicolous amphipod *Ampelisca abdita*, in U.M. Cowgill and L.R. Williams (eds), *Aquatic Toxicology and Hazard Assessment*, ASTM, Philadelphia, pp. 289-301.
131. Zajac, R.N. and Whitlach, R.B. (1989) Natural and disturbance induced demographic variation in an infaunal polychaete, *Nephtys incisa. Mar. Ecol. Prog. Ser.* **57**, 89-102.
132. Maurer, B.A. and Holt, R.D. (1996) Effects of chronic pesticide stress on wildlife populations in complex landscapes: processes at multiple scales. *Environ. Toxicol Chem.* **15**, 420-426.
133. Underwood, A.J. (1994) On beyond BACI: sampling designs that might increase reliability detect environmental disturbance. *Ecol. Appl.* **4**, 3-15.

12. THE USE OF BIOMARKERS IN VERTEBRATES

C.H. WALKER
35 Victoria Road
Mortimer, Berks, RG7 3SH, U.K

Abstract

In field studies biomarker assays can provide evidence of exposure of vertebrates to environmental chemicals, and sometimes also of consequent harmful effects. Appropriate combinations of assays can measure the sequence of changes which lead to the appearance of overt symptoms of toxicity; they can record changes at different organisational levels, ranging from the molecular, through the cellular, to effects upon the whole organism (physiological and behavioural). Assays that only require non destructive sampling have two major advantages, one scientific the other ethical. From the scientific viewpoint they can be based on serial sampling of individuals, thereby eliminating the problem of interindividual variations in the parameters that are measured. From the ethical point of view, there is growing opposition to testing procedures that cause suffering or death to vertebrate animals. Samples of blood, faeces or eggs can be taken without causing distress. Biomarker responses which are closely related to toxicity (eg those based upon the actual mechanism of toxicity) are of particular interest. They can give an integrated measure of the toxic effect of mixtures of compounds acting through the same mechanism (eg in the case of planar organohalogen compounds displaying Ah receptor - mediated toxicity), thus offering a way of approaching one of the most difficult problems in ecotoxicology. A major challenge is linking biomarker responses to consequent effects at the level of population and above. The success in linking eggshell thinning caused by DDE to the decline of certain raptor populations has demonstrated the potential of this approach. However much work needs to be done to make it more widely applicable. With the rapid advance of biochemical toxicology there is an excellent prospect for the development of new 'user friendly' assays to aid environmental risk assessment of vertebrates.

Keywords: Biomarkers, Vertebrates, Non-destructive sampling, Risk Assessment, Ah receptor, Acetylcholinesterase

D. B. Peakall et al. (eds.),
Biomarkers: A Pragmatic Basis for Remediation of Severe Pollution in Eastern Europe, 177–189.

12.1. Introduction

During the last 50 years great advances in the field of analytical chemistry have led to the detection and determination of a bewildering variety of environmental chemicals, many of them man made, often at very low concentrations. Consequently there is now a very large amount of data on levels of environmental chemicals in biota, surface waters, soils, and air. The problem is that the biological significance of these residues is but poorly understood. Little is known about the relationship between environmental concentrations of chemicals (individual compounds or mixtures), and consequent harmful effects at the levels of individual, population, community, or ecosystem. Very large safety factors have to be used when interpreting toxicity data for the purposes of environmental risk assessment [1]. Recently there has been rapidly growing interest in the application of the concept of biomarkers to the problem of environmental risk assessment [2-5].

Biomarkers are here defined as biological responses to environmental chemicals which give a measure of exposure and sometimes also of toxic effect. They are of considerable interest in relation to the assessment of harmful effects of chemicals upon vertebrates in the living environment [6], especially where they only depend upon non destructive sampling [7]. Non destructive biomarker strategies can sometimes be used as alternatives to traditional lethal toxicity testing when providing data for environmental risk assessment [8].

In the following account, biomarker strategies will be briefly reviewed, before describing a number of biomarker assays at the molecular, cellular or 'whole organism' level. Particular attention will be given to assays that provide a measure of toxicity, and to those which are non destructive. In the concluding section, the use of biomarker strategies to study the effects of pollutants upon individual vertebrates and upon vertebrate populations in the field will be reviewed.

12.2. Biomarker Strategies

The definition of biomarkers given earlier refers to responses of individual organisms to environmental chemicals at different organisational levels - molecular, cellular, and whole organism. However, the ultimate concern of ecotoxicologists is about effects at the levels of population, community, and ecosystem [1]. Establishing links between biomarker responses and consequent effects at the level of population and above is an issue of considerable interest and importance, and will be discussed later in this paper. For the moment it should be re-emphasised that current risk assessment practices are directed towards establishing the chances of sensitive species experiencing toxic effects when chemicals are released into the environment, and do not address the more difficult question of effects at the level of population and above. Biomarkers are of immediate relevance to environmental risk assessment [9].

When animals are continuously exposed to lipophilic organic chemicals in their food, or in ambient air or water, the tissue concentration of these chemicals will tend to increase with time until a steady state is reached. If the chemicals have toxic effect, there can be a progression with time from mild effects to severe effects and ultimately death [1]. At first when the tissue levels of the chemical are very low homeostasis will prevail, but with increasing concentration the animal will become stressed, and 'defence mechanisms' (eg the induction of detoxifying enzymes) will begin to operate. These defence mechanisms will bear an energy cost which may be of significance in relation to Darwinian fitness [10]. With a progressive increase in the tissue levels of the chemical, toxic symptoms will appear which are reversible, but will be followed by more severe symptoms and ultimately death. Different biomarker assays can monitor some or all of these stages. It is possible to use combinations of biomarkers to measure the sequence of changes that underly toxicity from initial molecular interaction at the target site, through consequent cellular disturbances, to physiological and behavioural effects [1]. A similar progression of biomarker responses can also be seen where different groups of animals are studied which have been exposed to differing doses of a chemical for a fixed period of time, as in the case of standard toxicity tests in the laboratory, or the sampling of wild animals along concentration gradients of pollutants [1,3].

The most valuable biomarkers are those which relate directly to toxicity, very obviously where they are part of the mechanism of toxicity itself. The inhibition of brain cholinesterase by organophosphorous and carbamate insecticides is one of the best known and most widely studied examples. With birds, inhibition of up to 40% can be tolerated, without any obvious indications of toxicity. Inhibition of 40-75% is associated with characteristic mild physiological and behavioural effects. Above 75% inhibition severe symptoms of neurotoxicity are seen, followed by death [11]. Thus the inhibition of brain acetyl cholinesterase can provide an index of the whole sequence of toxic changes. Unfortunately this assay depends on destructive sampling.

What are the characteristics of an ideal biomarker assay for the present purpose? As already explained, there are advantages in using biomarkers that provide a measure of toxicity as well as exposure; many currently used assays measure exposure alone. The virtues of non destructive biomarkers for vertebrates should also be emphasised [7]. Apart from the ethical issue, there are scientific advantages in using them. Serial sampling is possible (eg of blood) and this can resolve the problem of interindividual variation in the parameters upon which the assays are based (eg levels of proteins in blood). Also, where population studies are carried out , the experimenter should not reduce population numbers by sampling.

Other desirable characteristics of biomarker assays are sensitivity, specificity, stability, and simplicity. Sensitivity is desirable to detect early low level effects of chemicals. Specificity for a particular chemical can be very important in field studies, where it is necessary to distinguish between the effects of one chemical and others present at the same time. (Specificity with regard to test species is undesirable, because it reduces the applicability of an assay.) Where attempts are made to relate the presence of particular chemicals to harmful effects upon wild animals individually, or at the levels of population or above, then specific biomarker assays can provide the essential evidence for causality [1,6]. A biomarker response needs to be reasonably stable for it to be practically useful in certain field situations. Where pollution is episodic, and the pollutant unstable, biomarker responses can easily be missed if sampling occurs too long after exposure. At the more practical level, some systems used as biomarkers are unstable after sampling has occurred, even with low temperature storage (eg certain assays based upon cytochrome P_{450}). Finally assays need to be relatively simple and inexpensive if they are to be widely used by non-experts in the course of environmental risk assessment and field studies on chemical pollution. There is a considerable number of very promising biomarker assays which are only at the research stage, and can only be performed by a small number of specialised research laboratories. More will be said of these later.

A major advantage of biomarker assays that provide indices of toxicity is their ability to give an integrated measure of response to mixtures of compounds acting through a common mechanism of toxicity [12]. In other words it is possible to identify situations where there is potentiation of toxicity. An interesting development of this approach is the production of transgenic vertebrate cell lines which contain specific targets. An example of this is the use of hepatocytes which contain the Ah receptor, and can give an integrated measure of response to dioxins and planar PCBs [13,14]. Non specific assays that measure toxic response can be useful here - especially if used in combination with a more specific biomarker assay that can identify the causal agency.

The use of biomarker assays to measure or predict the harmful effects of chemicals in the field is of particular interest and importance. The potential of this approach has already been demonstrated. The attribution of population declines of raptors to pp'DDE has been confirmed by the demonstration of eggshell thinning in affected populations [1]. Similarly, declines of the dog whelk (*Nucella lapis*) have been attributed to, the action of tributyl tin on the grounds of evidence obtained using the 'Imposex' biomarker [1,3]. Unfortunately the biomarker approach is still at an early stage of development. Only a limited number of appropriate assays are generally available for use in connection with field studies. Progress depends on long term strategies and the investment of realistic sums of money.

Biomarker assays have been used under two distinctly different circumstances in the field. First, where investigating a pollution that has already occurred, and sampling organisms which have experienced high, low, or no exposure to a chemical. Sometimes it has been possible to take samples along a pollution gradient. Second by controlling exposure in one of two ways:

A By deploying 'clean' organisms into polluted and control areas and measuring responses.

B By releasing a chemical into the environment in a controlled field experiment and sampling organisms for assay.

The latter strategy has clear advantages over the former in the sense that there is greater control by the experimenter. In the case of option A, however, there is a need to know which chemicals need to be present in the environment are responsible for a particular effect. The test organism and the magnitude, timing and duration of dose can be predetermined, thus making it relatively easy to compare dose - response relationships observed in the laboratory with those in the field.

Whichever approach is used the intention is to identify biomarker responses to known levels of environmental chemicals, and then to relate these to consequent effects at the levels of population or above. These issues are discussed in more detail in a number of texts on biomarkers [2,3,4,5].

12.3. Biomarker Assays

In the following sections, biomarker assays will be described which are already in use, or are at an advanced stage of development. Destructive biomarkers will be briefly reviewed, and then non destructive ones will be considered in more detail, devoting separate sections to molecular, cellular and 'whole organism biomarkers.

12.3.1. DESTRUCTIVE ASSAYS

As already explained there are reasons for preferring non destructive assays to destructive ones. Unfortunately, because of the limited number of available assays, destructive assays sometimes represent the best or only choice for certain purposes. Some examples will now be given.

12.3.1.1 *Esterases*

Perhaps the most widely used biomarker assay in the present context is the inhibition of brain acetylcholinesterase by organophosphorous (OPs) and carbamate insecticides [15]. This has the advantage of representing the molecular mode of action of the insecticides in question. The degree of inhibition of the enzyme relates closely to the consequent manifestations of toxicity. Acetylcholinesterase and butyrylcholinesterase also occur in blood (see 3.2.1), but measurement of their activities can only provide biomarker assays of exposure in the great majority of cases. There is no general relationship between inhibition of the blood enzymes and those located in brain, although it may sometimes be known for particular compounds in particular species (see for example, 16,17). The assay has the advantages of being sensitive, relatively easy to perform, and relatively stable, as well as providing an index of sublethal effects. It has also been widely used to establish that organophosphorous compounds have been the causal agency when vertebrates have been lethally poisoned in the field.

12.3.1.2 *Binding to Ah receptor*

Assays which measure the induction of hepatic cytochrome P_{450} 1A1 have been widely employed in ecotoxicology. These are of two kinds - 1. Assays which measure an increase of activity catalysed by the enzyme (eg EROD activity) or 2. Assays which measure an increase in the quantity of protein (eg Western blotting) [3,18]. Induction of cytochrome P_{450} 1A1 is caused by a range of planar molecules notably polycyclic aromatic hydrocarbons (PAHs), dioxins, and coplanar polychlorinated biphenyls (PCBs). Induction occurs as a consequence of the inducing agent binding to the so-called 'Ah receptor' in the cytosol. When planar halogenated compounds interact with the Ah receptor there are certain toxic consequences (eg hormonal disturbances and embryo toxicity) as well as P_{450} 1A1 induction. Indeed P_{450} 1A1 induction has been used as an index of 'Ah receptor mediated toxicity' [3,19]. Thus, 1A1 induction can be regarded as a biomarker of toxic effect as well as a biomarker of exposure where it is clearly related to a toxic endpoint (eg embryonic mortality). The effects of combinations of PCBs, dioxins, and polychlorinated dibenzofurans (PCDFs) appear to be additive, and not synergistic. Because individual compounds differ in their binding affinities they are assigned 'toxic equivalency factors' (TEFs), expressed as fractions of the binding affinity of 2,3,7,8 tetrachlorodibenzodioxin (TCDD), which is given a value of 1. From the TEFs and the concentrations of compounds in tissues, TCDD 'toxic equivalents' are calculated, giving an integrated expression of toxic potential.

Although the exact mechanisms of toxic effects mediated by the Ah receptor have yet to be defined, this has nevertheless provided a valuable working concept in ecotoxicogy. A recent development has been the incorporation of the Ah receptor in transgenic mammalian cell lines which can be used to give a measure of the integrated effects of compounds which bind to the receptor. The CALUX system is an example [13]. The limitation here is that there is no simple extrapolation from the reponse of an *in vitro* system to toxicity *in vivo*. To date the measurement of IAI induction has depended very largely on the provision of samples of liver. The possibility of using skin instead will be discussed in a later section.

12.3.1.3 *Genotoxic biomarkers*

There has been much interest recently in the development of biomarkers for genotoxicity. To a large extent the succesful ones have measured DNA damage (eg adduct formation by ^{32}P postlabelling, and chromosome breakage by the comet assay). For a discussion of these see 3,20. In ecotoxicology they have been succesfully used on liver samples. However, there has also been some success in using blood samples for this type of assay which will be discussed in a later section. Assays for DNA damage are generally regarded as biomarkers of exposure only, because it has not been possible to predict consequent mutations from them. However, recent proposals for the use of PCR based DNA fingerprinting may provide a way of establishing mutagenic effects as well [21].

12.3.2. NON DESTRUCTIVE BIOMARKER ASSAYS.

12.3.2.1. *Molecular biomarker assays*

It is relatively easy to obtain blood samples from vertebrates, and there is already an impressive number of biomarker assays that can be performed upon blood (Table 1). Both cholinesterases and carboxylesterases are examples of 'B' type esterases that are found in blood. OPs act as suicide substrates towards them, forming stable phosphoryl adducts,and so causing virtually irreversible inhibition [14,22]. In mammals and birds, depression of blood 'B' esterase activities can last for several days after exposure to OPs, and considerably longer than this in one species of lizard [16,17]. Inhibition provides a valuable biomarker for exposure to OPs and carbamates in field studies, but does not give a reliable indication of toxic effect (see 3.1). A limitation of these assays is the variability of esterase levels in blood. Recently some ELISA assays have been developed, which facilitate the determination of specific activities of blood esterases, thereby circumventing this problem and permitting more sensitive determination of inhibition [23,24].

Anticoagulant rodenticides such as warfarin act as vitamin K antagonists, thereby inhibiting the completion of synthesis of clotting proteins in the liver. The precursors of clotting proteins are consequently released into blood instead of the fully synthesised proteins. Over a period of several days the levels of clotting proteins in blood fall, until clotting can no longer take place, and there is haemorrhaging. There are ELISA assays which can determine the levels of a clotting protein precursor in human blood [25]. However, it is not yet known whether the same assay is effective in wild vertebrates of interest in ecotoxicology. In principle this could provide an excellent biomarker of toxic effect for use in field studies. There is much concern about the possible effects of new rodenticides (eg brodifacoum and difenacoum) upon predators and scavengers (eg owls and corvids) which feed upon rodents.

Other blood biomarker assays that measure toxic effects are based upon the thyroxine (T4) antagonism of the PCB metabolite 4'OH,3,3',4,5' tetrachlorobiphenyl. This, and to a lesser extent other PCB metabolites, can compete with thyroxine for binding sites on, the blood. protein transthyretin. When thyroxine is displaced, thyretin dissociates from an associated retinol (vitamin A) binding protein. Consequently both thyroxine and retinol are lost from blood [26,27]. A number of toxic effects may result from this, including some typical symptoms of vitamin A deficiency. Three assays have been used to measure this toxic interaction — reduction of blood retinol; reduction of blood T4; and reduction of the number of free T4 binding sites on transthyretin. These assays have been succesfully used in a number of mammalian species, including seals. However, results with birds have beeen variable, some species apparently not showing the same mode of action as mammals.Lack of metabolic activation may be a contributory factor [27].

In medical toxicology there has been considerable success in developing assays for genotoxicity (see earlier discussion in 3.1 and 2,3). PAHs, for example, have been shown to form stable adducts with the DNA of white blood cells, using

TABLE 1. Biomarker assays for blood

Biomarker	Type of environmental chemical	Biomarker of toxic effect	Characteristics
Cholinesterase inhibition	Organophosphorous compounds (OPs), carbamate	No	Have been used as biomarkers of exposure. Variability of normal levels a problem. Recent immunochemical assays overcome this problem, but not available commercially.
Carboxylesterase inhibition	OPs and carbamates	No	Same comments as above except that they have not been widely used.
Increase in precursors of blood clotting protein (PIVKA)	Vitamin K antagonists such as Warfarin and related rodenticides	Yes	ELISA assays have been used in humans exposed to warfarin. Not known to what extent assay works in other species.
Fall in retinol (vit A) and thyroxine (T_4) levels, and reduction of available thyroxine binding sites of transthyretin.	Hydroxymetabolites of certain PCBs especially 4 OH' 3 3' 4 5' TCB.	Yes	Have been used successfully on mammalian species including seal. Results with birds variable. Assay of T_4 binding sites requires specialised laboratory.
Inhibition of ALA dehydratase	Lead	No	Good, specific test for exposure to lead. Has been widely used and is readily available.
Changes in porphyrin levels	A number of compounds, including halogenated compounds such as HCB, 3,3',4,4' TCB and TCDD (dioxin)	In some cases	Not very specific. Some potential for use in the field.
Formation of DNA adducts (white blood cells)	Polycyclic aromatic hydrocarbons (PAHs) and other environmental mutagens including oxyradicals.	Not at present state of knowledge.	Several techniques including ^{32}P post-labelling. HPLC/fluorescence and ELISA have been used in environmental studies. Possibility of detection by DNA fingerprinting.
DNA strand breaks	As above.	As above.	Alkaline unwinding assay and Commett assay have been used in environmental studies.
Formation of Haemoglobin adducts	As above	No	HPLC/fluorescence, GC/MS, and other complex assalytical techniques.
Vitellogenin production in male fish	Oestrogens	Yes	Determined by radioimmunoassay. Not specific for particular contaminants

After Walker 1998 [8]

techniques such as ^{32}P postlabelling [20]. Assays have also been developed for measuring DNA strand breaks caused by chemicals (comet assay, alkaline unwinding assay). As noted earlier these are used mainly on tissues such as liver. However, they may be applicable to tissues such as blood and skin obtainable by non destructive sampling [7].

Other biomarker assays that have been succesfully employed with blood, include aminolaevulinate dehydratase (ALAD) inhibition for lead, and changes in porphyrin patterns for a number of organohalogen compounds [7]. Both examples are biomarkers of exposure only.

Skin can be sampled non-destructively,and can be used for biomarker assays such as P_{450} induction and DNA adduct formation. In some recent work with dolphins, skin samples have been taken using a dart, and are being analysed for P_{450} induction as well as organochlorine residues [28].

The use of the eggs of birds, amphibians, fish and reptiles for biomarker assays can also be regarded as non-destructive [7]. The induction of P_{450} forms in bird embryos has been used as a biomarker assay in field studies [29].

Another approach to non destructive biomarker strategies is to analyse urine and faeces. Changes in the patterns of excretory products sometimes provide evidence for the toxic effects of environmental chemicals. A pertinent example is the determination of changes in porphyrin patterns by HPLC analysis following exposure to organohalogen compounds [30]. Recent work has highlighted the potential, of NMR analysis to detect changes in urinary components as a consequence of chemical toxicity [31].

The use of vertebrate cell cultures is a recent development in the context of biomarkers. An example of particular interest and importance addresses the problem of the combined effects of organohalogen compounds which cause Ah receptor mediated toxicity (see 12.3.1.2) [3,14]. Certain hepatoma cell lines contain the Ah receptor (eg mouse Hepa-1c 1c 7 and rat H 4 11e). Recently such cells have been transfected with reporter genes, as in the chemically activated luciferase gene expression (CALUX) system. Very small quantities of organohalogen compounds can cause the synthesis of luciferase via an Ah receptor-dependent pathway, leading to the emission of photons. The output of light can give an integrated measure of the interaction of a mixture of compounds with the Ah receptor.

Trout hepatocytes have also been used for biomarker assays [32]. Western blotting and incorporation of S^{35} methionine were used as assays. There was evidence for stress protein induction caused by four different pollutants. Two organic pollutants induced P_{450} 1A1 in a dose-dependent manner. Also primary cultures of trout hepatocytes have been used to measure the activity of environmental oestrogens in terms of vitellogenin release [33].

Cell systems hold considerable promise for the future and can provide biomarker assays based upon the mode of toxic action Indeed, the incorporation of particular receptors and detoxifying/activating enzymes into transgenic lines is an important part of this developing technology. Nevertheless there are problems in extrapolating from responses observed in cellular systems to the expression of toxicity *in vivo.*

12.3.2.2. *Changes in structure and function at the level of cell or organ*
Where samples of tissue can be obtained,structural changes may be observable by light or electron microscopy. Proliferation of endoplasmic reticulum, lysosomal damage, mitochondrial damage and hypertrophy of cells are all examples of chemically induced changes that can be observed in this way.

One of the most valuable biomarkers yet discovered in ecotoxicology is based on structural change. pp ' DDE, a persistent metabolite of the insecticide DDT causes shell thinning in certain species of birds [1]. This is due to a reduction in the transfer of Ca^{++} in the shell gland, probably because of the inhibition of Ca^{++} ATP ase [3,7]. Species such as the peregrine (*Falco peregrinus*), sparrowhawk (*Accipiter nisus*) and gannet (*Sula bassana*) all showed egg shell thinning related to environmental levels of pp' DDE.In some cases this caused a decline in reproductive success which led to population decline [1].

12.3.2.3 *Changes at the level of the whole organism*
Harmful effects upon the function of the whole organism can be measured using physiological or behavioural assays. Respiration, cardiac function, blood flow, and neuroactivity can all be monitored to give evidence of toxic effect.

In the case of neurotoxic compounds, there is particular interest in behavioural changes resulting from sublethal effects upon the nervous system. Many poisons, including the four major groups of insecticides (OC, OPs, carbamates, pyrethroids), act as neurotoxins. OPs, for example, can cause behavioural changes in birds at sub lethal levels [11]. Movement, feeding behaviour and singing can all be affected. It should be emphasised that effects like these may disturb feeding and reproduction, which can be more harmful at the level of population than lethal toxicity to a limited number of individuals.

12.4. Conclusions

The use of biomarker assays upon vertebrates,making appropriate comparisons between dose-response relatioships in the laboratory and in the field, provides a logical approach to the assessment of the environmental effects of pollutants which has already proven its worth. Such an approach could make a valuable contribution to the process of environmental risk assessment. Ideally the biomarker assays should provide a measure of toxicity, not just exposure, and should not require destructive sampling. Other desirable characteristics of assays are simplicity, stability, specificity, and sensitivity. No single assay is ever likely to have all these attributes, and much is to be gained by using carefully selected combinations of assays which will monitor different stages of the process of intoxication, at different levels of biological organisation. Where toxic responses are monitored, it is possible to measure the integrated effect of combinations of chemicals operating through a common mode of action.

A major challenge is extrapolating from biomarker responses in individual organisms, to effects at the level of populations and above. That such extrapolations are possible has already been established, albeit retrospectively, in two well documented studies - of egg shell thinning of certain raptors caused by pp' DDE, and of 'Imposex' in the dog whelk caused by TBT [1]. Where studies are undertaken on populations, the need for detailed planning and long-term strategies cannot be too strongly emphasised. There is also a strong case for the development of more 'user-friendly' biomarker assays to make this approach more widely applicable.

There is much current interest in the use of alternative methods to vertebrate toxicity tests in environmental risk assessment. The use of mammalian cell lines (including transgenic ones), invertebrates, microorganisms, QSARs and mesocosms have all received attention [34]. These are, however, all indirect and raise problems of extrapolation. The use of non-destructive biomarkers in live vertebrates provides the most direct solution to this dilemma

References

1. Walker, C.H., Hopkin, S.P., Sibly, R.M. and Peakall, D.B. (1996) *Principles of Ecotoxicology*, Taylor and Francis, London.
2. McCarthy, J.F., and Shugart, L.R. (eds.), (1990) *Biomarkers of Environmental Contamination*, Lewis, Boca Raton, Fl.
3. Peakall, D.B. (1992) *Animal Biomarkers as Pollution Indicators*,Chapman and Hall, London.
4. Huggett, R.J., Kimerle, R.A., Mehrle, P.M. and Bergman, H.L., (eds) (1992) *Biomarkers. Biochemical, Phsiological and Histological Markers of Anthropogenic Stress*, Lewis, Boca Raton, Fl.
5. Peakall, D.B. and Shugart, L.R. (eds.) (1993) *Biomarker Research and Application in the Assessment of Environmental Health*, Springer Verlag, Berlin.
6. Peakall, D.B. and Walker, C.H. (1994) The Role of Biomarkers in Environmental Risk Assessment (3) Vertebrates. *Ecotoxicology* **3** 173-179.
7 Fossi, M.C. and Leonzio, C. (eds.) (1994) *Non-destructive Biomarkers in Vertebrates,* Lewis, Boca Raton, Fl.
8. Walker, C.H. (1998) Biomarker Strategies to Evaluate the Environmental Fate of Chemicals, *Environmental Health Perspect.*, **106** (Suppl. 2), 613-620.
9. Walker, C.H. (1996) Biomarkers and Risk Assessment, *Pesticide Outlook,* 7, 7-12.
10. Depledge, M.H., and Fossi, M.C. (1994) The Role of Biomarkers in Risk Assessment (2) invertebrates. *Ecotoxicology* **3**, 161-172.
11. Grue, C.E., Hart, A.D.M. and Mineau, P. (1991) Biological consequences of depressed brain cholinesterase activity in wildlife in P. Mineau (ed.)

Cholinesterase-inhibiting Insecticides - Their Impact on wildlife and the Environment, Elsevier, Amsterdam, pp. 151-210.

12. Walker, C.H. (1998) The use of biomarkers to measure the interactive effects of chemicals. *Ecotoxicology and Environmental Safety*, **40**, 65-70.
13. Brouwer, A. (1996) Biomarkers for exposure (and effect) assessment of dioxins and PCBs in Use of Biomarkers in Exposure assessment, report R5. IEH report published by Medical Research Council, Leicester.
14. Garrison, P.K., Tullis, K., Aaarts, J.M.M., Brouwer, A., Giesy, J.P. and Dennison, M.S. (1996) Species specific cell lines as bioassay systems for the detection of TCDD-like chemicals. *Fundamental and Applied Toxicology,* **30**, 194-203.
15. Mineau, P. (1991) *Cholinesterase-inhibiting insecticides - Their Impact on Wildlife and the Environment*, Elsevier, Amsterdam.
16. Sanchez, J.C., Fossi, M.C. and Focardi, S. (1997) Serum B esterases as a non destructive biomarker in the lizard *Gallotia galloti* experimentally treated with parathion, *Environmental Toxicology and Chemistry*, **16**, 1954-1961.
17. Fossi, M.C., Sanchez, J.C. and Gaggi, C. (1995) The lizard *Gallotia galloti* as a biomarker of organophosphorous contamination in the Canary Islands, *Environmental Pollution*, **87**, 289-294.
18. Livingstone, D.R. and Stegeman, J.J. (1998) Forms and Functions of Cytochrome P_{450}, *Comparative Biochemistry and Physiology.* Special issue, in press.
19. Ahlborg, U.G., Becking, G.C., Birnbaum, L.S., Brouwer, A., Derks, H., Feeley, M., Golor, G., Hanberg, A., Larsen, J.C., Liem, A.K.D., Safe, S.H., Schletter, C., Waern, F., Younes, M. and Yrjanheikki, E. (1994) *Toxic Equivalency factors for Dioxin-like PCBs Chemosphere*, **28**, 1049-1065.
20. Shugart, L. (1994) Genotoxic responses in blood in M.C. Fossi. and C. Leonzio (eds) *Non- destructive Biomarkers in Vertebrates.* Lewis, Boca Raton, Fl.
21. Savva, D. (1996) DNA fingerprinting as a biomarker assay in ecotoxicology. *Ecotoxicology and Ecotoxicology News*, **3**, 110-114.
22. Thompson, H.M. and Walker, C.H. (1994) Blood esterases as indicators of exposure to organophosphorous and carbamate insecticides in M.C. Fossi and C. Leonzio (eds.) *Non-destructive Biomarkers in Vertebrates.* Lewis, Boca Raton, Fl.
23. Khattab, A.D., Walker, C.H., Johnston, G.O., Siddiqui, M.K. and Saphier, P.W. (1994). An ELISA assay for avian serum butyrylcholinesterase: a biomarker for organophosphates. *Environmental Toxicology and Chemistry*, **13**, 1661-1668.
24. Jackson J.B. (1994) The biochemical toxicology of serum carboxylesterases in the pigeon (*Columba livia*) PhD thesis, University of Reading.
25. Hamulyak, K., Thijssen, H.H.W. and Vermeer, C. (1994) Effects of coumanrin derivatives on coagulation proteins, *Human and Experimental Toxicology* **13**(a), 631.

26. Brouwer, A., Reijnders, P.J.H. and Koeman, J.H. (1989) PCB contaminated fish induce vitamin A and thyroid deficiency in the common seal (*Phoca vitulina*), *Aquatic Toxicology*, **15**, 99-105.
27. Brouwer, A., Murk,A.J. and Koeman, J.H. (1990) Biochemical and physiological approaches in ecotoxicology, *Functional Ecology*, **4**, 275-281.
28. Aguilar, A., and Borrell, A. (1994) Assessment of Organochlorine Pollutants in cetaceans by means of skin and hypodermic biopsies in M.C. Fossi and C. Leonzio (eds.), *Non-destructive Biomarkers in Vertebrates*, Lewis, Boca Raton, Fl.
29. Tillett, D.E., Ankley, G.T., Verbrugge, D.A., Giesy, J.P., Ludwig, J.P. and Kubiak, T.J. (1991) H411E rat hepatoma cell bioassay-derived 2,3,7,8-tetrachlorodibenzo-p-dioxin equivalent in colonial fish-eating waterbird birds eggs from the Great Lakes, *Archives of Environmental Contamination and Toxicology*, **21**, 91-101.
30. De Matteis F. and Lim, C.K. (1994) Porphyrins as non-destructive bioindicators of exposure to environmental pollutants, in M.C. Fossi and C. Leonzio (eds.), *Non-destructive biomarkes in Vertebrates*, Lewis, Boca Raton, Fl.
31. Nicholson, J.K. (1994) N.M.R.spectroscopy and pattern recognition as approaches to the detection of novel molecular markers of toxicity, *Human and Experimental Toxicology*, **13**(Abstract), 630.
32. Grosnik, B.E. and Goksoyr, A. (1996) Biomarker protein expression in primary cultures of salmon hepatocytes exposed to environmental pollutants, *Biomarkers*, **1**, 45-54.
33. Sumpter, J.P. and Jobling, S.,(1995) Vitellogenesis as a biomarker for oestrogenic contamination of the aquatic environment, *Environmental Health Perspectives*, **103**, 173-178.
34. Walker, C., Kaiser, K., Klein, W., Lagadic, L., Peakall, D., Sheffield, S., Soldan, T. and Yasuno, M. (1998) Alternative Testing Methodologies for Ecotoxicology, *Environ. Health Perspectives*, **106** (Suppl. 2), 441-451.

1. REPORT OF THE WORKING GROUP ON THE KATOWICE ADMINISTRATIVE DISTRICT, POLAND: A REVIEW OF RESEARCH DONE TO DATE, AND RECOMMENDATIONS FOR FUTURE RESEARCH

S. W. KENNEDY (Rapporteur), S. GODZIK (Chair) K. DMOWSKI, R. HANDY, A. KEDZIORSKI, P. KRAMARZ, L. MANUSADZIANAS AND A. MURK

1.1. Summary

The Katowice Administrative District (KAD) is the oldest and largest industrial region in Poland, and the air, soil and water has been heavily contaminated with many pollutants from factories in the area. Pollutants of major concern are polycyclic aromatic hydrocarbons (PAHs), other organic compounds, and heavy metals. There is also concern that other, not yet identified, compounds might be having effects on biota. Biomarkers have been used with some success in the KAD to document the effects of pollutants. This chapter provides a brief review of the biomarker research on invertebrate and vertebrate species that has been carried out in the KAD, and makes recommendations for enhancing the use of biomarkers for future environmental monitoring programs.

1.2. Introduction

The Katowice Administrative District (KAD) is the oldest and largest industrial region in Poland. All power stations of a capacity of more than 6500 megawatts are located in this region and since it contains 10% of the Polish population there is a considerable input of pollutants from individual households and district heating units. The KAD is the sole source of hard coal and cadmium, and a major source of lead and zinc for the entire country. Details of the types of pollutants in the KAD are contained in the paper by Godzik and co-workers (this volume). Historical records on the location of old smelters, municipal waste sites, chemical plants and other anthropogenic sources are either unavailable or difficult to obtain. Information is particularly difficult to obtain for factories that were dismantled many years ago. Even for factories dismantled in recent years detailed information concerning the degree and spatial distribution of pollutants over the area is often not available.

The pollutants that have caused the greatest concern are air pollutants, especially sulphur dioxide. More recently some attention has been given to ozone.

D. B. Peakall et al. (eds.),
Biomarkers: A Pragmatic Basis for Remediation of Severe Pollution in Eastern Europe, 191–210.

Widespread damage to the forests of the region have been caused by air pollution. Polycyclic aromatic hydrocarbons (PAHs) and other organic compounds are also important concerns. Major sources of PAHs are cokeries and factories that process tar products. Various chemical factories have polluted the environment with other organic compounds. Recent increases in the numbers of automobiles have increased the amount of pollution from this source.

Monitoring of air quality in the KAD started in 1975 and a Regional Environmental Network was established in 1993. Automatic recording apparatus to measure concentrations of SO_2, NO, NO_2 and total suspended particulates were set up. In recent years, there have been enhanced efforts to determine the effects of pollutants on the biochemistry and physiology of vertebrates, invertebrates and plants.

In this chapter we briefly review the biomarker research on *invertebrate* and *vertebrate* species that has been carried out in the KAD, and make recommendations for enhancing the use of biomarkers for future environmental monitoring programs. *Plant* biomarkers are discussed elsewhere in this book (Godzik *et al.*; Ernst).

1.3. Biomarkers Used to Date

1.3.1. VERTEBRATES

1.3.1.1. *Biomarkers of Genetic Damage*

Investigators from the Institute of Oncology in Gliwice, in collaboration with scientists from Finland, Sweden, Norway and the United States have carried out several studies that used biomarkers and *in vitro* bioassays to determine human exposure to PAHs in the KAD. They found higher levels of DNA adducts in white blood cells of coke oven workers and in residents from towns close to cokeries compared to residents of rural Poland [1]. Other studies used a battery of biomarkers to measure molecular and genetic damage in peripheral blood from residents of the KAD to assess exposure and effects of PAHs [2,3,4]. The following assays were carried out: DNA and protein adducts, sister chromatid exchange, micronucleus formation, DNA strand breaks and DNA repair capacity. Most of these biomarkers were indicative of exposure to PAHs, and results showed the usefulness of using various biomarkers to help increase understanding of the effects of PAHs on human health. The same team of investigators have found associations between ambient air pollution and DNA damage in Polish mothers and newborns in the city of Krakow, an industrialized city in close proximity to the KAD [5]. To our knowledge, biomarkers of genetic damage have not been studied in wildlife species in the KAD, and are strongly recommended.

1.3.1.2. *Other Biomarkers*

Several studies in the KAD have used biomarkers to attempt to determine the effects of pollutants on wildlife species. For example, Dmowski and Kozakiewicz found elevated levels of metallothionein (MT) in kidneys of small insectivorous mammals and rodents collected in the vicinity of the Miasteczko Slaskie zinc smelter [6]. MT content in kidneys of Bank voles (*Clethrionomys glareolus*) and mice (several species of *Apodemus*) ranged from 20-42 μg/g (wet weight basis) while the concentration in the Common screw (*Sorex araneus*) was 175 μg/g. Wlostowski [7] studied Bank voles (*Clethrionomys glareolus*) from the Rybnik coal region and from a relatively uncontaminated site. Intensive binding of Cd by MT in livers from the KAD site, especially in livers of immature animals, was observed.

Over-wintering voles from the KAD had significantly elevated tissue respiration rate in comparison with voles collected from the uncontaminated site. However in other studies carried out in the late 1970s, shortly after the Katowice steelwork was opened, no significant changes in the tissue respiratory metabolism of hens (*Gallus domesticus*) or voles collected in the vicinity of the plant were found [8]. Hemoglobin concentration was lower in Great Tit (*Parus major*) nestlings collected near the Olkusz zinc smelter compared to levels in this species of bird in a relatively uncontaminated areas of Sweden [9].

Jankowski *et al.* [10] found that heavy metals appeared to cause accumulation of dolichol in magpie (*Pica pica*) livers. Dolichol is a mixture of linear polyisoprenoid alcohols which can accumulate in organs of vertebrates with age or upon exposure to some environmental pollutants and have a negative effect on cellular membrane functions. Significantly lower activity of phosphofructokinase and pyruvate kinase, important enzymes of the glycolytic pathway, were reported in muscle, brain, intestine and liver of adult field mice (*Aopdemus agrarisus*) collected from a polluted site (Sosnowiec) in comparison with mice collected from a reference site (Rudziniec). However, in the same study no differences in activities of these enzymes were found in pigeons and the trend was the opposite in frog muscle in animals collected from the same two locations [11].

Elevated levels of zinc protoporphyrin in children and workers exposed to lead have been reported [12].

1.3.2. INVERTEBRATES

Studies in the KAD to determine the effects of pollutants on the biochemistry and physiology of invertebrates have, in general, been similar to those carried out in Western European countries and North America. Due to the extreme nature of the ecological problems in the KAD, investigators had opportunities to investigate the influence of high concentrations of multiple stresses on the biochemistry and physiology of various organisms. Investigations over such large gradients of exposure have not been carried out to our knowledge in Western Europe and North America. However, there were technical limitations, and more advanced and sophisticated techniques used in other countries often could not be applied in the KAD.

1.3.2.1. *Changes in Respiratory Metabolism Rate*

Early studies carried out on selected invertebrates (earthworm, *Lumbricus terrestris*; grasshopper, *Chorthippus sp.*; coccinellid beetle, *Coccinella septempunctata*) in close proximity to the Katowice steel factory revealed changes in respiratory metabolism (eg., decreased oxygen uptake that was followed by a mild elevation above control in subsequent years). This phenomenon occurred mainly in earthworms, and to some degree, in grasshoppers. It was also reported that inhibition of metabolic rate was negatively correlated with the distance from the steel factory [8]. Higher oxygen uptake was found in *L. terrestrius* collected in the vicinity of the plant than in animals collected from a reference location [8] and similar differences were reported in a study carried out 7 years later [13]. Increased respiration rate was also reported in in spiders exposed to industrial pollution, and these animals also contained relatively high levels of lipids [14]. In contrast, a decrease in the rate of oxygen uptake was reported in aphids, (*Macrosiphoniella*) exposed to SO_2 or acid rain [15]. Two species were studied (*M. oblonga* and *M. artemisiae*) and inhibition of oxygen uptake was greater in the former species. Similar effects were observed in other species of insects exposed to various gaseous (e.g. SO_2 and NO_x) or dust pollutants (steel-work dust fallout). Migula [16] reported that small doses of atmospheric pollutants caused elevated oxygen consumption in chewing and sap-sucking species, indicating possible involvement of the *hormesis* phenomenon.

1.3.2.2. *Adenylate Nucleotides and the AEC Index*

Adenylate energy charge (AEC; AEC = (ATP + 0.5 ADP)/(ATP + ADP + AMP)) was proposed by Atkinson [17] as a general measure of the response of an organism to environmental stressors. This index and a total pool of adenine nucleotides were suggested as measures of metabolic homeostasis in invertebrates, particularly those inhabiting aquatic biotopes [18,19]. AEC values between 0.7 and 1.0 reflect energy balance of an organism, while lower values indicate a shift of adenine nucleotides from ATP to ADP and AMP; i.e. enhanced ATP hydolysis. AEC values below 0.5 indicate near-to-death energy status.

Biochemical analysis of honey bees collected from several localities in the KAD that differed in levels of air pollutants (Pb, Cd, Cu, SO_2, dust and dust fallout) revealed differences in adenylate concentrations. Adenylates were lower in bees from more polluted sites, but this decrease did not correlate with dust fallout or heavy metal concentration [20]. However, heavy metals were associated with species-specific responses in adenylate concentration in caterpillars [16]. A decreased level of adenylates was reported in foraging ants inhabiting polluted sites contaminated with Cd and Hg compared to ants collected from relatively clean sites [16,21]. In another study, the cantharid beetle (*Rhagonycha fulva*) and a species of aphid (*Aphis fabae*) collected in the KAD had lower levels of adenine nucleotides than the same species collected from a reference site, but this was not observed for other insects investigated in this study [22], and the AEC index for all the species ranged between 0.8 - 0.9, suggesting a good energy balance.

1.3.2.3. *Metabolic Enzymes*

Species-specific differences in enzymes involved with carbohydrate metabolism in response to air pollution have been reported in two aphid species (*M. oblonga and*

M. artemisiae). Decreased glucose-6 phosphate isomerase (GPI), isocitrate dehydrogenase (IDH) and malate dehydrogenase (ME) activity appeared to be associated with high SO_2 or acid rain concentration in these species [15]. Similarly, decreased activity of glycolytic and amino acid-metabolizing enzymes which appeared to be caused by atmospheric pollutants were observed in another species of aphid (*Acyrthosiphon pisum*) [23]. In contrast, significant stimulation of the several of the same enzymes occurred in holometabolous moths (*Leucoma salicis* and *Euproctis chrysorrhoea*) kept in similar conditions [23]. Different metabolic responses to pollution were also obtained in recent studies carried out on a few other species of invertebrates [11]. Higher activitities of phosphofructokinase and pyruvate kinase were reported in grasshoppers exposed to high levels of pollutants, while in the slug (*Arion rufus*) from this site the activities of both enzymes were lower. The two sites differed in the concentrations of various metals and tar compounds. Because no correlation was found between any specific pollutant and enzyme activity, such assays can only be regarded as non-specific biomarkers. Further research in the laboratory to determine the effects of specific pollutants and complex mixtures of pollutants on metabolic enzymes in invertebrates are recommended.

1.3.2.4. *Detoxifying Enzymes*

Recently, there has been increased attention by Migula and colleagues [24,25] and other investigators to changes in the activity of detoxifying enzymes in invertebrates in response to pollution. In several studies, enzymes involved with phase I (e.g. carboxylesterases; CarE) and phase II metabolism (e.g. glutathione S-transferase; GST) have been examined. In addition, enzymes involved with oxidative stress (superoxide dismutase (SOD) and catalase (CAT)) have been studied in various species. In spiders (*Pardosa palustris, Linyphia triangularis, Metellina segmentata, Araneus diadematus*), higher activity of GST and SOD was found in individuals exposed to various contaminants (mainly heavy metals) compared to individuals from a less contaminated site. Thus, GST and SOD appeared to be non-specific biomarkers of exposure. Species-specific differences occurred with CarE activity: the activity was higher in *P. palustris* but lower in other species inhabiting polluted biotopes in comparison with individuals collected from a reference site outside of the KAD. These differences appeared to be associated with life style of the species: *P. palustris* is a non web-building and detritivore spider, while the others are typical web predators [24]. Different findings were observed for grasshoppers (*Chorthippus brunneus*). GST activity was significantly lower in insects from areas within the KAD contaminated with heavy metals compared to GST activity in individuals obtained from sites outside the industrialized region [25,26]. However, in grasshoppers collected form a heavily polluted site (Katowice-Szopienice) that were exposed to various doses of Cd in the laboratory, increased activity of GST, CarE and peroxidase were observed (Augustyniak, M.Sc. thesis, University of Silesia, Katowice). Reasons for differences in apparent responses to pollutants among species are not known, and systemic studies to try to identify causal relationships are recommended if such biomarkers are to be applied to future investigations.

1.3.2.5. *Heat Shock Proteins*
Heat shock proteins (hsp) are a group of highly conserved proteins which are induced in cells upon exposure to various stresses, including heat, metals, some organic pollutants, ultraviolet radiation, injury and disease. Induction of hsp represents a general mechanism of cellular response to protein-disintegrating stressful situations [27]. Stress proteins play a protective role in cells exposed to stress, but are also constituitively expressed as heat shock cognates (hsc) under normal conditions [28]. Hsp expression has been suggested as a biomarker of stress caused by toxic pollutants for aquatic [27] and terrestrial [29] invertebrates.

Pyza and colleagues [30] measured hsp 70 proteins in centipedes (*Lithobius mutabilis*) collected from sites close to a zinc and lead smelter (Olkusz) and an uncontaminated site. The sites were in the Krakow District, adjacent to the KAD. There was no correlation between stress protein levels and exposure to pollutants, leading the authors to suggest that this biomarker has limited or no value in this species of invertebrate in this location. They noted that one must be cautious when attempting to use hsp 70 as a biomarker because its expression varies with the animal's stage of development [31], age [32] and season [33]. These cautionary remarks are in general agreement with those made in a recent review article on stress proteins as biomarkers [34] where contradictory studies were noted, and areas of neglected research were reviewed. Overall, it was concluded that the selection of heat shock proteins was scientifically sound, but that further evaluation is required.

1.4. Recommended Biomarkers and *In Vitro* Bioassays

A wide range of biomarkers and *in vitro* bioassays have been developed to determine the effects of environmental pollutants, and some of these have not yet been used in the KAD. Here we indicate methods that could be applied to determine biochemical and physiological effects of pollutants on wild species of vertebrates and invertebrates in this district of Poland. Potentially useful plant biomarkers are discussed elsewhere in this volume (Godzik *et al.*; Ernst). It is beyond the scope of this chapter to discuss *all* possible assays, but we hope that the information provided will be helpful to investigators who wish to strengthen future biomonitoring studies in the KAD.

1.4.1. POLLUTANTS OF KNOWN CONCERN

1.4.1.1. *PAHs*
Some PAHs are genotoxic, and there are several biomarkers that can be used to measure exposure and effects. Available biomarkers include both direct measurement of DNA or protein damage (adducts) and measurement of secondary biological effects (mutations and cytogenetic damage). Adducts can be detected by

^{32}P-postlablelling, HPLC, immunochemical or mass spectrometric methods [35-38]. Various methods for measuring mutations and cytotoxic damage are available [39]. DNA strand breaks can be detected using alkaline unwinding assays [40]. Several of these biomarkers have been used for determining human exposure to PAHs in the KAD (discussed above) and elsewhere [40]. Some of these biomarkers have also been used with success to determine DNA alterations in fish and other vertebrates in other areas of the world [41,42]. We recommend using the following biomarkers for wild vertebrate species expected to be exposed to PAHs in the KAD: DNA adducts, DNA strand breaks and measurement of cytochrome P4501A content and activity. Exposure to PAHs should be assessed, and one approach that should be considered is the measurement of fluorescent aromatic compounds (FACs) in the bile [42].

1.4.1.2. *Other Organic Compounds*

Major organic pollutants known to be of concern in the KAD are benzene, toluene, phenol and formaldehyde. In acute exposure, tissue residue levels of these compounds may give an indication of exposure, but absorption of these compounds from mixtures of pollutants may vary. For example, benzene toxicity may be reduced by toluene [43]. In addition, metabolism during chronic exposure may make interpretation of tissue residues difficult. However, the presence or absence of metabolites can serve as biomarkers. For example, urinary hippuric acid is a preclinical biomarker of toluene exposure in humans [44]. Trans, trans muconic acid is a metabolite of benzene which can be measured in body fluids [45]. Perhaps existing veterinary samples could be analyzed retrospectively for this purpose in cattle to produce spatial and/or temporal maps of regional exposure. Blood and urine could also be collected from freshwater fish [46] for analysis of metabolites. Phenol can be determined directly in tissue homogenates [47]. Aldehyde dehydrogenase is essential to formaldehyde metabolism [48] and perhaps this enzyme might serve as a biomarker of formaldehyde exposure.

There are several histological methods which, when combined with haematology, can indicate the presence and effects of some organic compounds. For example DNA damage in peripheral lymphocytes and liver tissue in combination with anaemia may indicate benzene exposure [43]. Mixtures containing low levels of benzene and phenol produce non-pathological hepatocyte proliferation in rats [49]. Kraft mill effluents typically contain high levels of phenolic compounds, which produce lipidosis and macrophage aggregration in the liver of fish [50]. Phenoxyl metabolites from benzene, phenol itself, and other solvents can cause lipid peroxidation [51], but so do heavy metals [52]. For carcinogenic compounds generally, the frequency of hepatic tumours and the type of tumour may indicate chronic exposure and provide some indication of the type of compound [50,53]. Tumour scores from fish at different locations can certainly give an indication of surface water contamination.

Some organic pollutants and redox active metals generate free radicals which cause lipid peroxidation. Thiobarbituric acid reactive substances (TBARS) is the end product of lipid peroxidation [48] and provides a useful biomarker [52,54].

Conversely, the utilization of anti-oxidant defense mechanisms may indicate the presence of oxygen radicals. For example, reduction of alpha tocopherol can occur during Cu exposure [54].

1.4.1.3. *Metals*

Knowledge of metal concentrations in biological tissues often provides limited information on their effects, but we strongly recommend measurement of metal concentrations in parallel with biomarker assays to help establish cause-effect linkages. Semi-quantitative multi-element analysis using inductively coupled plasma mass spectroscopy (ICP-MS) can be used to identify the main contaminants and/or to prioritize further investigation. Multi-element analysis can be achieved in small samples using inductively coupled plasma emission spectrophotometry (ICP-EAS), or by repeat analysis for individual metals using flame or graphite furnace atomic absorption spectrophotometry (FAAS or GFAAS). These techniques, particularly the multi-element approaches, are relatively inexpensive and can be automated for routine analysis. These procedures require ion-clean laboratory techniques and a reliable source of reference material. In some situations it may be useful to pretreat samples with metal chelators. This approach can be used to differentiate between the toxic effects of metals and organic compounds in environmental samples. It involves the chelation of toxic metals in the sample (e.g. water, soil-extract) prior to assessment of toxicity [55]. If the addition of chelators eliminates the toxic effects, then a metal(s) rather than an organic pollutant may be suspected as the causative agent. This approach is especially useful when facilities for metal analysis are not available, or in the early stages of an investigation to identify contaminants. If a metal is suspected, then more specific toxicity assays may be applied to follow up the investigation.

Biomarkers available for the metals of concern in the KAD include:

Metallothionein: Induction of MT is a well known biomarker of exposure to Cd, Zn, Hg, and Cu [56-59] and has been used for a limited number of studies in the KAD (discussed above). MT is a metal chelator, and probably reduces the toxicity of metals in the tissues where it occurs. Since several metals will induce MT, this assay can indicate an overall stress response to multi-element exposures. One should be aware that other factors, including certain hormones, second messengers and other chemicals can induce MT synthesis [60,61]. Therefore, appropriate chemical analysis should be carried out in parallel with measurement of this biomarker. Early methodologies used pulse polarography to quantify MT [56], but the recent development of rapid photometric protocols provides a practical advantage for the investigator [62].

δ-Aminolevulinic Acid Dehydratase (ALAD): ALAD is found in red blood cell membranes and is involved in haemoglobin synthesis. ALAD activity is inhibited by Pb and determination of ALAD activity in serum samples can indicate exposure to this metal. ALAD activity has been correlated with serum lead concentrations in fish [63,64] and birds [65,66].

Ion Flux Measurements: This approach relies on the fact that toxic metals sometimes compete with other essential metals for uptake. For example, Zn can affect Ca uptake [67] or inhibit the membrane proteins involved in the regulation of the major plasma ions (e.g. Cu inhibition of Na,K-ATPase [68]). These effects have traditionally been measured by determination of serum mineral levels, or even whole body mineral content [69]. However, ion flux measurements are much more sensitive, and may be selective. For example, micromolar concentrations of Cd disturbs Ca fluxes but not Na balance [70]. Wood [71] reviewed the application of ion flux measurements to quantify biological effects in fish and highlighted the advantages of flux techniques over traditional blood and tissue sampling approaches. These are: (*i.*) 10-400 fold greater sensitivity, (*ii.*) non-invasive, (*iii.*) simplicity and applicability to field work, (*iv.*) measurements made at the site of toxic action. Ion flux measurements are often sensitive to micromolar changes in toxic metal concentration [67] and respond to the biologically available fraction of the metal. Wood [71] illustrates the application of the *in vivo* approach using live animals to test for aquatic pollution. Ion flux measurements can also be made from purified membrane vesicle preparations of mammalian, invertebrate and fish tissues respectively [72,73,74]. Therefore, it might be possible to make ion flux ecotoxicity assessments *in vitro*.
P-type ATPase Activity: This method is similar to the ion flux approach, but measures the activity of the enzymes responsible for ion transport, rather than the ion fluxes themselves. The main advantage of this approach is that the enzyme inhibition may be the mode of toxic action which ultimately contributes to mineral imbalances, cardiovascular collapse and death [75]. Inhibition of ATPase activity is indicative of metal exposure [68,76], and perhaps exposure to some organic chemicals [77]. The interpretation of ATPase assays are therefore best achieved in concert with trace metal determination in the tissue. A variety of colorimetric assays exist to determine ATPase activities in tissue homogenates and purified membrane preparations [78].

Histological Biomarkers: The goal of this approach is identification of the main categories of pollutants using a suite of histological observations on a variety of internal organs [50]. This approach requires considerable expertise in histopathology to develop a suite of identifying lesions which may be scored as evidence for a particular pollutant. Nonetheless, the approach may help to determine whether the causative agents are metals or organic compounds. For metals, the presence of granular deposits in the liver may be indicative of metal intoxication [79], and such observations combined with routine histological staining for the presence of metals can differentiate between animals from clean reference sites and those from contaminated areas. In addition, histochemical stains for enzymes particularly affected by metal exposure may be helpful (e.g. for Cu glutathione transferase).

1.4.2. POLLUTANTS OF POSSIBLE CONCERN

1.4.2.1. *Dioxins, PCBs and other Halogenated Aromatic Hydrocarbons*

Halogenated aromatic hydrocarbons (HAHs), such as dibenzo-*p*-dioxins (PCDDs), dibenzofurans (PCDFs) and polychlorinated biphenyls (PCBs) are persistent environmental contaminants which have the potential to cause a wide range of toxic and biological effects in humans and wildlife. There has been some analysis of HAH concentrations in the KAD, but there do not appear to have been many studies to determine the effects of these contaminants on biota. Chemical residue analysis is essential for determining levels of HAHs in the environment, but often provides limited information on the toxic potency of complex mixtures of these chemicals because (*i.*) there are hundreds of isomers and congeners, with toxic potencies that range over several orders of magnitude, (*ii.*) environmental mixtures of HAHs are extremely complex, and (*iii.*) there are large differences in sensitivity to HAHs among different species of animals. Therefore, we recommend the application of biomarkers and cell culture bioassays to determine the extent of possible problems due to HAHs in the KAD. One approach would be to exploit the fact that there is substantial evidence that many of the effects of HAHs are mediated by the aryl hydrocarbon receptor (AHR). In recent years, several bioassays which utilize cultured cells for measuring AHR-dependent effects of HAHs extracted from environmental samples (eg. soil, water, air or biological tissue) have been developed. Kennedy and colleagues have developed avian hepatocyte bioassays for measuring potency to induce cytochrome P4501A (CYP1A) and porphyria, two responses which are mediated by the AHR [80,81,82]. Other laboratories have developed bioassays that measure CYP1A induction in primary and continuous cultures of rodent, fish and human cells which respond to HAHs with AHR-dependent induction [83-86]. These bioassays should also be considered for use in the KAD.

Measurement of CYP1A content and catalytic activity in fish and birds have been useful biomarker assays for HAH exposure in wildlife species and should be considered for application to future studies in the KAD. However, the presence of AhR-dependent induction of CYP1A has not been confirmed in aquatic invertebrates [87]. We recommend the concomitant measurement of chemical residues because there are natural and anthropogenic chemicals that can affect CYP1A. In some species of birds, porphyrin concentrations may be a useful biomarker of exposure to PCBs or other PAHs [88].

1.4.2.2 *Endocrine Disrupting Compounds*

One of the most important and controversial issues in environmental toxicology today concerns the potential effect of anthropogenic chemicals on the development, growth and reproduction of wildlife species and humans. There is considerable evidence that some environmental pollutants can modulate or mimic the action of hormones, leading to adverse effects in some species of wildlife. Such pollutants, which include organic chemicals and metals, are commonly referred to as "endocrine disrupting chemicals" (EDCs), or "endocrine modulating chemicals".

There is clear evidence that wildlife species have been affected by EDCs in some areas of the world [89,90] but the chemicals responsible for causing such effects are often not known. There is considerable international debate as to whether EDCs are affecting human reproduction and development. Investigators who are working in the Katowice Area should ensure that they remain up-to-date with ongoing research on the effects of EDCs. Exposure to estrogens and its mimics, particularly alkylphenolic compounds, causes the synthesis of vitellogenin and several studies have showed the usefulness of vitellogenin as a biomarker of exposure to estrogen-like compounds [90].

1.5. General Recommendations

Enhanced application of biomarkers and *in vitro* bioassays could be used to augment existing monitoring programs in the KAD ,and help identify and prioritize the extent of problems. The value of using biomarkers is that they can give important information of the effects of pollutants. Judicious use of biomarkers would complement certain existing programs that only measure the levels of specific pollutants. In the KAD an increased use of biomarkers can help define areas which are most at risk from known and unsuspected pollutants. Biomarkers can be used to:

1. Complement existing biomonitoring programs,
2. Identify currently unknown hot spots which require urgent attention,
3. Determine which organisms and populations are at risk,
4. Determine the severity of the effects,
5. Determine how widespread the effects are,
6. Monitor the effectiveness of clean-up and remediation measures.

Demonstration of cause-effect linkages for contaminants will be just as important as determining that effects of pollutants in some areas of the KAD are minimal. Both situations can be established with the judicious use of biomarkers. A variety of biomarkers and bioassays will be needed, not only because existing chemical residue databases are incomplete, but also because of the variety of biological problems within the district. Some biomarkers are less expensive to carry out than chemical residue analysis and should be considered in situations where labour costs are relatively low.

There are environmental monitoring programs in the KAD and programs using biomarkers have been useful for identifying many problems. We recommend incorporation of the biomarkers indicated above into future research. In addition we recommend that investigators consider taking pro-active measures to contact environmental scientists from North America and areas of Europe that are less polluted than the KAD. Many of these scientists are very willing to share their ideas and new methods.

We propose the use of a risk-based approach for environmental assessment which incorporates existing legislative ecotoxicity tests from the European

Community (EC 1992 - Annex to EC directive 92/69/EEC) and guideline ecotoxicity protocols recently validated by inter-laboratory testing (e.g. via the Paris Commission), and incorporation of biomarkers at appropriate stages of the risk assessments. We also recommend incorporation of the toxicity identification and evaluation (TIE) approach outline in several US EPA documents (e.g. Marine Toxicity Identification Evaluation, Phase I Guidance Document, 1996) into future assessments. In addition, we propose:

1. The development and implementation of a comprehensive screening program, preferably at the national level. This would include contaminant identification in all environmental compartments, and screening for biological effects using biomarkers and bioassays,
2. Enhanced site-specific risk assessments at the major industrial sites based on US EPA guidelines or similar methods,
3. Enhanced research to identify hotspots, and assessment of the effects of complex mixtures of pollutants on wildlife species,
4. Enhanced efforts to determine the levels of PCBs, dioxins and pesticides in wildlife and humans. When more information is obtained on the levels of these types of pollutants in the KAD, appropriate biomarkers should be used to determine their effects,
5. Establishment of a regional and/or national specimen bank.

References

1. Hemminki, K., Grzybowska, E., Chorazy, M., Twardowska-Saucha, K., Sroczynski, J.W., Putman, K.L., Randerath, K., Phillips, D.H., Hewer, A., Santella, R.M., Young, T.L. and Perera, F.P. (1990) DNA adducts in humans environmentally exposed to aromatic compounds in an industrial area of Poland, *Carcinogenesis* **11**, 1229-1231.

2. Perera, F.P., Hemminki, K., Grzybowska, E., Motyklewicz, G., Michalska, J., Santella, R.M., Young, T.-L., Dickey, C., Brandt-Rauf, P., DeVivo, I., Blaner, W., Tsai, W.-Y. and Chorazy, M. (1992) Molecular and genetic damage in humans from environmental pollution in Poland, *Nature* **360**, 256-258.

3. Grzybowska, E., Hemminki, K. and Chorazy, M. (1993) Seasonal variations in levels of DNA adducts and X-spots in human populations living in different parts of Poland, *Environ. Health Perspect*, **99**, 77-81.

4. Øvrebø, S., Fjeldstad, P.E., Grzybowska, E., Kure, E.H., Chorazy, M. and Haugen, A. (1995) Biological monitoring of polycyclic aromatic hydrocarbon exposure in a highly polluted area of Poland, *Environ. Health Perspect.* **103**, 838-843.

5. Whyatt, R.M,, Santella, R.M., Jedrychowski, W., Garte, S.J., Bell, D.A., Ottman, R., Gladek-Yarborough, A., Cosma, G., Young, T.L., Cooper, T.B., Randall, M.C., Manchester, D.K., Perera, F.P. (1998) Relationship between ambient air pollution and DNA damage in Polish mothers and newborns, *Environ. Health Perspect.* **106 Suppl 3,** 821-826.

6. Dmowski, K. and Kozakiewicz, M. (1993) Contamination of shrews (*Sorex sp.*) in the surroundings of the Miateczko Slaskie zinc smelter, Report of Warsaw University Grant BW-1104.

7. Wlostowski, T. (1986) Detoxification of cadmium ions by metallothionein in the liver of free-living Bank vole, *Acta Theriologica* **31**, 523-535.

8. Pytasz, M., Migula, P. and Krawczyk, A. (1980) The effect of industrial pollution on the metabolism rate in several animal species from the ironworks in the "Katowice" region. (in Polish) *Acta Biologica* (*Katowice*) **8**, 81-103.

9. Nyholm, N., Sawicka-Kapusta, K., Swiergosz, R. and Laczewska, B. (1995) Effects of environmental pollution on breeding birds in southern Poland, *Air and Soil Poll.* **85,** 829-834.

10. Jankowski, W., Dmowski, K. and Mankowski, T. (1990) Environmental pollution amplifies doichol accumulation in Magpie livers, Abstract in the *20th FEBS meeting*, Aug. 19-24, Budapest.

11. Rakoszek, K. (1996) Comparative analysis of phosphofructokinase and pyruvate kinase activities in selected animals from industrially polluted areas. (in Polish) M.Sc. thesis, University of Silesia, Katowice.

12. Grabecki, J., Panasiuk, Z. and Jarkowski, M. (1991) Assessment of exposure to lead of children living in the province of Katowice. Some results of biological monitoring. Abstract in *Int. East-West Symposium on contaminated areas in Eastern Europe: Origin, monitoring, sanitation*, Nov. 25-27, Gosen/Berlin.

13. Pokora, I. (1986) Oxygen metabolism of earthworms (*Oligochaeta, Lumbricidae*) from "Katowice" steelwork surroundings (in Polish), M.Sc. thesis, University of Silesia, Katowice, 53 pp.

14. Zimakowska-Gnoinska, D. (1981) The effect of industrial pollution on bioenergetic indices and on chemical composition of polyphagous predators - *Araneae*, *Pol. Ecol. Studies* **71**, 61-76.

15. Krawczyk., A. (1985) Aphid sensitivity to sulphur dioxide. Part III. Effect of SO_2 on respiratory metabolism of *Macrosiphoniella oblonga* (MORDV.) and

Macrosiphoniella artemisiae (B. DE F.) (in Polish), *Acta Biologica* (Katowice), **17**, 130-141.

16. Migula, P. (1989) Bioenergetic indices as indicators of environmental contamination in insects, *Nutritional Eocl. Ins. & Env* 157-166.

17. Atkinson, D.E. (1977) *Cellular energy metabolism and its regulation*, Academic Press, New York, 293 pp.

18. Giesy, J.P., Duke,C., Bingham,R.D., Dickson,G.W. (1983) Changes in phosphoadenylate concentration and adenylate energy charge as an integrated biochemical measure of stress in invertebrates, *Toxicol. Environ. Chem.* **6**, 259-295.

19. Haya, K., Waiwood, B.A., (1983) Adenylate energy charge and ATPase activity: Potential biochemical indicators of sublethal effects caused by pollutants in aquatic animals, in: J.O. Nriagu, (ed.), *Aquatic Toxicology*, J. Wiley and Sons Inc., pp. 308-333.

20. Migula,.P., Biñkowska,.K., Kafel,.A, Kedziorski,A., & Nakonieczny, M. (1989) Heavy metal contents and adenylate energy charge in insects from industrialized region as indices of environmental stress, in J. Bohac and V.Ruzicka (eds.), *Proc. Vth Int, Conf. Bioindicatores Deteriorisationis Regions*, Ceské Budejovice,. pp. 340-349.

21. Migula, P. and Glowacka, E. (1996) Heavy metals as stressing factors in red wood ants, *Fresenius J. Anal. Chem.* **354**, 653-659.

22. Migula, P., Dolezych, B., Kielan, Z., Laszczyca, P. and Howaniec, M. (1990) Responses to stressors in animals inhabiting industrially polluted areas (in Polish), in S. Godzik (ed.) Zagroaenia i stan Srodowiska przyrodniczego rejonu Slasko-krakowskiego. Wydawnictwo SGGW-AR, Warszawa, pp: 108-129.

23. Migula, P., (1985) *Sensitivity of some insects to industrial gaseous and dust pollutants and their tolerance of environmental thermal conditions (in Polish) Sci. Papers of the University of Silesia,* ***No. 765*** *Katowice. 115pp.*

24. Wilczek, G. and Migula, P. (1996) Metal body burdens and detoxifying enzymes in spiders from industrially polluted areas, *Fresenius J. Anal. Chem.* **354**, 643-647.

25. Augustyniak, M. and Migula, P. (1996) Patterns of glutathione S-transferase activity as a biomarker of exposure to industrial pollution in the grasshopper *Chorthippus brunneus (Thunberg). St. Soc. Sci. Toruñ* **4** (3) 9-15.

26. Purszke, A. (1997) GST pattern in grasshoppers *Chorthippus brunneus* under different environmental pollution (in Polish). University of Silesia, Katowice, 50 pp.

27. Sanders, B.M. (1993) Stress proteins in aquatic organisms: An environmental perspective, *Crit. Rev. Toxicol.* **23**, 49-75.

28. Ashburner, M. (1982) The effects of heat shock and other stress on gene activity: An introduction, in M.J. Schlesinger, M. Ashburner and A. Tissieres (eds.), *Heat Shock from Bacteria to Man*, Cold Spring Harbor Press, New York, pp. 1-9.

29. Kohler, H.R., Rahman, B. Graff, S., Berkus, M. and Triebskorn, R. (1996) Expression of the stress-70 protein family (Hsp70) due to heavy metal contamination in the slug, *Deroceras Reticulatum*: An approach to monitor sublethal stress conditions, *Chemosphere* **33**, 1327-1340.

30. Pyza, E., Mak, P., Kramarz, P. and Laskowski, R. (1997) Heat shock proteins (HSP70) as biomarkers in ecotoxicological studies, *Ecotoxicol. Environ. Saf.* **38**, 244-251.

31. Feder, M.E., Cantano, N.V., Milos, L., Krebs, R.A. and Lindquist, S. L. (1996) Effect of engineering *Hsp*70 copy number on Hsp70 expression and tolerance of ecologically relevant heat shock in larvae and pupae of *Drosophila Melanogaster*, *J. Exp. Biol.* **199**, 1837-1844.

32. Williams, J.H., Farag, A.M., Stansbury, M.A., Young, P.A., Bergman, H.L. and Petersen, N.S. (1996) Accumulaton of Hsp70 in juvenile and adult rainbow trout gill exposed to metal-contaminated water and/or diet, *Environ. Toxicol. Chem.* **15**, 1324-1328.

33. Tursman, D., Duman, J.G. and Knight C.A. 1994) Freeze tolerance adaptations in the centipede, *Lithobius foficatus. J. Exp.Zool. 268, 347-353.*

34. De Pomerai, D. (1996) Heat-shock proteins a bimarkers of pollution, *Hum. Exp. Toxicol.* **15**, 279-285.

35. Santella,R.M. (1997) DNA damage as an intermediate biomarker in intervention studies, *Proc. Soc. Exp. Biol. Med.* **216**, 166-177.

36. Helbock, H.J., Beckman, K.B., Shigenaga, M.K., Walker, P.B., Woodall, A.A., Yeo, H.C., Ames, B.N. (1998) DNA oxidation matters: The HPLC-electrochemical detection assay of 8-oxo-deoxyguanosine and 9-oxo-guanine, *Proc. Natl. Acad. Sci.* USA **95**, 288-293.

37. Varanasi, U., Reichert, W.L., Stein, J.E. (1989) ^{32}P-postlabeling analysis of DNA adducts in liver of wild English sole (*Parophrys vetulus*) and winter flounder (*Pseudopleuronectes americanus*), *Cancer Res.* **49**, 1171-1177.

38. Qu, S.-H., Bai, C.-L. and Stacey, N.H. (1997) Determination of bulky DNA adducts in biomonitoring of carcinogenic chemical exposures: features and comparison of current techniques, *Biomarkers* **2**, 3-16.

39. Farmer, P.B. *et al.* [>24 authors] (1996) Biomonitoring human exposure to environmental carcinogenic chemicals, *Mutagenesis*, **11**, 363-381.

40. Shugart, L.R. (1988) An alkaline unwinding assay for the detection of DNA damage in aquatic organisms, *Marine Environ. Res.* **24**, 321-325.

41. Shugart, L.R., Bickham, J., Jackim, G., McMahon,G., Ridley.W., Stein, J. and Steiner, S. (1992) DNA alterations, in R.J. Huggett, R.A. Kimerle, P.M. Mehrle and H.L. Bergman (eds.), *Biomarkers: Biochemical, Physiological and Histological Markers of Anthropogenic Stress*, Lewis Publishers Inc., Boca Raton, pp. 125-153.

42. Stein, J.E., Collier, T.K., Reichert, W.L., Casillas, E., Hom, T. and Varanasi, U. (1992) Bioindicators of contaminant exposure and sublethal effects: Studies with benthic fish in Puget Sound, Washington, *Environ. Toxicol. Chem.* **11**, 701-714.

43. Plappert, U., Barthel, E., Seidel, H. J. (1994) Reduction of benzene toxicity by toluene, *Environ. Mol. Mutagenesis* **24**, 283-292.

44. Ukai, H., Takada, S., Inui,S., Imai,Y, Kawai, T., Shimboo, S. and Ikeda, M. (1994) Occupational exposure to solvent mixtures - effects on health and metabolism. *Occup. Environ. Med.* **51**, 523-529.

45. Gobba, F., Rovesti, S., Borella, P., Vivoli, R, Caselgrandi, E., and Vivoli, G., (1997) Inter-individual variability of benzene metabolism to trans,trans-muconic acid and its implications in the biological monitoring of occupational exposure, *Sci. Total Environ.* **199**, 41-48.

46. Curtis, B. J. and Wood, C. M. (1991) The function of the urinary bladder *in vivo* in the freshwater rainbow trout, *J. Exp. Biol.* **155**, 567-583.

47. McCahon, C. P., Barton, S. F., and Pascoe, D. (1990) The toxicity of phenol to the freshwater crustacean *Asellus aquaticus* (L.) during episodic exposure - Relationship between sub-lethal responses and body phenol concentrations, *Arch. Environ. Contam. Toxicol.* **19**, 926-929.

48. Timbrell, J. A. (1994) *Principles of Biochemical Toxicology* (2nd edition), Taylor and Francis, London.

49. Constan, A.A., Yang, R. S., Baker, D. C., and Benjamin, S. A. (1995) A unique pattern of hepatocyte proliferation in F344 rats following long-term exposures to low-levels of a chemical mixture of groundwater contaminants, *Carcinogenesis* **16**, 303-310.

50. Teh, S. J., Adams, S. M., and Hinton, D. E. (1997) Histopathologic biomarkers in feral freshwater fish populations exposed to different types of contaminant stress, *Aquatic Toxicol* **37**, 51-70.

51. Stoyanovsky, D.A., Goldman, R., Claycamp, H.G. and Kagan, V.E. (1995). Phenoxyl radical-induced thiol-dependent generation of reactive oxygen, species-implications for benzene toxicity. *Arch. Biochem. Biophys.* **317**, 315-323.

52. Baker, R.T.M., Handy, R.D., Davies, S.J., Snook, J.C. (1998) Chronic dietary exposure to copper affects growth, tissue lipid peroxidation, and metal composition of the grey mullet. *Chelon labrosus*, *Marine Environ. Res.* **In press.**

53. Metcalfe, C.D., Cairns, V.W. and Fitzsimons, J. D. (1988) Experimental induction of liver tumours in rainbow trout (*Salmo gairdneri*) by contaminated sediment from Hamilton Harbour, Ontario. *Can. J. Fish. Aquatic Sci.*, 45, 2161-2167.

54. Baker, R.T.M., Martin, P. and Davies, S.J. (1997) Ingestion of sub-lethal levels of iron sulphate by African catfish affects growth and tissue lipid peroxidation, *Aquatic Toxicol.* **40**, 51-61.

55. Hockett, J. R. and Mount, D. R. (1996) Use of metal chelating agents to differentiate among sources of acute aquatic toxicity, *Environ. Toxicol. Chem.* **15**, 1687-1693.

56. Roch, M., Noonan, P. and McCarter, J. A. (1986) Determination of no effect levels of heavy metals for rainbow trout using hepatic metallothionein., *Water Res.* **20**, 771-774.

57. Hogstand, C. and Haux, C. (1991) Binding and detoxification of heavy metals in lower vertebrates with reference to metallothionein, *Comp. Biochem. Physiol.* **100C**, 137-141.

58. Olsson, P. E. (1996) Metallothioneins in fish: induction and use in environmental monitoring. In Toxicology of Aquatic Pollution (E W Taylor editor). pp.187-203. Cambridge, University Press.

59. George, S.G. and Olsson, P.-E. (1994) Metallothioneins as indicators of trace metal pollution, in K.J.M. Kramer (ed.), *Biological monitoring of coastal waters and estuaries*, CRC Press Inc., Boca Raton, pp. 151-178.

60. Hamer, D. (1986) Metallothionein, *Ann. Rev. Biochem.* **55**, 913-951.

61. Palmiter, R.D. (1993) Molecular biology of metallothionein gene expression, in K.T. Suzuzuki, N. Imura, M. Kimura (eds), Metallothionein III: Biological roles and medical implicaions, Birkhauser Verlag, Basil, pp. 63-79.

62. Viarengo, A., Ponzano, E., Dondero, F., and Fabbri, R. (1997) A simple spectrophotometric method for metallothionein evaluation in marine organisms: an application to Mediterranean and Antarctic molluscs, *Marine Environ. Res.* **44**, 69-84.

63. Hodson, P.V., Blunt, B.R. and Spry, D.J. (1978) Chronic toxicity of waterborne and dietary lead to rainbow trout (*Salmo gairdneri*) in Lake Ontario water, *Water Res.* **12**, 869-878.

64. Larsson, Å, Haux, C. and Sjöbeck, M. (1985) Fish physiology and metal pollution: results and experiences from laboratory and field studies, *Ecotoxicol. Environ. Safety* **9**, 250-281.

65. Pain, D.J. (1989) Haematological parameters as predictors of blood lead and indicators of lead poisoning in the black duck (*Anas rubripes*), *Environ. Pollut.* **60**, 67-81.

66. Scheuhammer, A.M. (1989) Monitoring wild bird populations for lead exposure, *J. Wildl. Manage.* **53**, 759-765.

67. Hogstand, C, Wilson, R. W., Polgar, D. and Wood C. M. (1994) Effects of zinc on the branchial calcium uptake in freshwater rainbow trout during adaptation to waterborne zinc, *J.Exp. Biol.* **186**, 55-73.

68. Li, J., Lock, R. A. C., Klaren, P. H. M., Swarts, H. G. P., Schuurmsns Stekhoven, F. M. A. H., Wendelaar Bonga, S. E. and Flik, G. (1996) Kinetics of Cu^{2+} inhibition of Na^{+}/K^{+}-ATPase, *Toxicol. Lett.* **87**, 31-38.

69. Gippo, R. S. and Dunson, W. A. (1996) The body ion loss biomarker. 1. Interactions between trace metals and low pH in reconstituted coal mine-polluted water, *Environ. Toxicol. Chem.* **15**, 1955-1963.

70. Verbost, P. M., van Rooji, J., Flik, G., Lock, R. A.C., and Wendelaar Bonga S. E. (1989) The movement of cadmium through freshwater trout branchial epithelium and its interference with calcium transport, *J. Exp. Biol.* **145**, 185-197.

71. Wood, C. M. (1992) Flux measurements as indices of H^+ and metal effects on freshwater fish, *Aquatic Toxicol.* **22**, 239-264.

72. Bers, D.M. (1979) Isolation and characterization of cardiac sarcolemma, *Biochem. Biophys Acta*, **555**, 131-146.

73. Duerr, J.M. and Ahearn, G.A. (1996) Characterization of a basolateral electroneutral Na^+/H^+ antiporter in Atlantic lobster (*Homarus americanus*) hepatopancreatic epithelial vesicles, *J. Exp.Biol.* **199**, 643-651.

74. Klaren, P.H.M., Wendelaar Bonga, S. E. and Flik, G. (1997) Evidence for P_2-purinoceptor-mediated uptake of Ca^{2+} across a fish (*Oreochromis mossambicus*) intestinal brush border membrane, *Biochem. J.* **322**, 129-134.

75. Hogstand, C., and Wood, C. M. (1998) Toward a better understanding of the bioavailability, physiology, and toxicity of silver in fish: Implications for water quality criteria, *Environ. Toxicol. Chem.* **17**, 547-561.

76. Beckman, B.R. and Zaugg, W.S. (1988) Copper intoxication in Chinook Salmon (*Oncorhynchus tshawytscha*) induced by natural springwater: effect on gill Na^+K^+-ATPase, hematocrit, and plasma glucose, *Can. J. Fish. Aquatic Sci.* **45**, 1430-1435.

77. van Wezel, A.P., Schmitz, M.G.J. and Tielens, A.G.M. (1997) Acetylcholinesterase and ATPase activities in erythrocyte ghosts are not affected by 1,2,4-trichlorobenzene: implications for toxicity by narcotic chemicals, *Environ. Toxicol. Chem.* **16**, 2347-2352.

78. Esmann, M. (1988) ATPase and phosphatase activity of Na^+K^+-ATPase: Molar and specific activity, protein determination, *Methods in Enzymology* **156**, 105-115.

79. Lanno, R.P., Hicks, B. and Hilton, J.W. (1987) Histological observations on intrahepatocytic copper-containing granules in rainbow trout reared on diets containing elevated levels of copper, *Aquatic Toxicol.* **10**, 251-263.

80. Kennedy, S.W., Jones, S.P. and Bastien, L.J. (1995) Efficient analysis of cytochrome P4501A catalytic activity, porphyrins and total proteins in chicken embryo hepatocyte cultures with a fluorescence plate reader, *Anal. Biochem.* **226**, 362-370.

81. Kennedy, S.W., Lorenzen, A. Norstrom, R.J. (1996) Chicken embryo hepatocyte bioassay for measuring cytochrome P4501A-base 2,3,7,8-tetra-chlorodibenzo-*p*-dioxin equivalent concentrations in environmental samples, *Environ. Sci. Technol.* **30**, 706-715.

82. Kennedy, S.W., Lorenzen, A. Jones, S.P. Hahn, M.E. and Stegeman, J.J. (1996) Cytochrome P4501A induction in avian hepatocyte cultures : A promising approach for predicting the sensitivity of avian species to toxic effects of halogenated aromatic hydrocarbons, *Toxicol. Appl. Pharmacol.* **141**, 214-230.

83. Anderson, J.W., Rossi, S.S. Tukey, R.H. Vu, T., and Quattrochi, L.C. (1995) A biomarker, P450RGS, for assessing the induction potential of environmental samples, *Environ. Toxicol. Chem.* **14**, 1159-1169.

84. Hahn, M.E., Woodward, B.L., Stegeman,J.J. and Kennedy, S.W. (1996) Rapid assessment of induced cytochrome P4501A protein and catalytic activity in fish hepatoma cells grown in multiwell plates: Response to TCDD, TCDF, and two planar PCBs, *Environ. Toxicol. Chem.* **15**, 582-591.

85. Murk, A.J., Legler, . Denison, M.S., Giesy, J.P. van de Guchte, C., Brouwer, A. (1996) Chemical-activated luciferase gene expression (CALUX): A novel *in vitro* bioassay for Ah-receptor active compounds in sediments and pore water, *Fund. Appl. Toxicol.* **33**, 149-160.

86. Tillitt, D.E., Giesy, J.P. and Ankley, G.T. (1991) Characterization of the H4IIE rat hepatoma bioassay as a tool for assessing toxic potency of planar halogenated hydrocarbons in environmental samples, *Environ. Sci. Technol.* **25**, 87-92.

87. Hahn, M.E. The aryl hydrocarbon receptor: A comparative perspective, *Comp. Biochem. Physiol.* **121C**, in press.

88. Kennedy, S.W., Fox, G.A., Trudeau, S. Bastien, L.J. and Jones, S.P. (1998) Highly carboxylated porphyrin concentration : A biochemical marker of PCB exposure in herring gulls, *Mar. Environ. Res.* **46**, 65-69.

89. Colborn, T. and Clement, C., eds. (1992) *Chemical-Induced Alterations in Sexual and Functional Development: The Human/Wildlife Connection*, Princeton Scientifc Publishing Co., Inc., Princeton New Jersey.

90. Jobling, S., Nolan, M., Tyler, C.R., Brighty,. G. and Sumpter, J.P. (1998) Widespread sexual disruption in wild fish, *Environ. Sci. Techol.* **32**, 2498-2506.

2. REPORT OF THE WORKING GROUP ON THE KATOWICE ADMINISTRATIVE DISTRICT ASSESSING ENVIRONMENTAL QUALITY AND EFFICACY OF REMEDIAL ACTIVITIES IN THE UPPER VISTULA BASIN OF POLAND, USING BIOMARKERS

TRACY K. COLLIER[1] (Rapporteur), KRYSTYNA GRODZINSKA[2] (Chair), LAURENT LAGADIC[3],
JAN FREDRIK BØRSETH[4], JESUS GONZALEZ-LOPEZ[5], ALBERT LEBEDEV[6], MIRDZA LEINERTE[7]
[1]National Oceanic and Atmospheric Administration, National Marine Fisheries Service, Northwest Fisheries Sçience Center, Environmental Conservation Division, Ecotoxicology Program, 2725 Montlake Blvd. E., Seattle, WA 98112, USA; [2]W. Szafer Institute of Botany, Polish Academy of Sciences, Lubicz 46, PL-31-512, Krakow, Poland; [3]Institut National de la Recherche Agronomique (INRA), Station Commune de Recherche en Ichtyophysiologie, Biodiversit et Environnement (SCRIBE), Equipe Ecotoxicologie Aquatique, Campus de Beaulieu, F-35042 Rennes Cedex, France; [4]Rogaland Research Station, Stavanger, Norway; University of Granada, Department of Microbiology, E-18071 Granada, Spain; [5]Lomonosov Moscow State University, Department of Chemistry, Leninskje Gory, RU-119898 Moscow, Russia; [6]Ministry of Environmental Protection and Regional Development, Rupniecibasela 25, LV-1045 Riga, Latvia.

2.1. Summary

The Upper Vistula Basin, in Southern Poland, is a diverse area, with a number of environmental problems. The working group discussed three primary ecosystem types within this area: forest catchments, urban areas, and retention reservoirs. Environmental conditions and current monitoring efforts were well presented for the forest catchments and urban areas; much less information was presented for retention reservoirs. Accordingly, the working group was able to make specific recommendations for incorporation of biomarkers in any programmes to ascertain the efficacy of remediation efforts in the first two ecosystem types listed above. Those recommendations are given at the end of this chapter. The working group recommended further assessment of the environmental quality of the region's retention reservoirs, as an obligatory first step in determination of the suitability of similar biomarker approaches. Overall, the working group was favourably impressed with the ongoing and planned environmental assessment efforts of scientists in this region.

D. B. Peakall et al. (eds.),
Biomarkers: A Pragmatic Basis for Remediation of Severe Pollution in Eastern Europe, 211–227.

2.2. Introduction

A detailed description of the Upper Vistula Basin (UVB), including its geology, climate, land use patterns, and general environmental quality, is given in a previous paper in this volume (Grodzinska, 1998). A brief summary of some salient features of this region is given here. The Vistula River is the largest in Poland, with a drainage basin of nearly 170,000 km^2. The upper portion of this drainage, or the Upper Vistula Basin, comprises slightly more than 50,000 km^2, or approximately 15% of the area of Poland. This area is more heavily forested than much of the rest of Poland, and is also very rich in natural mineral resources. In the western portion of the basin deposits of hard coal, zinc and lead ores, salt, gypsum, and mineral aggregates are found, while the eastern portion contains oil, natural gas, and sulphur deposits, along with natural mineral aggregates. However, the far eastern portion of the basin has no significant mineral resources. Largely as a consequence of this distribution of resources, the western portion of the UVB is the most heavily industrialized, with lesser industrial activities in the eastern portion.

Environmental quality of this region has deteriorated substantially over the past decades, and there is considerable pollution of air, soil, and water, as well as flora and fauna. There are also in existence several ongoing monitoring efforts aimed at assessing the environmental quality of the region, or certain portions of it, and changes in the status of environmental quality (e.g. trends determination). The working group assigned to this area was therefore charged with evaluating the ongoing monitoring efforts in order to determine if 1) they appear adequate for the environmental problems of most concern; 2) current programs should or could be modified to better provide information on the issue of environmental quality in the UVB; 3) there are substantial contaminant classes which are not being assessed by current approaches; and 4) what types of biomarker measurement are recommended overall by the working group which can provide the best assessment of current environmental quality and also provide a basis for determining the efficacy of remedial actions which may be undertaken in the UVB.

Because of the large area and diversity of ecosystems and pollutants involved, the working group proposed that the ecosystems of concern be somewhat prescribed, based largely on the expertise of Polish scientists as to the highest priority areas for study. Accordingly, we have structured this chapter along the following lines:

- Evaluation of current efforts to monitor environmental quality of forest catchments, focused on the Ratanica forest catchment, and recommendations for alterations to current efforts;
- Evaluation of current efforts to monitor environmental quality of urban areas, focused on the city of Krakow, and recommendations for alterations to current efforts;
- Suggestions for monitoring environmental quality of aquatic ecosystems, focusing on retention reservoirs in the UVB.

2.3. Forest catchments

2.3.1. DESCRIPTION OF THE CATCHMENT

The Ratanica forest catchment is located in Southern Poland (49°51'N, 20°02'E) in the western part of the UVB, 40 km south of the city of Kraków and in the vicinity of a large water reservoir providing the city with drinking water (see figure in Grodzinska, this volume). The entire catchment covers 241.5 ha, with the upper portion (88 ha) of the catchment being forested. The lower portion is comprised of meadows and intensively fertilized fields (Grodzinska and Weiner 1994; Grodzinska and Laskowski 1996; Grodzinska 1998 this volume).

The bedrock of the Ratanica catchment is typical for this part of Poland and represents Carpathian flyash (largely sandstone) of tertiary origin. Over this bedrock there is a layer of podsolized soils (according to FAO classification the soils represent Orthic Podsols and Haplic or Stagnic Luvisols). The soils in the catchment are strongly acidic, especially in the upper part of the soil profile, where pH averages 3.9. At depth the soil pH gradually increases, reaching 6.4 at a depth of 150 cm.

The vegetation of the Ratanica catchment is typical of the Carpathian Foothills. The forests are of anthropogenic character, dominated by coniferous species (*Pinus sylvestris* and *Larix decidua*). The main deciduous tree species is beech (*Fagus sylvatica*).

The Ratanica catchment has been subject to moderate yet chronic industrial emissions transported from distant regions for more than 40 years. In addition, it is influenced by local emissions from farming activities and combustion products derived from a number of domestic sources.

2.3.2. CURRENT MONITORING EFFORTS

Studies of environmental quality in the Ratanica catchment began in the 1980s. Five years of research (1986-1990) have been presented in numerous papers and a special volume of *Ekologia Polska*: (Grodzinska and Weiner, 1993). More extensive biogeochemical and biomarker studies, focused on the forested part of the catchment, were carried out from 1991 through 1995 (Grodzinska and Laskowski 1996). These results are summarized below.

The 1991-1995 studies included assessments of air pollution, water chemistry and water quality, chemical monitoring of leaf litter, and biomarker studies. Air pollution monitoring was conducted at one site, where measurements were made of SO_2, NO_2, suspended dust, and ozone. Water chemistry was monitored at 29 sites, and included measurement of 14 ions in bulk precipitation, throughfall, stemflow, soil water, and stream water. The measured ions were: NH_4^+, NO_3^-, SO_4^{2-}, Cl^-, PO_4^{3-}, Ca^{2+}, Mg^{2+}, Na^+, K^+, Mn^{2+}, Zn^{2+}, Cu^{2+}, Pb^{2+} and Cd^{2+}, as well as pH and electrical conductivity. Twelve elements (N, S, P, Na, K, Ca, Mg, Mn, Zn, Cu, Pb, and Cd) were measured in leaf litter, also at 29 sites. The biomarker measurements included: 1) the use of species and community indices as indicators of environmental quality (e.g. changes in species composition, distribution, and abundances of lichens; changes in species composition of stream algae); 2) indicators of ecological processes (e.g. rates of litter decomposition, defoliation of Scots pine and discolouration of pine needles); and 3) accumulation of pollutants in biological matrices (e.g. the accumulation of heavy metals by mosses and several genera of fauna, including *Lumbricidae, Chilopoda, Arachnida, Staphylinidae, Carabidae, Insectivora, Rodentia*, and the accumulation of S-compounds in pine needles).

There were several findings from this series of studies, as summarized below and in the paper by Grodzinska (this volume).

- High inputs of nitrogen (mainly in the form of NH_4^+), calcium and sulphur are characteristic of this forest catchment.
- S, Na, K, Cu, and Zn are transported to the forest floor mainly by water (throughfall, stemflow), whereas Ca, Mn, and P are transported via organic matter (litterfall). N, P, Ca, K, Mg, and Mn fluxes showed a significant annual pattern of inputs that was consistent for the five years of the study.
- The significant contribution of NH_4^+ ion in the throughfall and stemflow fluxes appear to be connected with intensive agriculture in surrounding fields.
- During the study period there was a decrease in the inputs of heavy metals (Cd, Pb, Zn), S and N. Compared to the studies in the 1980s the influx of heavy metals has declined by 50% while sulphur and nitrogen have decreased by approximately 35% and 20%, respectively. These declines appear to be connected with declines in industrial emissions.
- There is considerable evidence that the forest ecosystem in the Ratanica catchment is under stress. Pine trees (*Pinus sylvestris*) showed substantial adverse effects in the form of defoliation, crown damage, and altered needle growth and structure. The heavy metal content in *Pleurozium schreberi* was 3 times higher than in this same species from relatively clean areas in Poland. Also, changes in species composition of lichens was noted, with a shift towards species which are considered to be resistant to the effects of air pollution, and there appeared to be inhibition of litter decomposition processes.

Overall, because of constant inputs of pollutants and comparatively slow reductions of toxic elements from the catchment, it is believed that this ecosystem is facing continued threats to its sustainability. Moreover, the reductions noted in heavy metal fluxes are not thought to result from improved pollution control measures, but rather from reduced industrial activity, which will likely increase with improving economic conditions. Accordingly, monitoring of this ecosystem should continue.

2.3.3. ADJUSTMENTS AND ADDITIONS TO CURRENT EFFORTS

2.3.3.1 *Terrestrial and Aquatic Invertebrates*

On the basis of preliminary faunistic inventories, invertebrates may be useful for assessing both individual health and the impact of contaminants on essential ecological processes such as organic matter recycling. Criteria for selecting such bioindicator species and recommendations for their use in soil ecotoxicology are available (Edwards *et al.*, 1996; Migula, 1996). It is important to consider position in food webs when choosing species.

Accumulation and Effects of Metals in Invertebrate Species. The use of predatory or herbivorous insect species for investigations of contamination by metals is at an early stage of development (Migula *et al.*, this volume). Forthcoming investigations should clarify their utility for monitoring metal contamination in soils. Acetylcholinesterases (AChE) and nonspecific cholinesterases (ChE) appear to be potential tools for assessing the effects of metallic ions such as arsenic, copper and mercury in invertebrates (Bocquené *et al.*, 1997; Mineau, 1991). In the Ratanica forest catchment, acidic rains are likely to increase the transfer of metals from the soil to the stream (Grodzinska, this volume). Cholinergic effects of some pollutants have been associated with behavioural changes in stream-dwelling insects with possible subsequent impact on community (Lagadic *et al.*, 1994), AChE may constitute a relevant biomarker of such a transfer of metals to water. Biomarkers related to energy metabolism also appear to be potent tools for the assessment of joint effects of pH and metals in stream invertebrates. Energy re-allocation between maintenance, growth and reproduction has been reported for *Asellus aquaticus* (Maltby, 1991) and *Cambarus robustus* (Daveikis and Alikhan, 1996) exposed to acid effluents containing heavy metals, with potential changes in community structure and ecosystem functioning (Lagadic, this volume).

Detrivorous invertebrates such as isopods (*Oniscus sp.*, *Porcellio sp.*) are promising for the biomonitoring of metals. Extensive studies demonstrated that they bioaccumulate metals which elicit specific physiological responses (Hopkin, 1989). Earthworms also play a functional key-role in soil ecosystems. They ingest enormous quantity of soil from which they extract their food but can also be contaminated by soil pollutants. It has been reported that bioconcentration factors (BCFs) for cadmium, lead and zinc in moles have been shown to be closely related to metal concentrations in earthworms but not in soils (Ma, 1987). Additional discussion of the application of biomarkers to earthworms can be found in Weeks *et al* (this volume).

Accumulation and Effects of Organic Contaminants in Invertebrate Species. Isopods (*Porcellio* sp., *Trachelipus rathkei*) have been shown to accumulate PAHs (van Brummelen *et al.*, 1991; Faber and Heijmans, 1996) with subsequent induction of CYP1A-like monooxygenase activity. Interference of organic pollutants with the cytochrome P450 monooxygenase system may result in changes in the metabolism of steroids with possible consequences for growth and reproduction (van Brummelen and Stuijfzand, 1993; Faber and Heijmans, 1996).

In earthworms it is possible to measure the effects of both inorganic and organic contaminants by determining changes in the permeability of the lysosomal membrane of earthworm coelomocyte cells using the neutral red retention assay as described by Weeks & Svendsen (1996). This assay makes use of the fact that healthy lysosomes take up and retain the dye (neutral red) indefinitely, whereas dye in lysosomes in coelmocyte cells taken from "stressed" earthworms, gradually leaks into the surrounding cytoplasm. This response has been quantified in both the laboratory and field using various model organic and inorganic contaminants (see for example, Svendsen *et al.*, 1996, Svendsen & Weeks, 1997a,b). This marker is non-specific, responding equally sensitively to organic or inorganic contamination; however, if used in combination with an earthworm immunocompetence marker (Svendsen *et al.*, 1997), such as total immunoactivity of the coelomocytes, then it is possible (with caution) to be more specific as to the likely nature of contamination.

2.3.3.2. *Plants*

Generally it was felt that the measurements currently being made in terrestrial plants (e.g. heavy metal content, needle damage, impaired crown growth) are providing useful information, and should be continued. This would especially apply to areas undergoing remediation activities. It would be desirable to include measurements of nitrogen and ozone pollution in appropriately sensitive plant species. No data were presented on the results from assessing species diversity in freshwater algae, thus no recommendations are made in regard to this type of measurement.

2.4. Urban areas

2.4.1 DESCRIPTION OF THE CITY OF KRAKÓW

The city of Kraków covers 330 km^2, and this historical urban complex is on the UNESCO World Cultural Heritage List. With a population of nearly 1 million, it is one of the most polluted cities in Poland, even though within its borders are several parks and small forests. The major sources of air pollution in Kraków region include thousands of individual coal-fired stoves and small local boiler houses, the Nowa Huta steel mill (the largest metallurgical complex in Poland), the large number of other industrial plants inside and outside Kraków region, and high levels of automobile and truck traffic. During the 1970s and 1980s Kraków was classified as an area of ecological disaster. During the past 10 years this situation has improved due largely to decreased emissions resulting from depressed economic conditions. Nonetheless, the concentrations of pollutants in the air, and rates of dust deposition, consistently exceed standards set by the Polish government.

2.4.2. CURRENT AND PAST MONITORING EFFORTS

An ecological monitoring program was started in the Kraków region in 1981. The primary objective of the program has been to identify and document the current status of air quality in the region, and to assess effects of such air pollution on levels of contamination of vegetation in the area. Within the city of Kraków, this effort included monitoring as follows.

Air quality monitoring was carried out at 19 stations, making the following measurements: quantity of dustfall and chemical dust composition (Cd, Pb, Cr, Ni, Zn, Fe), concentration of suspended particles; SO_2, NO_x, O_3 and heavy metals in aerosols.

The chemistry of atmospheric precipitation was studied at three stations: pH, N, NO^{3-}, SO_4^{2-}, PO_4^{3-}, Cl^-, Ca^{2+}, Mg^{2+}, Na^+, K^+, Zn^{2+}, Cu^{2+}, Pb^{2+} and Cd^{2+} were measured.

Soil monitoring was carried out at 12 stations. Measured in soils were the following: physical composition, pH, concentration of heavy metals (Cd, Pb, Cr, Ni, Zn, Cu, Fe, S); monitoring of vegetation included measurements of the concentrations of heavy metals (Cd, Pb, Cr, Ni, Zn, Cu, Fe, S) in vegetables (lettuce, parsley, carrot) growing in allotment gardens.

There have been other monitoring initiatives, including an automatic air quality monitoring system begun in 1991 with assistance from the USA. This program involves the continuous measurement of concentrations of suspended particulate matter, SO_2, NO_2, CO, O_3, and chemical measurements of dust, at seven fixed stations and one mobile station. Meteorological parameters are also measured. Since 1996, the concentrations of dioxins and dibenzofurans have been determined in suspended particulate material at four locations within the city. These programs are funded through the Voivoidship Environmental Protection and Water Management Fund, and coordinated by the Voivoidship Inspectorate for Environmental Protection. The ecological monitoring program is conducted by several institutions, with scientific supervision by Professor Grodzinska.

Results from these efforts have shown that air quality in the city is generally poorer than mandated under Polish national standards (Turzanski and Wertz, 1997). For example, in 1996, mean annual levels of SO_2, CO, and suspended particulate material were all higher than the standards. During winter months, levels were highest. While concentrations of heavy metals measured in dust particles were generally below permissible standards, mean annual concentrations of benzo[a]pyrene (BaP) were 5 ng per m^3, which exceeded levels allowed under the standards. In the winter, levels of BaP were as high as 52 ng per m^3. While mean levels of dust deposition were well below permissible levels, there was high deposition of sulfates, calcium, chloride, nitrogen, fluorine, and heavy metals from the atmosphere, and precipitation was shown to be highly acidic (mean pH of 4.42).

Results of soil monitoring show that while there is substantial heavy metal contamination of soils within the city, especially in the city center and near the steel mill and busy roadways, levels have decreased substantially between 1984 and 1995 (Grodzinska et al, 1994-1995; Grodzinska 1996). The heavy metal contamination of vegetation in allotment gardens was most pronounced in lettuce and parsley leaves, and lower levels were detected in carrots and parsley roots. While levels of metals in vegetation have also generally declined over the past decade, the levels of cadmium in all vegetables exceeded Polish standards during every year from 1984 through 1995 (Grodzinska et al, 1994-1995; Grodzinska 1996).

Additional monitoring efforts, for which the working group did not have summaries of results, included monitoring of the pH of tree bark and levels of heavy metals in tree rings of 100-year old beech during the 1970s and 1980s (Grodzinska, 1977; Kazmierczakowa *et al.*, 1984); assessment of health status of the human population (i.e. cancer, respiratory and cardiac disease, allergic reactions) during the 1980s (Guminska and Delorme, 1990); and measurement of heavy metals in human hair, blood, and teeth during the 1990s (Zielonka and Wodzien, 1993; Zachwieja *et al.*, 1997; Karolak, personal communication; Zarzecka and Krywult, personal communication).

2.4.3 ADJUSTMENTS AND ADDITIONS TO CURRENT EFFORTS

2.4.3.1 *Biomarkers in plants*

The group felt that appropriate biomarkers measurements in plants could build on the considerable air quality dataset available. These could include the following: accumulation of sulphur and nitrogen in pine needles; accumulation of heavy metals in tree bark; species composition of lichen assemblages, to include distribution and abundance of environmentally sensitive and resistant species; accumulation of salt in vascular plants (e.g. *Sambucus nigra*, *Tilia* sp); distribution and abundance of native flora known to be sensitive to ozone pollution. Examination of these biomarkers, together with correlation with air quality measures, would validate their use as surrogate indicators of air quality, and could be directly useful in substantiating improvements of air quality.

2.4.3.2 *Biomarkers in soil invertebrates*

The invertebrate species previously proposed for the monitoring of contaminants in the Ratanica forest catchment could equally be used in the urban-industrial Krakow region. However, particular attention should be devoted to monitoring the contamination of vegetables by cadmium in allotment gardens, as this represents an assessment of risk to human health. In addition to earthworms, snails and slugs might be appropriate for the monitoring of cadmium in such a situation. Slugs have recently been proposed as the model invertebrate in a so-called 'slug-watch' monitoring programme (Marigomez *et al.*, 1996).

Invertebrates may provide reliable markers for monitoring the effects of organic pollution in urban areas using either bioassays or direct measurements of biomarkers. Tests designed to detect mutations in *Drosophila melanogaster* somatic and germinal cells may be used to assess the genotoxic effects of air pollutants (Lee *et al.*, 1983; Frölich and Würgler, 1990; Vogel *et al.*, 1991; Batiste-Alentorn *et al.*, 1995). As they reach the aquatic environment, urban pollutants can affect invertebrates which may thus provide valuable biomarkers for both PAHs and metals. The polytene chromosomes of the salivary gland of chironomid larvae are a promising tool to monitor genotoxic contaminants (Bentivegna and Cooper, 1993) and metals (Michailova and Belcheva, 1990; Aziz *et al.*, 1991) in water. In heavily polluted aquatic environments, chironomid larvae often present morphological abnormalities affecting mainly mouthparts and antennae (Warwick, 1988; van Urk and Kerkum, 1987; Dickman *et al.*, 1992; Janssens de Bisthoven *et al.*, 1992).

2.5. Retention reservoirs

It seems clear from the available data that a major water quality issue for the UVB is input of nitrogen and phosphorus from heavily fertilized fields, as well as from sewage effluents, resulting in eutrophication. This is supported by the finding that almost all (98+%)waters in this basin fail microbial standards. Thus biomarkers of nutrient enrichment of freshwater ecosystems should be recommended. These could include the following: concentration of chlorophyll "a"; composition of phytoplankton; presence of blue-green algae; concentration of blue-green algae ("blooms"); presence of bottom (benthic) invertebrates; composition of benthic invertebrates; concentration of benthic invertebrates; presence of attached algae; and macrophyte abundance. Many of these indicators are discussed in Cimdins (1995).

There was also discussion of the likelihood that there could be substantial contamination of this region's aquatic sediments by organic contaminants, such as organochlorines and PAHs, in association with industrial activities. The working group concluded that there needs to be an assessment of this potential problem, to allow a decision to be made on whether to pursue development of a biomarker program to support remedial actions. There are two recommended ways to do this. One is to use screening biomarkers for organic contaminants (such as measuring hepatic CYP1A or body burdens, or bile metabolites [fluorescent aromatic compounds (FACs)] in vertebrates, or body burdens in invertebrates), while another is to perform appropriate chemical analyses of sediment samples. While these recommendations technically encompass more than the use of classic biomarkers, it was felt that these approaches are necessary as an initial assessment of the scope of any potential contamination of the retention reservoirs. A comprehensive example of the application of both tissue residue analysis and biomarker measurement to the assessment of the impacts of contaminants on fish is given in Collier et al. (1998a, b). The technical memorandum (Collier *et al.*, 1998a) includes detailed descriptions of methods, both for laboratory assays and field sampling, including statistical considerations in the evaluation of data.

The application of screening biomarkers to the biota in the retention reservoirs should be done by including both benthic and water column feeding organisms in the study. This would increase the possibility of being able to discriminate between effect of ongoing contamination of the water and 'old sins' represented by contaminants in the sediment. Another distinction could be made between phytoplankton grazers and carnivorous organisms. Phytoplankton grazers might better reflect the influence of heavy metals while carnivores could be better suited for the bioaccumulative organic contaminants. In order to optimize the interpretation of biomarker screening data, emphasis should be put on coordination of the sampling program both for chemical and biomarker measurements. Suitable screening biomarkers for organics could be hepatic CYP1A and bile FACs in fish and lysosomal response and tissue residues in molluscs and crustaceans. Heavy metal exposure could be screened by ICP-MS analyses directly of fish bile and by also measuring body burden in the invertebrates.

It should be noted that many of the biomarkers which have been identified in vertebrates can potentially be measured in invertebrates, the notable exceptions being biomolecules, structures or processes which have no equivalents in the latter (e.g. thyroid hormones, endoskeletal structures, kidney or lung functions, etc) though structural and functional homologies allow common interpretations of direct individual effects of exposure (Lagadic and Caquet, in press). The many biomarkers that can be measured in both vertebrates and invertebrates have been extensively reviewed (McCarthy and Shugart, 1990; Peakall, 1992; Huggett *et al.*, 1992; Lagadic *et al.*, 1997; Lagadic *et al.*, in press). They allow direct assessement of a wide range of pollutants in natural ecosystems where invertebrates display a number of characteristics that make them ecologically relevant (Lagadic, this volume). In particular, invertebrates are most useful to evaluate the distribution and bioavailability of pollutants in various compartments of the ecosystems. Thus, the levels of bioavailable pollutants in aquatic sediments are potentially to be more reliable when directly obtained from burrowing benthic invertebrates than from bottom-feeding fish that eat them (Steingraeber and Wiener, 1995; Finley, 1985), though measurement of body burdens in more mobile fish species may allow for an integration of a wider area.

Beside metals and other inorganic compound found in high concentrations in the Dobczyce reservoir (Grodzinska, this volume), pollutants of concern are also those which can affect the endocrine control of reproduction in animals. Organic pollutants acting as endocrine disruptors (pesticides, industrial effluents and wastes) may threaten human health (Safe, 1994; Kavlock *et al.*, 1996). Some species of aquatic invertebrates have been identified as target animals for endocrine-disruptor pollutants. Comparative studies on the effects of chemicals on reproductive capacity and steroid metabolism in *Daphnia magna* have shown that short-term exposure to toxicants that impair reproduction also affects steroid metabolism (Baldwin *et al.*, 1995; Parks and LeBlanc, 1996). Changes in steroid metabolism, which may result from dysfunction of cytochrome P450 monooxygenase system, can therefore be considered as an early-warning indicator of reproductive toxicity. This is further supported by extensive investigations of the effects of tributyltin (TBT) on the reproduction and some other metabolic processes of many marine mollusc species (Hawkins *et al.*, 1994; Lagadic, this volume). Biomarkers of endocrine disruption are comparatively well established in fish species, such as assays for vitellogenin or zona radiata proteins in male or immature fish [see Arukwe and Gokoyr (1998) for a discussion of these approaches], or the assessment of reproductive function and success in maturing female fish (Johnson *et al.*, 1998). Such assessment may be warranted in fish species from contaminated portions of retention reservoirs.

To provide definitive data on the qualitative and quantitative composition of chemical contaminants in the sediments, it is recommended that sediments of the retention reservoirs of this region should be analyzed for heavy metals (more than the 5-6 elements included in previous analyses), PAHs, phenols, organochlorines including PCBs, PCDDs, and PCDFs, and pesticides. Methods and instrumentation used in these analyses should be standardized with other laboratories, in order to allow direct comparison to pre-existing data from other regions.

2.6. Recommendations

The following recommendations are made:

- Current monitoring efforts in the Ratanica forest catchment should continue, as a model for forest catchments throughout the UVB

- Within this monitoring effort, measurements of biomarkers in soil invertebrate species should be conducted. Discussion of the types and purposes of such measurements are to be found in Migula (this volume) and Lagadic (this volume).

- The current programme for monitoring of biomarkers in plants within the forest catchment is largely appropriate; however measurements of the effects of nitrogen and ozone pollution should be added.

- Within the Krakow urban region, there is an excellent dataset on air quality; this dataset would be very well augmented by addition of appropriate biomarker measurements in plants, as described in that section of the chapter.

- It is recommended that biomarkers of contaminant effects on soil invertebrates be measured with the Krakow urban region as well, similar to measures used within the forest catchment, with the addition of measures specifically for use in allotment gardens.

- Human health monitoring within Krakow is called for; however little data were available to the working group. Recommendations as outlined in the working group report by Kennedy et al (this volume) may be appropriate.

- Within the retention reservoirs, there were not enough data available to the working group to allow recommendations of many specific biomarker approaches. However, it is probably appropriate to monitor plant communities within the retention reservoirs to examine for impacts of nutrient enrichment.

- The group does recommend that initial screening and analyses of the spatial extent and severity of chemical contamination be conducted. This would be especially useful if chemical analyses of sediments were performed, and certain so-called screening biomarkers were measured in biota.

• The utilization of biomarker approaches to issues of contaminant impacts within the retention reservoirs will be dependent on the results of these intial screeenings for degree of contamination. The application of biomarkers of endocrine disruption should also be considered.

References

Arukwe, A., and Goksøyr, A. (1998). Xenobiotics, xenoestrogens, and reproduction disturbances in fish. *Sarsia* **83**, 225-241.

Aziz, J.B., Akrawi, N.M. and Nassori, G.A. (1991) The effect of chronic toxicity of copper on the activity of Balbiani rings and nucleolar organizing region in the salivary gland chromosomes of *Chironomus ninevah* larvae, *Environ. Pollut.* **69**, 125-130.

Baldwin, W.S., Milam, D.L., LeBlanc, G.A. (1995) Physiological and biochemical perturbation in *Daphnia magna* following exposure to the model environmental estrogen diethylstilbestrol, *Environ. Toxicol. Chem.* **14**, 945-952.

Batiste-Alentorn, M., Xamena, N., Creus, A., Marcos, R. (1995) Genotoxic evaluation of ten carcinogens in the *Drosophilia melanogaster* wing spot test, *Experientia* **51**, 73-76.

Bentivegna, C.S. and Cooper, K.R. (1993) Reduced chromosomal puffing in *Chironomus tentans* as a biomarker for potentially genotoxic substances, *Environ. Toxicol. Chem.* **12**, 1001-1011.

Bocquené, G., Galgani, F. and Walker, C.H. (1997) Les cholinestérases, biomarqueurs de neurotoxicité, in L. Lagadic, T. Caquet, J.-C. Amiard and F. Ramade (eds.), *Biomarqueurs en Ecotoxicologie - Aspects Fondamentaux*, Masson, Paris, 209-239.

Cimdins, P., (ed.) (1995) Manual of Practical Hydrobiology. University of Latvia, Vide, Riga, Latvia.

Coles, J.A., Farley, S.R. and Pipe R.K. (1995) Alteration of the immune response of the common marine mussel *Mytilus edulis* resulting from exposure to cadmium, *Dis. Aquat. Org.* **22**, 59-65.

Collier, T.K., Johnson, L.L., Stehr, C.M., Myers, M.S., Krahn, M.M., and Stein, J.E. (1998a). Fish Injury in the Hylebos Waterway of Commencement Bay, Washington. U.S. Dep. Commer., NOAA Tech. Memo. NMFS-NWFSC-36, 576 p.

Collier, T.K., Johnson, L.L., Stehr, C.M., Myers, M.S., Krahn, M.M., and Stein, J.E. (1998b). A comprehensive assessment of the impacts of contaminants on fish from an urban waterway. *Mar. Environ. Res.* **46**, 243-247.

Daveikis, V.F. and Alikhan M.A. (1996) Comparative body measurements, fecundity, oxygen uptake, and ammonia excretion in *Cambarus robustus* (Astacidae, Crustacea) from an acidic and a neutral site in northeastern Ontario, Canada, *Can. J. Zool.* **74**, 1196-1203.

Dickman, M., Brindle, I. and Benson, M. (1992) Evidence of teratogens in sediments of the Niagara river watershed as reflected by chironomid (Diptera: Chironomidae) deformities, *J. Great Lakes Res.* **18**, 467-80.

Edwards, C.A., Subler, S., Chen, S.K. and Bogomolov, D.M. (1996) Essential criteria for selecting bioindicator species, processes, or systems to assess the environmental impact of chemicals on soil ecosystems, in N.M. van Straalen and D.A. Krivolutsky (eds.), *Bioindicator Systems for Soil Pollution*, NATO ASI Series, Kluwer Academic Publisher, Dordrecht, 67-83.

Faber, J.H. and Heijmans, G.J.S.M. (1996) Polycyclic aromatic hydrocarbons in soil detritivores, in N.M. van Straalen and D.A. Krivolutsky (eds.), *Bioindicator Systems for Soil Pollution*, NATO ASI Series, Kluwer Academic Publisher, Dordrecht, 31-43.

Finley, K.A. (1985) Observations of bluegills fed selenium-contaminated *Hexagenia* nymphs collected from Belews Lake, North Carolina, *Bull. Environ. Contam. Toxicol.* **35**, 816-825.

Frölich, A. and Würgler, F.E. (1990) Drosophila wing-spot test improved detectability of genotoxicity of polycyclic aromatic hydrocarbons, *Mutation Res.* **234**, 71-80.

Grodzinska, K. (1977) Acidity of tree bark as a bioindicator of forest pollution in southern Poland. *Water, Air, and Soil Pollution* **8**, *3-7.*

Grodzinska, K. (ed.) (1996) *Monitoring ekologiczny wojewodztwa krakowskeigo w latach 1993-1995* [*Ecological monitoring of Krakow voivodship in 1993-1995*], Biblioteka Monitoringu Srodowiska, Krakow.

Grodzinska, K., Godzik, B., and Szarek, G. (1994-1995) Krakowskie ogrody dzialkowe - skazcnie warzyw I gleby metalami ciezkimi [Allotment gardens in Cracow - vegetables and soil contamination by heavy metals*], Foila Geographica, series Geographica - Physica* **26-27**, 113-138.

Grodzinska, K. and Laskowski, R. (eds.) (1996) *Ocena stanu srodowiska I procesow zachodzacych w lasach zlewni potoku Ratanica (Pogorze Wielicklie, Polska poludniowa) [Environmental assessment and biogeoichemistry of a moderately polluted Ratanica catchment, Southern Poland*], Biblioteka Monitoringu Srodowiska, Warszawa.

Grodzinska, K. and Weiner, J. (eds.) (1993) Watershed processes and vegetation in the region of chronic atmospheric pollution (Carpathian Foothills, S Poland), *Ekol. pol.* **41**, 285-450.

Guminska, M., and Delorme, A. (eds.) (1990) *Kleska ekologiczna Krakowa* [*Ecological disaster of Krakow*], Polski Klub Ekologiczny, Krakow.

Hawkins, S.J., Proud, S.V., Spence, S.K. and Southward, A.J. (1994) From the individual to the community and beyond: water quality, stress indicators and key species in coastal systems, in D.W. Sutcliffe (ed.), *Water Quality and Stress Indicators in Marine and Freshwater Ecosystems: Linking Levels of Organisation (Individuals, Populations, Communities)*, Freshwater Biological Association, Ambleside, 35-62.

Hopkin, S.P. (1989) *Ecophysiology of metals in terrestrial invertebrates*, Elsevier Applied Science Publishers Ltd, London.

Huggett, R.J., Kimerle, R.A., Mehrle Jr., P.M. and Bergam, H.L. (eds.) (1992) *Biomarkers. Biochemical, Physiological, and Histological Markers of Anthropogenic Stress*, SETAC Special Publications Series, Lewis Publishers, Boca Raton, FL.

Janssens de Bisthoven, L.G., Timmermans, K.R. and Ollevier, F. (1992) The concentration of cadmium, lead, copper and zinc in *Chironomus* gr. *thummi* larvae (Diptera, Chironomidae) with deformed *versus* normal menta, *Hydrobiologia* **239**, 141-9.

Johnson, L.L., J.T. Landahl, L.A. Kubin, B.H. Horness, M.S. Myers, T.K. Collier, and J.E. Stein. (1998). Assessing the effects of anthropogenic stressors on Puget Sound flatfish populations. *J. Sea Res.* **39**, 125-137.

Kavlock, R.J., Daston, G.P., DeRosa, C., Fenner-Crisp, P., Earl Gray, L., Kaattari, S., Lucier, G., Luster, M., Mac, M.J., Maczka, C., Miller, R., Moore, J., Rolland, R., Scott, G., Sheehan, D.J., Sinks, T. and Tilson, H.A. (1996) Research needs for the risk assessment of health and environmental effects of endocrine disruptors: a report of the U.S. EPA-sponsored workshop, *Environ. Health Perspect.* **104**, suppl 4, 715-740.

Kazmierczakowa, R., Grodzinska, K. and Benarz, Z. (1984) Content of heavy metals in xylem of 100-year old beech (Fagus sylvatica L.) in Southern Poland. *Bull. Pol. Acad. Sci. Biol.* **32**, 329-338.

Lagadic, L. and Caquet, T. (1998) Invertebrates in testing of environmental chemicals: are they alternatives? *Environ. Health Perspect.* **106** (Supplement 2), 593-611.

Lagadic L, Caquet T and Ramade F. (1994) The role of biomarkers in environmental assessment. (5) Invertebrate populations and communities. *Ecotoxicology* **3**, 193-208.

Lagadic, L., Caquet, T., Amiard J.-C. and Ramade F. (eds.) (1997) *Biomarqueurs en Ecotoxicologie - Aspects Fondamentaux*, Masson, Paris.

Lagadic, L., Caquet, T., Amiard J.-C. and Ramade F. (eds.) (in press) *Utilisation de Biomarqueurs pour la Surveillance de la Qualité de l'Environnement.* Tec & Doc Lavoisier, Paris.

Lee, W.R., Abrahamson, S., Valencia, R., von Halle, E.S., Würgler, F.E., and Zimmering, S. (1983) The sex-linked recessive lethal test for mutagenesis in *Drosophila melanogaster* - A report of the United -States Environmental Protection Agency Gene-Tox program, *Mutation Res.* **123**, 183-279.

Ma, W. (1987) Heavy metal accumulation in the mole, *Talpa europea*, and earthworms as an indicator of metal bioavailability in terrestrial environments, *Bull. Environ. Contam. Toxicol.* **39**, 933-938.

Maltby, L. (1991) Pollution as a probe of life-history adaptation in *Asellus aquaticus* (Isopoda), *Oikos* **61**, 11-18.

Marigomez, I., Soto, M., and Kortabitarte, M. (1996). Tissue-level biomarkers and biological effect of mercury on sentinel slugs, *Arion ater. Arch. Environm. Contam. Toxicol.* **31**, 54-62.

Maryanski, M., Niklinska, M., Kramarz, P., Laskowski, R. and Nuorteva, P. (1995) Zinc and copper concentrations in in litter invertebrates of different trophic levels and taxonomic positions. *Archiwum Ochrony Srodowiska* **3-4**, 207-212.

McCarthy, J.F. and Shugart, L.R. (eds.) (1990) *Biomarkers of Environmental Contamination.* Lewis Publishers, Boca Raton, FL.

Michailova, P. and Belcheva, R. (1990) Different effect of lead on external morphology and polytene chromosomes of *Glyptotendipes barbipes* (Staeger) (Diptera: Chironomidae). *Folia Biol. (Krakow),* **38**, 83-88.

Migula, P.J. (1996) Constraints in the use of bioindicators and biomarkers in ecotoxicology, in N.M. van Straalen and D.A. Krivolutsky (eds.), *Bioindicator Systems for Soil Pollution*, NATO ASI Series, Kluwer Academic Publisher, Dordrecht, 17-29.

Mineau, P. (ed.) (1991) Cholinesterase-inhibiting Insecticides. Their Impact of Wildlife and the Environment. Elsevier, 348 pp.

Parks, L.G. and LeBlanc, G.A. (1996) Reductions in steroid hormone biotransformation/elimination as a biomarker of pentachlorophenol chronic toxicity, *Aquat. Toxicol.* **34**, 291-303.

Peakall, D.B. (1992) Animal biomarkers as pollution indicators. Chapman and Hall, London, UK.

Safe, S.H. (1994) Polychlorinated biphenyls (PCBs): environmental impact, biochemical and toxic responses, and implications for risk assessment. *Crit. Rev. Toxicol.* **24**, 87-149.

Starmach, J. (ed.) (1998) (in press *Zbiornik dobczycki - zagrozenie srodowiska naturalnego w zlewnirzeki Raby powyzej zbiornika I ich wppyw na jakosc wody [Dobczyce reservoir - threats to the natural environment in Raba river catchment above and letter water reservoir and their effect water reservoir quality*/, Malopoiski Instytut Samorzadu Terytorialnego I Administracji, Krakow.

Steingraeber, M.T. and Wiener, J.G. (1995) Bioassessment of contaminant transport and distribution in aquatic ecosystems by chemical analysis of burrowing mayflies (*Hexagenia*), *Regul. Riv. Res. Manage.* **11**, 201-209.

Turzanski, K.P., and Wertz, J. (eds.) (1997*) Raport o stanie srodowiska w wojewodztwie krakowskim w 1996 roku [Report on environmental status of Krakow voivodship in* 1996]. Biblioteka Monitoringu Srodowiska , Krakow.

Van Brummelen, T.C. and Stuijfzand, S.C. (1993) Effects of benzo(a)pyrene on survival, growth and energy reserves in the terrestrial isopods *Oniscus asellus* and *Porcellio scaber*. *Sci. Total Environ.*, Suppl. (1993), 921-930.

Van Brummelen, T.C., Verweij, R.A. and Van Straalen, N.M. (1991) Determination of benzo(a)pyrene in isopods (*Porcellio scaber* lattr.) exposed to contaminated food. *Comp. Biochem. Physiol.* C **100**, 21-24.

Van Urk, G. and Kerkum, F.C.M. (1987) Chironomid mortality after the Sandoz accident and deformities in *Chironomus* larvae due to sediment pollution in the Rhine, *Aqua* **4**, 191-6.

Vogel, E.W., Nivard, M.J.M. and Zijlstra J.A. (1991) Variation of spontaneous and induced mitotic recombination in different *Drosophila* populations - A pilot study on the effects of polyaromatic hydrocarbons in 6 newly constructed tester strains. *Mutation Res.* **250**, 291-298.

Warwick, W.F. (1988) Morphological deformities in Chironomidae (Diptera) larvae as biological indicators of toxic stress, in M.S. Evans (ed.) *Toxic Contaminants and Ecosystem Health: A Great Lakes Focus*, John Wiley, New York, 281-320.

Zachwieja, Z., Chlopicka, J., Barton, H., Schlegel-Zawadzka and Folta, M. (1997) The cadmium content in the hair of children as an indicator of environmental pollution. *Polish J. Enviromental Studies,* **6** Suppl. 202-205.

Zielonka, E. and Wodzien, M. (1993) Blood cadmium and lead levels among school children in Cracow. *Folla Med. Cracow.* **34**, 85-86

3. REPORT OF THE WORKING GROUP ON NORTHERN BOHEMIA

L.SHUGART (Rapporteur) J. CIBULKA (Chair) L. BORUVKA, E. CURDOVA, L. HOLSBEEK, A. KAHRU, S. KOTELEVTSEV, W. KRATZ, J. KUBIZNAKOVA, and C. WALKER.

3.1. Summary

The overall purpose of this chapter is to provide information that will be helpful in the design and implementation of a biomarker-based program that is applicable to environmental problems associated with pollution in Northern Bohemia of the Czech Republic. As participants in the preparation of this chapter, it was our task to sort through existing knowledge on biomarkers and pollution (especially as found in the study area) and condense it in such a fashion that information relevant to the topic of this book would be presented in a clear, concise and readily understandable manner. This task was accomplished by arranging the material in the following manner. First, an attempt was made to describe the Northern Bohemian region. Emphasis was placed on those physical features that identify the region and particular attention was given to environmental problems that have arisen as a result of pollution, as well as other factors that have contributed to the problem. Current activities were identified that are a direct result of the problem, such as environmental legislation, corrective or preventive measures, monitoring programs and studies on health issues. Second, information was prepared on the implementation of a biomarker program that describes several classes of biomarkers appropriate to the types of pollutants that exist in the environment in the Northern Bohemian region. Because of the complexity of the problem of environmental pollution in this region (i.e., extent and types of toxic substances), an integrated approach was emphasized. To be effective, biomarker data must be integrated with results from field studies, bioassays and chemical analysis. Pertinent recommendations are detailed that must be addressed. Finally, several important documents, that were either presented or discussed at the meeting of this NATO Advanced Research Workshop and are relevant to this study area, are included as Appendixes to this chapter.

3.2. Description of Study Area

Northern Bohemia stretches along the Northwest borders with Germany and Poland between the cities of Kadan and Liberec on the Southwest and Northeast, respectively and comprises about one sixth of the Czech Republic (Figure 1, map). It has an area of approximately 16,000 km^2 with over 1.5 million inhabitants living in about 2,500 towns and villages. A National Park and several Protected

D. B. Peakall et al. (eds.),
Biomarkers: A Pragmatic Basis for Remediation of Severe Pollution in Eastern Europe, 229–277.

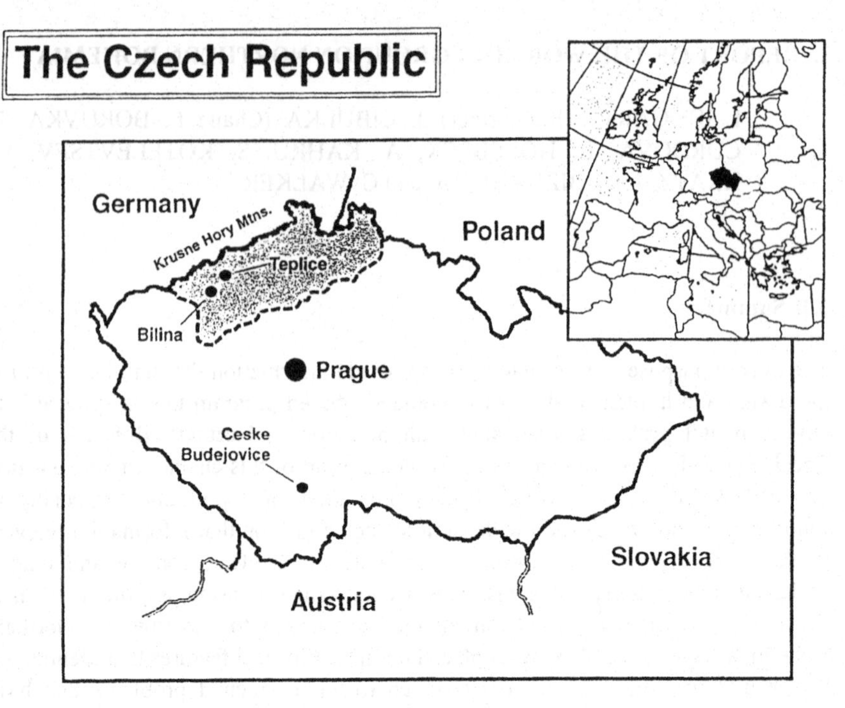

Figure 1. *Map of The Czech Republic (shaded area: Northern Bohemia)*

Landscape Areas currently exist in this region of the Czech Republic [1]. Current land use is given in Table 1.

The basic unit of the geological structure is represented by the Bohemian Massif, which is the easternmost part of the European Hercynides. The regional geology of the Bohemian Massif is very diverse. The bedrock is predominantly granite with metamorphic rock containing precious ores, whilst sedimentary basins contain large deposits of brown coal, lignite and clays. Thus, various soil substrates and different soils have developed in this relatively small area [2]. The physical-geographical situation of the study area is very complicated. Climatic factors affect the natural weathering rate that may also be accelerated by acidification of clay and bedrock. Geological processes have played important roles also. During the Alpine geological process, the southern half of Krusne Hory Mountain system declined and the lignite basin arose. Its recent morphological appearance is the result of the long term process of sedimentation of clays and sands. These lignite deposits are the richest ones in Middle Europe [3]. Layers have varying thickness and lie quite close to the surface. This fact enabled their surface exploitation, mainly after World War II.

TABLE 1. Land Use in Northern Bohemia[1]

Type of Landscape	Area	
	[km^2]	[%]
Urban agglomeration	787	5
Deteriorated landscape (dumps, open pits, etc.)	240	1
Recreation and park areas	169	1
Agricultural land	9093	54
Forests	5222	31
Non-forested area	1189	7
Surface water	94	>1
Total:	16794	100

[1] Czech Republic Program/Coordination of Information on the Environment

The area of interest belongs to a moderate climatic region that is the consequence of frequent changing of a wide spectrum of weather conditions. The period of plant growth lasts from 110 to 170 days per year and the annual mean precipitation is very variable (400-1800 mm). Altitude varies from 200 m to 1600 m above sea level. At altitudes less than 700 m, moderately warm and very wet areas can be found and above 700 m, cold areas predominate.

In the Bohemian basin, orographical conditions strongly affect the winds which blow mainly from west to east over most of the whole year. Temperature inversions are more frequent and vertically developed on the southern side of the Krusne Hory Mountains and in the Bohemian basin they are more frequent in the cold half of the year. A mean depth of an inversion layer is usually about 300 m but more vertically developed inversions can occur where the top of the inversion layer is above the altitudes of border mountain massifs of the Czech Republic. These vertically developed and very intense inversions are mainly connected with anticyclones situations which can last for several days. A temperature difference between the top and bottom of inversion layer is 8-12°C but extreme values up to 20°C can be reached. The maximum value of days with inversion occurrence inside the Bohemian basin is 115, corresponding value for the outside area is 59. Air pollution in this region is strongly affected by temperature inversions. Ground concentrations of pollutants in the air reach their extreme values during their occurrence.

3.3. Contamination Problems

3.3.1. HISTORICAL AND CURRENT INDUSTRIAL ACTIVITIES

In historical times this part of middle Europe was bounded by heavily forested mountains. Today, the Northern Bohemian region can be characterized as a highly industrialized area with prevailing mining, chemical and refinery industries [4]. A typical feature of this area is a high concentration of coal burning power plants (7 of them) producing electricity not only for local industrial and household consumption but also supplying energy to a substantial part of the Czech Republic (Figure 2).

Since the Middle Age there has been a long-term history and tradition of mining activities in the Northern Bohemian region for Au, Ag, Pb and U as well as iron and coal. A few geochemical anomalies with enhanced content of such elements as As, Be, Pb, U, Co, Zn, Ra and Rn have also been mined.

During the Second World War, the resources in this region were exploited to provide raw materials and cheap electricity to support an expanding heavy industry. Before 1989, more lignite was extracted than anywhere else in Europe, with an output of 200-220 million tons per year, representing 25% of all European production. This has resulted in a countryside devastated by the scars left by large scale surface mining. The environmental and health consequences of this situation have been studied there by many institutions.

3.3.2. INDUSTRY ASSOCIATED POLLUTION

The most important environmental problems in the Northern Bohemian region are associated with: a) emission of sulphur dioxide and solid particles by power plants operating on brown coal; b) devastation of the surrounding landscape by coal strip mining; c) damage of forests due to atmospheric pollution; d) pollution from chemical industry; e) pollution from commercial truck operation; f) mining and export of stone, and; g) agricultural recession with concomitant demographic changes and environmental damage.

Environmental pollution associated with known industries in Northern Bohemia are listed in Table 2. Changes in the consumption of electricity, extraction of

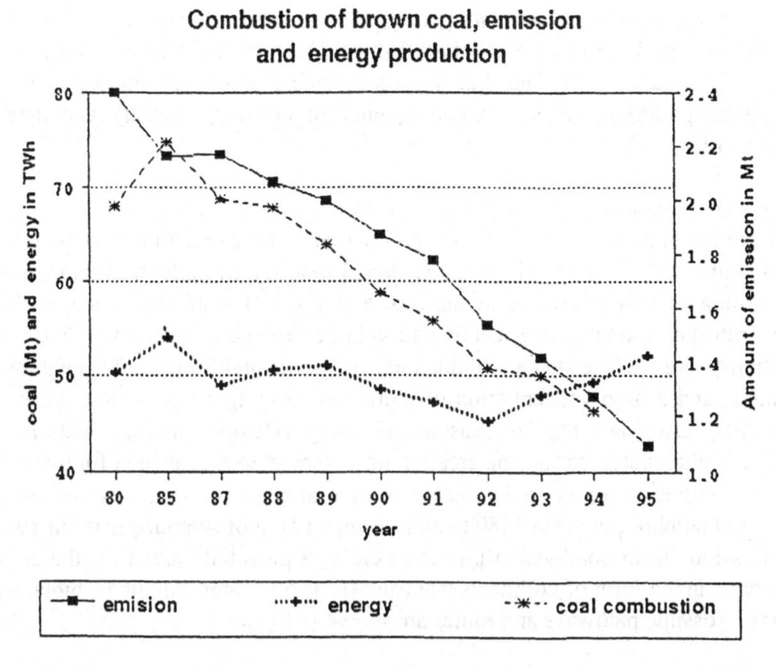

Figure 2. *Consumption of electricity, extraction of lignite and amount of emission in Czech Republic during 1980 -1995.*

lignite and amount of emission during 1980-1995 are show in Figure 2. The main environmental problems in the Northern Bohemian region arise from atmospheric pollution. It must be stressed that the mines do not cause the majority of the atmospheric pollution which is a consequence of chemical industry and power plant activities.

3.3.2.1. *Air Pollution*

Coal has been the backbone for power production and the smelting of ores; it has been a source of raw material for the chemical industry and a home-heating fuel. The burning of it has resulted in massive emissions of sulphur dioxide, carbon oxides, nitrogen oxides, trace metals and volatile organic compounds (VOCs) to the atmosphere. Due to a combination of unfavorable natural geographic conditions and a high concentration of industries emitting air pollutants (gaseous and solid), there are regular seasons of smog (October through February). Temperature inversion causes the creation of a closure above the area followed by rising concentrations of air contaminants. Furthermore, depositions have increased from 1 g of sulphur per m^2 in 1880 to an extreme of 50 g of sulphur per m^2 in 1992. The emissions from combustion processes create a potential hazard for the entire ecosystems in the form of pollutants that can affect soil, water, buildings, biota, and humans. Possible pathways and routes are shown in Figure 3.

TABLE 2. Industrial Pollution in Northern Bohemia

Industry	Environmental pollution
Power Plants	Air pollution ($SO2$, NOx) Heavy metals incl. Cd, Hg & As Particulate matter Solid waste
Refineries	Air pollution (phenols, PAHs & sulphur) Solid waste Waste water
Pulp & Paper Mills	Waste water (organics incl. phenols & dioxins)
Chemical Plants	Air pollution (organics) Waste water (organics)

Accessible data on the emission of environmental pollutants, particularly organics, are sparse. Changing industrial processes complicate the problem. Since 1974 available information about the sources and magnitudes of emission of all pollutants have been recorded in the Register of Emissions and Sources of Atmospheric Pollutants (REZZO). Management of the database is the responsibility of the Czech Hydro-Meteorological Institute under the control of the Air Protection Division of Czech Environmental Inspection Office (ORGREZ). Table 3 gives the emissions of pollutants to the atmosphere in the Czech Republic during 1994. About 70% of the total SO_2 emission can be attributed to activities in

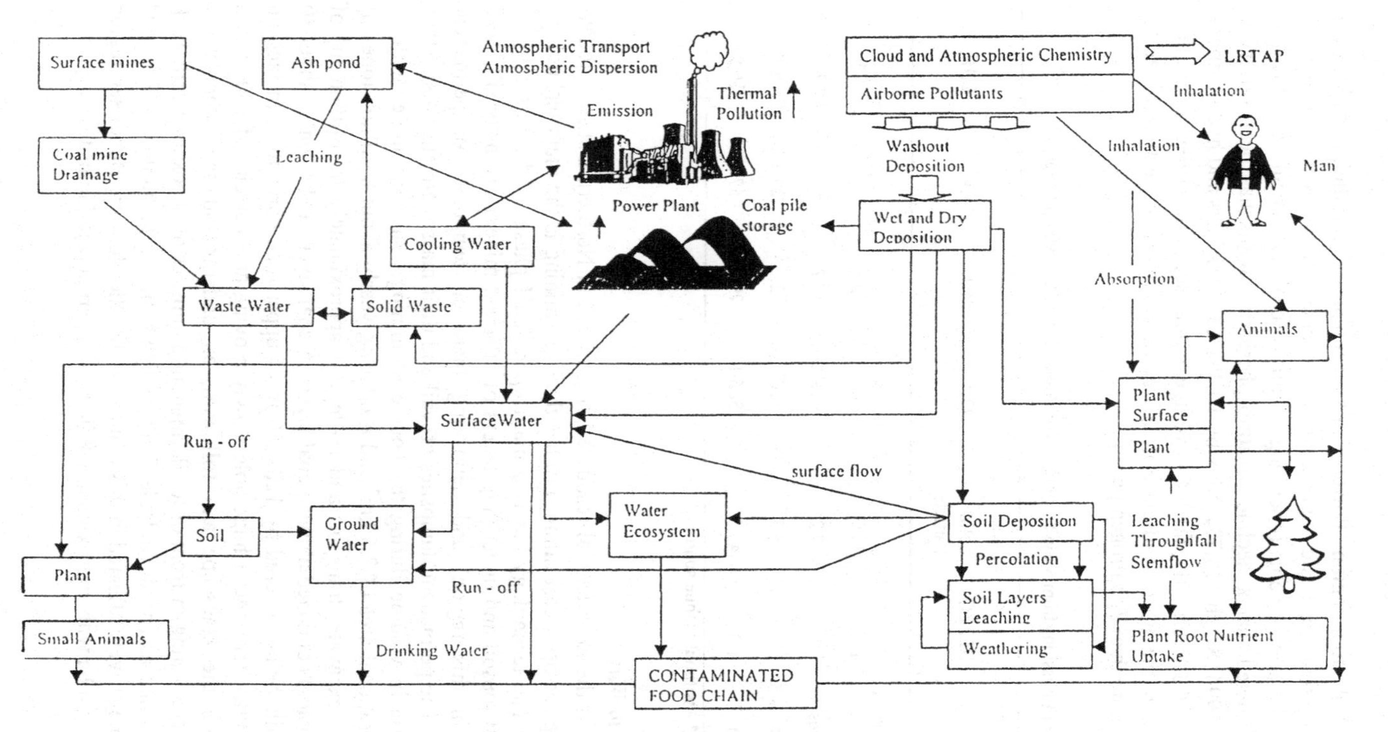

Figure 3. *Possible environmental risk from coal-fired power plant [52, with permission]. Pathways and transport of contaminants to various media (LRTAP: long range transport air pollutants).*

the Northern Bohemian region. Power plants, heat plants, strip mines and chemical industries represent the basic sources of local contamination while the biggest polluters are the power plants (operating on brown coal) and chemical industries. It should be noted that a decline of atmospheric pollutants has been observed compared to 1994: SO_2 by 38.7%, NOx by 10.4%, CO by 25.2% and hydrocarbons (CxHx) by 17.9%. Currently, the total average annual deposition fluxes of sulphur and nitrogen in the open landscape in Northern Bohemia are estimated at 5g sulphur and more than 2g nitrogen per m^2.

TABLE 3. Total Emissions from Main Stationary Pollution Sources for 1994 (kt/year)

Source	REZZO	Solid Particles	O_2	NO_x	CO	C_xH_x
Large Sources	1	35.0	567.7	75.6	19	7.9
Medium Sources	2	7.2	8.4	1.7	9	2.1
Small Sources*	3	5.3	10	2	29.2	6.5
Total Emission	1-3	47.5	586.1	79.3	57.2	16.5

Small Sources* - without mobile sources

3.3.2.2. *Solid Waste*

Brown coal is the main natural resource being mined in Northern Bohemia today. Strip mining (or open cast mining) rather than pit mining is the main technology employed. The technology has two main operational phases. The first stage is stripping off topsoil and underlying layers of different minerals (mostly silt, sand and clays in this region) and depositing them on temporary or permanent dumpsites. The permanent dumpsites prevail and their masses are enormous since rock increase in volume during the course of mining. Just one mine, with its mining operational area of 2.5 km^2 and producing yearly about 8 million tones of coal, has to excavate, transport and re-deposit approximately 60 million m^3 of minerals (deads) contained above a coal layer. A coal layer can be "only" about 30 m thick while being covered by 120–150 m of upper rock horizons. In fact, an operating mine represents a huge hole slowly moving at a velocity of 100 m in a year through the landscape and following an underground coal seam while destroying everything in front of it. Permanent dumpsites of deposited sterile rock surround a mine as an unavoidable feature of this mining technology. The total area being negatively affected in this manner within the Northern Bohemian region is about 26,000 ha with approximately 8,000 ha as permanent dumpsites.

3.3.3. AFFECTED HABITATS

The landscape of Northern Bohemia has been heavily damaged by past and current anthropogenic activities and the visual effects are quite evident. In this region many different types and kinds of habitat exist that may be affected (damaged) by industrial pollution. Table 4 identifies five major habitats that are susceptible to industry-specific pollution (see Table 2 for specific pollutants). It is obvious from Table 4 that pollution associated with power plants has the greatest potential to affect all the major habitats. This is because much of the pollution generated by power plants is distributed via the atmosphere to all the habitats in the region.

The magnitude of the risk involved in the operation of a hypothetical 500 MW lignite-fired power plant can be inferred from the list of toxic elements that are released to the environment as depicted in Table 5. Organic pollutants such as polycyclic aromatic hydrocarbons (PAHs) have not been included in this list.

TABLE 4. Potential for Habitat Damage by Industry-specific Pollution

Habitat	Industry			
	Power Plants	Refineries	Pulp and Paper Mills	Chemical
Agriculture	3	1	NE	2
Aquatic	3	2	3	3
Forest	3	1	1	2
Permanent grass lands	3	1	NE	2
Urban	3	1	1	3(*)

Habitat damage: 1 - small; 2 - moderate; 3 - considerable; NE - not evaluated; (*) - odor

An example of the effect of atmospheric pollution on a particular habitat is the destruction of forests. Even though forest decline is a complex problem, air pollution coupled with acute synergistic impact of extreme climatic conditions are believed to be responsible for the extensive deforestation of spruce monocultures and the acidification of soils especially in the mountainous regions. Symptoms of forest decline and damage were noted in the 1930's and significant damage was recorded in the period 1966-1977. By the 1980's, forest stands in the Jizerske Hory Mountains were almost completely destroyed and intensive damage was noted to the spruce forests in the Krkonose mountains. Since the 1980's the remaining spruce forest stands have stagnated while young stands have grown relatively well. However, all spruce stands in Northern Bohemia showed severe damage after the winter of 1996.

Another major problem for habitats in Northern Bohemia arose because of the increased mobility and leaching of toxic elements by acid rain from hazardous waste storage sites, strip bedrock and deads from mining activities. The flux of H^+ achieves values >250 mg per m^2 per year and a number of heavy metals (Pb, Cd and Ni) are found in the environment at higher than normal concentrations.

Furthermore, the increased bioavailability of the toxic elements Al, As and Be in the soil and surface waters of the region pose a serious threat not only to terrestrial and aquatic ecosystems but also the region's inhabitants (Figure 3). Additional important information concerning environmental pollution of soils in Northern Bohemia is provided by Dr. Lubos Boruvka of the Czech University of Agriculture in Prague and can be found in the Appendix 2.8.1 of this chapter. Czech legislation concerning soil pollution is discussed there. The main soil groups of the Northern Bohemian region are briefly characterized in relation to their vulnerability and sensitivity to pollution and acidification.

TABLE 5. Flow of Toxic Elements Through 500 MW Lignite-fired Power Plant

Element	Content g/t coal	Flow T/yr	Fly Ash µg/kg	Escape %	MIC µg/m³	Factor Toxicity	Risk Factor
As	6.7	20.1	8-183	.6	3	2	17
Be	5.5	16.5	10-17	1.5	0.01	550	825
Cd	2.8	8.5	2.7-8.2	2.6	3	0.93	2.4
Co	11.4	34.2	70-171	(15)	2	(5.7)	(86)
Cr	82	246	70-171	3.7	1.5	54.7	202.3
Cu	120	360	908-3300	18.1	2	60	1086
Pb	20	60	89-187	3.6	0.7	28.6	103
Mn	540	1620	425-3135	1.7	10	54	92
Hg	0.1	0.3	0.1-6.1	29	0.3	0.03	9
Ni	92	273	45-260	1.9	1	92	175
Se	(4)	(12)	nd	(21)	2	2	42
Tl	12	36	5-12	0.68	(2)	6	4.1
V	(25)	75	28-260	4.4	2	12.5	55
Zn	160	480	670-4100	22.7	50	3.2	73
S	16000	$4.8X10^4$	2500	95	150	106.7	10133

Remarks: nd - not determined; ( ) - extrapolated value; Factor Toxicity = Content divided by MIC - Maximum Immission Concentration to air environment; Risk Factor = Escape times Factor Toxicity; Values Factor toxicity according to [51].

3.4. Remedial Activities in the Study Area

The material in this section is discussed mainly in relation to the mining industry in the Czech Republic. As such it is not intended to be all inclusive, but rather to serve as a general guide to on-going remedial activities in this country and how these activities are influenced by the new legislation that is being enacted.

3.4.1. RELEVANT NEW ENVIRONMENTAL LEGISLATION

As a consequence of political and economic changes in the Czech Republic after November 1989, the government is hoping to join the European Community (EC). One of the basic conditions for being accepted as a new EC member is to implement harmonized legislation with the rest of Europe. That is why an extensive collection of new and farsighted laws have been implemented in the Czech Republic from 1990. The preamble to the Act On The Environment [5] states: "The Federal Assembly of the Czech and Slovak Federal Republic, recognizing the fact that human beings, together with other organisms, are an intrinsic part of nature, considering the natural interdependence of human beings and other organisms, respecting humanity's right to transform nature in accordance with the principle of permanent and sustainable development, aware of its responsibility for maintaining a favorable environment for future generations, and emphasizing the right to a favorable environment as one of humanity's principal rights, has approved this Act."

A new feature concerning the environmental laws (Czech National Council Act on Environmental Impact Assessment, Act No. 244/1992 Coll. [6]) is a requirement for all subjects to perform environmental evaluation of their current or planned operations (environmental impact assessment, EIA). Such evaluation can be prepared in various ways defined in the laws (see below). Biological monitoring of affected landscape is always a compulsory part of such documentation. The EIA documentation must be prepared only by a licensed expert specialized in the environmental problem areas and accredited for this activity by the Ministry of the Environment.

The Act on the Environment (17/1992 Coll.) defines duties in §17 with respect to the protection of the environment for all who use landscape and natural resources. Enclosure No. 1 is a list of activities amenable to this Act. Pit and strip mining are listed there (par. 3.1.). Enclosure No. 2 defines a structure (table of contents) of the compulsory documentation and evaluation of planned activities with respect to the environment.

The Act on Environmental Impact Assessment (244/1992 Coll.) defines a procedure for environmental evaluation of new constructions and technologies. It has several enclosures describing its scope. Mining activity is again listed there but it is applicable only if a mine is opened as a brand new operation. The Ministry of the Environment does not apply this legislation to ongoing mining activities within currently defined mining areas. This is a very important official statement.

Apart from these major environmental laws, there are a number of valid measures regulating all aspects of strip mining. For example, there is a measure requiring that all mines save and deposit a part of their financial profits in their special bank accounts to be used in future years for landscape revitalization. Eventually coal mining will be stopped or substantially slowed down and the only mine activity will be closure and subsequent landscape reparation. A mining permit is usually issued for a period of five years and must be regularly updated. The issuing of a permit allowing prolongation is based on the evaluations of many technical and economic documents including the environmental report described above. Since mine management lacks the ability to perform scientific biological monitoring for preparing the report, they usually look for professional cooperation with specialized institutions or universities. For a specific example of such a cooperative venture, the reader is referred to the report of Drs. Jiri Cibulka and Eva Curdova in Part A of this book. The material in this report is taken in part from the presentation Dr. Cibulka gave at the opening Plenary session of this NATO ARW.

3.4.2. CORRECTIVE AND PREVENTIVE MEASURES IN THE REGION

The basic reasons for performing corrective and preventative measures in the region are the following: a) regional industrial activity cannot be immediately stopped since it is currently vital for our state economy; b) strip mining can be limited since there will be future alternative sources of energy (e.g. nuclear power plant near Temelín, Ceske Budejovice, South Bohemia); c) mine operations must continue to a limited extent to accumulate sufficient financial resources for subsequent landscape rehabilitation.

The main corrective measure being used in the region is the application of a scrubber technology on several local power plants. Currently, stripping off sulphur oxides, ash and other air pollutants using effective scrubbers and stack filters has had a very positive effect on the atmosphere quality. There is a distinct trend of decreasing atmospheric pollution. However, this technology is rather costly and it will take several more years to equip all significant stacks and chimneys with these special scrubbers.

Coal strip mining in Northern Bohemia has declined during the past years since it is considered as unacceptable for the future. Mining should totally stop there by the year 2030 and even now all mines focus their attention on correction of environmental damage caused by them in the past decades. Current efforts are mainly concentrated on landscape re-creation and revitalization employing sophisticated, scientific recultivation methods (complex process of artificial soil profile creation and subsequent tree planting). In reality it would take decades for natural revitalization processes to transform sterile heaps of rock into green hills, fields or forests. A special treatment must be applied to form an "artificial" topsoil layer on their surface that will provide water and nutrition for plants. Due to a lack of natural topsoil for this dumpsite recultivation, special procedures must be used for incorporating rock material into fertile soil profiles. Such pedological measures make it possible to use dumpsites as agricultural plots or, more often, plant them with various plant species (grasses, deciduous and coniferous trees, etc.). The main types of recultivated areas are classified as agricultural, forest, and recreation type. There could be a mixed variation. Artificial water reservoirs are always part of a new landscape.

There is a significant non-governmental (NGO) environmental ("green") movement (Duha – Rainbow or Deti Zeme – Children of the Earth, etc.) working within the region and closely watching over all kinds of industrial activities there. As far as the Bílina Mines are concerned, it must be stressed that there is good cooperation between the mine management and the representatives of the movement. There are cases of successful joint rescue actions (flora and fauna rescue or emergency relocation, etc.) and of financial support to the movement. Also, all relevant planned activities of the mine are usually discussed with the movement.

3.4.3. MONITORING PROGRAMS

The Ministry of Agriculture, the Ministry of Health and the Ministry of the Environment have been performing an all state monitoring program and have made an inventory of the majority of all known contaminants. The emission monitoring network consists of 27 automated stations for emissions, 24 manual stations belonging to the State Hydrometeorological Institute, 58 stations belonging to the State Hygienic service and 20 stations belonging to the Research Institute of Crop Productions.

The State Veterinary Administration monitors organic contaminants in the atmosphere, water and soil. Results are published annually and are available to the public. Because polychlorinated biphenyls (PCBs) are toxic and persist in numerous environmental compartments, careful attention is devoted to monitoring their contents in surface water, plant and animal products [7].

The banking of environmental samples, in particular biological materials, is not routinely performed in the Czech Republic. The main reason is the lack of finance to operate such a costly institution for several decades [8]. However, there are local efforts to keep certain environmental samples (e.g. historical soil samples) for long periods of time under defined and carefully conditions.

3.4.4. HUMAN HEALTH

The most important feature associated with the unfavourable environmental conditions in Northern Bohemia is the shortening of human life span and life expectancy. It is a fact that the Czech women and men have shorter lives by approximately 5 and 7 years, respectively, compared to the average values of the developed world. This phenomenon is even more obvious among the Northern Bohemia population. Other explanations can contribute to this phenomenon. A reason for shorter lives could be attributed (partly) to unhealthy nutritional and living styles and habits that are widely spread among the Czech population (heavy smoking, fatty food, alcohol abuse, lack of exercise, a low standard of awareness concerning preventive health care and hygiene, etc.) in combination with relatively good average salary in labor professions. These habits are very typical for some social groups in Northern Bohemia. Sociologists and health experts have been studying this aspect and there is also foreign cooperation on the problem.

The "Teplice Program" is an extensive research program that monitors human health in the city of Teplice which is situated in one of the mining districts of Northern Bohemia [9]. This program is well funded until 1999. The approach is multifaceted and covers atmospheric monitoring, atmospheric genotoxicity and human reproduction. In the last category, particular emphasis is placed on pregnancies, immunity and health of children, sperm quality of men, endocrine functions, mortality, sociological aspects, risk assessment and good laboratory practice. The city of Prachatice in Southern Bohemia is used as a control since this region is considered one of the cleanest in the Czech Republic. An important component of the Teplice Program is the Biomarkers of Exposure to Mutagens and Carcinogens Project. Studies within this project have been conducted to simultaneously evaluate personal exposures to air pollution in the form of respirable particles containing PAHs and biomarkers of exposure, biological effective dose, genetic effects, and metabolic susceptibility [10,11].

3.5. Implementation of Biomarker Program

3.5.1. PURPOSE AND USE

It is an undeniable fact that Northern Bohemia is one of the most polluted regions in the Czech Republic. Fortunately, the government is working to reduce environmental pollution. Having made this commitment, the government is now faced with the problem of assessing the health risks that follow from exposure to environmental pollutants as well as from any remedial activities undertaken.

Before a study concerned with environmental pollution is initiated, an objective or need for the study must be established. Generally the objective is determined by the answers to certain questions (e.g., Has exposure occurred? What is the extent of exposure? What are the effects?) and the results of the proposed study should provide information needed to answer these questions. Often the objectives of the study are derived from regulatory issues. For example, an effective risk management policy (the process of deciding what can and should be done to mitigate or control risk) concerned with hazard identification or hazard assessment must incorporate risk assessment (the objective characterization and evaluation of knowledge about toxic effects). Biomarkers have the potential to improve the accuracy, reliability, and scientific basis for quantitative risk assessment [12-14]. Biomarkers are a direct measurements of biological events or responses that result from exposure to xenobiotics and reflect molecular and/or cellular alterations that occur along the temporal and mechanistic pathways connecting ambient exposure to a toxicant and the eventual effect [14].

Several components within the realm of risk assessment are amenable to the use of biomarkers. First, biomarkers may be helpful in hazard assessment by identifying toxic reactions and populations at risk. Second, in the arena of exposure assessment, biomakers can help (a) establish the fact of exposure; (b) reduce exposure misclassification; (c) identify exposed population (screening); (d) reconstruct exposure assessment; and (e) provide an integrative measure of exposure. Finally, in dose-response analysis, biomarkers may serve as surrogate measures of the key variables that are difficult, disruptive, or expensive to measure.

3.5.2. STRATEGY

3.5.2.1. *Introduction.*

In May of 1991 an important NATO Advanced Research Workshop was held at the Netherlands Institute for Sea Research, Texel, The Netherlands on the topic "Biomarkers". This activity brought together a group of international scientists to join in intensive, high-level discussions on the use of biomarkers to assess environmental health. The results of the workshop were published as a NATO ASI Series publication (Biomarkers: Research and Application in the Assessment of Environmental Health, edited by David B. Peakall and Lee R. Shugart, vol. 68, 1993 [15]). Many important topics in this book are discussed from a practical point of view. Included are topics on the strategy for the development and application of a biomarker-based biomonitoring program as well as the identification of problems associated with the implementation of the study and the interpretation of results. This seminal work should be consulted before the initiation of any important study that plans to incorporate biomarkers as tools for the measurement of biological events or responses that are thought to be related to exposure to environmental contaminants.

Regardless of the specific objective or motivation for a study that incorporates biomarkers, many elements of the study design will remain constant. This concept, which is the basis for an effective biomarker strategy, is shown in Figure 4 (taken from Peakall and Shugart [15]) and depicts a logical pathway for defining the essential elements of the study.

3.5.2.2. *Integrated Approach*

Several useful questions should be addressed when applying biomarkers to problems of environmental pollution: What adverse changes have occurred and what is the extent and severity of these changes? What adverse change can be attributed to the pollutant(s) at the site under investigation? What adverse change has the potential for abatement as a result of clean-up? These are very broad questions and to be answered adequately or addressed successfully, a Biomarker Program should integrate all relevant information and data. These included data from several sources such as field studies, bioassays (section 2.5.4.2. of this chapter), as well as biomarkers.

Field surveys (including ecological profiling) provide the most direct way to demonstrate adverse change and in most cases reflect the biological integration of all stresses, however, they can only establish correlation between effects and presumed cause. Bioassays can be employed to evaluate toxicity of environmental samples and in some instances may be the only way to experimentally establish a causal relationship between a contaminant and an observed effect. Biomarkers can provide evidence of exposure to sublethal concentrations of stressors, such as chemicals, and may provide a link to subsequent adverse consequences for the organism, population and ecosystem.

3.5.3. BIOMARKER SELECTION

Ideally, a biomarker should be biologically relevant, sensitive and specific. In addition, it should be readily accessible, inexpensive, and technically feasible. This combination of requirements is rarely achieved, and some tradeoff is inevitable in order to obtain useful data in a timely manner [14, 15].

The selection of appropriate biomarkers (prioritization process) occurs only after considerable information has been obtained about the area of study. For example, the "check list" found in Figure 4 indicates that species selection should precede biomarker selection. One reason is that species indigenous to the habitats present in the study area will often dictate the appropriate types of biomarkers that are applicable in that species for the classes of pollutants under investigation. Species selection in a polluted forest habitat may serves as an example. Conifers constitute more than 54% of the total tree species in the Northern Bohemian region and suffer the most visible stress from pollution. Thus the Norway Spruce (*Picea abies*) should be an excellent species for a biomarker study of atmospheric pollution. Other potentially useful forest species would be birds, small mammals, insects, lichens, soil organisms because soil acts as a sink for air pollutants, and indigenous plants. The appropriateness of any species to the study would however depend upon its compatibility with the biomarkers proposed. Other important factors to be considered are the availability of the species and the objective of the study.

Conversely, previous studies should not be overlooked as they may contain important information on species selection. For instance, for the past several years in the Czech Republic biological monitoring of atmospheric pollution and the assessment of air quality has been pursued using bioindicators. Organisms employed for various types of atmospheric pollutants are listed in Table 6 and could provide the starting point for the preparation of an initial list of appropriate species. Furthermore, the reader is referred to Appendix 2.8.2. of this chapter for a more detailed discussion and examples of the successful use of indigenous species as bioindicators in mining operations in Northern Bohemia.

Biomarker selection is determined by the class and type of pollutant(s) under investigation. In this regard a useful exercise that will aid in the selection process is the preparation of a lists of the major pollutants that occur and/or are anticipated to occur in the study area. For example, from the information contained in Tables 2 and 4, as well as data collected from ongoing monitoring studies (section 3.4.3. of this chapter), it can be assumed that most habitats in Northern Bohemia will be heavily contaminated with pollutants from energy plants and to a lesser extent pollution from chemical plants and refineries. Thus a wide range of pollutants will be encountered in any study area, and will include many potentially toxic chemicals/agents such as heavy metals, contaminated particulate matter, S02 and NOx as well as organics (PAHs, PCBs, etc.). A detailed list of biomarkers that

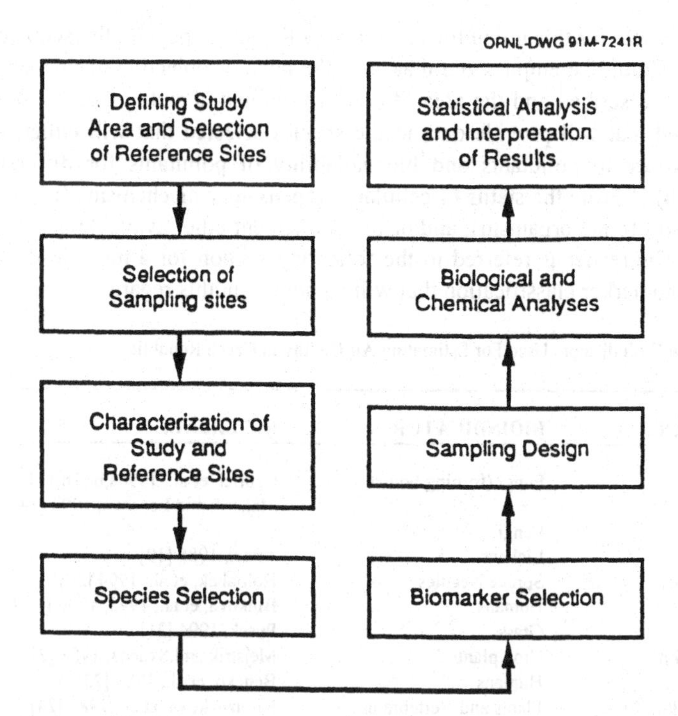

Figure 4. *General elements of a biomarker program [15, with permission].*

would be appropriate for this ubiquitous roster of pollutants is beyond the scope of this chapter. It should be emphasized that once the need or objective of the study has been clearly described and the area of concern defined, a suite of biomarkers can be identified that will provide data in the species selected that will either: a) document exposure to pollutants and bioavailability of pollutants for different trophic levels; b) indicate the status of cellular compensatory mechanisms (i.e., the potential for harm to the organism); and/or c) complement other available data and observations. The reader is referred to the following section for a more in-depth discussion on biomarker classification that will be helpful in this regard.

TABLE 6. Review Bioindicators Used For Estimating Air Quality in Czech Republic

AIR POLLUTANT	BIOINDICATOR	REFERENCE
Heavy metals	Fungi (fruiting bodies)	Gabriel, et al., 1996a,b [16,17]
		Mejstrik and Lepsova, 1993 [18]
	Fungi	
Radionuclides	Lichens	Andel, 1985 [19]
Persistent organics	Spruce Needles	Holoubek, et al., 1994 [20]
PAHs	Humans	Binkova, et al., 1995, 1996 [10,11]
Acid pollutants	Grass	Pysek, 1994 [21]
Co, Cd, Cr, Ni, Zn	Crop plants	Mejstrik and Svacha, 1988 [22]
Be	Humans	Bencko, et al., 1980 [23]
Al, Be, Cd, Cu, Pb, Zn	Plants and Vertebrates	Kubiznakova, et al., 1989 [24]
Fluoride	Humans	Valach and Sedlacek, 1989 [25]
		Molokov, et al., 1985 [26]
	Common Vole	Kierdorf, et al., 1996 [27]
	Red Deer	
Pb, Cd, Cl	Spruce Needles	Mankovska, 1978 [28]
Pb	Ants	Stary and Kubiznakova, 1987 [29]
Pb-traffic pollution	Winged Insects, Butterflies	Stary, et al, 1989 [30]
Air Pollution	Beetles	Bohac, 1995 [31]

3.5.4. IMPORTANT CLASSES OF BIOMARKERS

3.5.4.1. *Introduction*

In general, the biological responses (biomarkers) elicited in an individual organism after exposure to a chemicals or other substances (i.e., physical agents such as x-rays or ultraviolet radiation) constitutes a biomarker. The following discussion will focus on biological responses attributable to chemicals, since they are the main source of pollution in Northern Bohemia.

In 1987, the National Research Council (NRC) of the National Academy of Sciences USA issued a report examining the current state of the science for the use of biomarkers in environmental health research and risk assessment [32]. The NRC report provided a framework for the classification of biomarkers used for human health studies. Categories of exposure, effect and susceptibility were identified. These assignments are not mutually exclusive but do indicate the application of biomarkers in the exposure/disease paradigm set forth in the NRC report. The recent biomarker studies on the effect of air pollution on human health in Northern Bohemia (Teplice Program) subscribe to this paradigm [10,11]. The model has been expanded and modified [14,33] and even a new scientific journal (BIOMARKERS) accepts submissions based on the NRC classification of biomarkers. Many studies concerned with environmental species have recognize this classification [34], but other ways of categorizing biomarkers have been proposed. Hugget, et al, [35] classified biomarkers of anthropogenic stress (mainly chemicals) based on the biological level at which the response is measured (i.e., biochemical, physiological and histological). Here, classification is related to metabolic pathways, induction of resistance mechanisms, and analyses that employing physiological and biochemical criteria [36]. This, in turn, has led to greater emphasis on understanding sublethal effects to pollution at the level of biological organization where exposure can be adequately described and assessed.

3.5.4.2. *Molecular Markers of Toxicity*

Molecular and biochemical responses useful as biomarkers can usually be placed in one of several general categories that depend upon the endpoint to be examined [36,37]. These are: a) protein induction or expression (e.g., cytochrome P450 enzymes, metallothionein, stress proteins); b) DNA damage (e.g., adducts, strand breakage, clastogenic events); and c) excretory products (e.g., metabolites of detoxification, products of cellular catabolism). An example of each category will be given, but the reader is referred to the scientific literature for a more detailed discussion [33, 35-37].

a) Cytochrome P450 Enzymes: These constitute a family of haemproteins that are the major monooxygenase detoxification system of phase I metabolism in living organisms [38] including both eukaryotes and prokaryotes. Numerous cytochrome P450 genes code for different versions of the enzymes. The P450 1A1 enzyme is induced by a range of planar lipophylic compounds including PAHs, coplanar PCBs, polychlorinated dibenzodioxins (PCDDs) and polychlorinated dibenzofurans (PCDFs). The induction of hepatic microsomal P450 1A1 provides the basis for biomarker assays that have been widely used in vertebrates both in the laboratory and the field. The biomarker assays depend upon either an increase in enzyme activity (e.g., the EROD assay) or an increase in the enzyme protein (Western blotting or ELISA assay). The magnitude of the induction can provide a measure of the degree of interaction of an inducing compound(s) with the aryl hydrocarbon hydroxylase receptor (Ah-receptor) in the cytoplasm of the exposed cell. Where there is evidence for strong induction of P450 1A1 in the presence of high levels of PAHs, biomarkers for genotoxicity (i.e., structural damage to DNA-see below) should also be performed. Biomarker assays based on P450 1A1 induction have been used to measure exposure of birds, fish and mammals and because of the specificity of the inducer, the response can provide an integrated measure of exposure to a mixture of inducing compounds. Furthermore, since this biomarker response has also been related to a number of other toxic end points, it can provide an index of Ah-receptor mediated toxicity [37].

b) DNA damage: Substances capable of modifying the genetic material of living organisms are termed genotoxicants. Many of these are anthropogenic chemicals (PAHs, PCBs, heavy metals) and are widely distributed in our environment. Genotoxicants can interact with DNA and potentiate structural alterations in this important cellular macromolecule. The number of structural alterations per genome is often quite small and therefore sensitive analytical methods are required to detect the unique types that occur. Specific types of damage that occur to the structure of DNA can be used as end points for assessing exposure to genotoxicants. Persistence of damage may potentiate deleterious responses at other levels of biological organization which in turn become end points of health-related effects.

Structural modifications of DNA are probably the best understood genotoxic event for which analytical techniques with the appropriate selectivity are available and can take the form of adducts (where the chemical or its metabolite becomes covalently attached to the DNA), of strand breakage, or of chemically modified bases [39,40]. The most common methods employed for the detection of DNA adducts are 32P-postlabeling and immunoassays using adduct-specific antibodies [41]. The former is the most suitable technique for the detection of DNA adducts induced by unknown environmental and occupational chemicals or mixtures[42] and has been used extensively in biomarker studies with mammals (including humans [10,11]) and animals in the field [40] (including aquatic organisms [43]).

Excretory Products: Small molecules such as amino acids, nucleosides, organic acids, porphyrins, and other chemicals are often found associated with physiological fluids (i.e., urine, bile, milk, or blood). Their presence or amounts can be indicative of cellular detoxification processes or dysfunctions. The metabolism of many chemicals (e.g., PAHs) results in the formation of numerous conjugated and unconjugated metabolites that are excreted. These metabolites can be used as biomarkers of exposure. In addition, many metabolite markers are present in physiological fluids under normal conditions, but altered profiles of these compounds can be indicative of a toxic response. Porphyrins are the intermediate metabolites of heme biosynthesis. Disturbances in the heme biosynthetic pathway can lead to an excess of porphyrins in the blood and other tissues. Analysis of accumulated and excreted porphyrins is important in the diagnosis of porphyrias and lead poisoning [33]. For example, exposure to halogenated aromatic hydrocarbons can cause deregulation of heme biosynthesis and lead to massive increases in liver concentrations of uroporphyrin and heptacarboxylporphyrin and cause clinical symptoms similar to those found in an inherited form of porphyria.

3.5.5. SUPPLEMENTAL INFORMATION

As implied in the section on "Integrated Approach" (section 3.5.2.2. of this chapter), it is not realistic to assume that a successful environmental monitoring program can rely on biomarker data exclusively. Supplemental investigations that provide important information concerning the status of the environment should be sought. There are numerous methods and approaches that may be applicable. Those listed below have been tested and found reliable and their implementation into or concomitant with an ongoing biomaker program should greatly enhance the chance for success.

3.5.5.1. *Integrated Field Methods in Soil Ecology*

Because soil can become the repository of a considerable portion of the pollution dispersed into the environment (see Appendix 3.7.1. of this chapter), it is understandable that knowledge of the interaction and impact that pollution has on soil organisms and soil processes is an important issue in environmental health. For example, the function of soil organisms in maintaining forest ecosystems could be exploited as an ancillary technique along with field and biomarker studies to determine the impact of pollution in forest habitats. New methods amenable to examining structural parameters of ecological communities present in the soil as well as functional processes of these communities is detailed by Dr. Werner Kratz in Appendix 3.8.3. of this chapter.

3.5.5.2. *Bioassays*

In general, the application of the biomarker approach entails the measurement of biological responses in indigenous species in the field. The actual determination of the biological response is quite often conducted in the laboratory (i.e., the measurement of the catalytic activity of the P450 1A1 protein in liver extracts or the detection of certain classes of DNA adducts in extracted DNA). This is in contrast to the determination of toxicity as studied in a laboratory setting using bioassays or ecotoxicity tests. Here, bioassay refers to a specific procedure performed to determine the strength, nature, constitution, or potency of a sample from the degree of response elicited in the test system (organism, cell culture, etc.). Ecotoxicity tests (see below) are various implementations employing lethal and non-lethal response end points with diverse exposure durations [44].

An in vitro system, the H4IIE rat hepatoma cell bioassay [45] has been established to characterize the overall potency of planar halogenated hydrocarbons (PHH) in extracts from environmental samples. This cell line has an aryl hydrocarbon hydroxylase activity that is inducible by 2,3,7,8-tetrachlorodibenzo-p-dioxin and other PHHs such as PCBs. A strong correlation has been demonstrated between in vitro induction of this activity and many of the toxic attributes associated with environmental PHHs. A similar system is the chicken embryo hepatocyte bioassay [46]. Details for this bioassay can be found in the chapter on the Pollution Problems of the Vistula River in Poland (Working Group 2 report, this book).

3.5.5.3. *Ecotoxicity Testing*

In Northern Bohemia, it is anticipated that soil, sediment and wastewater (leachate from solid waste deposits of burned coal, deads from open-pit mining, etc.) will contain heavy metals, PAHs, PCBs and other toxic pollutants. Furthermore, these pollutants will almost surely occur as complex mixtures in the samples to be tested. This will complicate the determination of overall toxicity because of antagonistic and/or synergistic interactions between pollutants. Toxicity tests that use simple living organisms can circumvent this problem because they are usually not specific towards certain chemicals, but test for the overall toxicity of the sample. Appropriate manipulation of the test system (i.e., addition of EDTA to eliminate the effect of heavy metals) may discern classes of chemicals responsible for the toxic response. At sites under investigation where previous analytical data on the occurrence of these pollutants is absent or sparse, toxicity screening can be incorporated into a biomarker program to provide rapid and useful supplemental information. Vice versa, positive results from toxicity screening could be verified by chemical analysis which in turn would suggest the biomarker(s) appropriate for the toxic chemical(s) present in the study area (section 3.5.3. of this chapter).

There are currently many toxicity tests commercially available but they all are aiming at aquatic toxicity. Those that use daphnia, rotifers, and algae are relatively laborious to perform because of required extensive preparation and long periods of exposure (24-72 hours) for analysis. However, with the CerioFASTTM toxicity assay, exposure times for Ceriodaphnia have been shorten to 1 hour [47].

The MicrotoxTM is one of the most widely used assay for toxicity testing and incorporates a luminescent bacterium [48]. The test is most suitable for screening purposes as it is rapid, cost-effective and sensitive. The test has been used successfully to screen the toxicity of a large number of single chemicals and metallic salts and although developed for the analysis of water or extracted samples, it has been adapted as a solid-phase assay. The assay measures the effects of contaminants on light production of bioluminescent bacterium, Vibrio fischeri (formally referred to as Photobacterium phosphoreum). The inhibition of luminescence represents inhibition of electron transport systems, basic processes found in all organisms.

The genotoxicity of an environmental sample can be measured using the SOS Chromotest [49]. This test involves the incubation of the sample with a genetically engineered strain of Escherichia coli, where the production of β-galactosidase is under the control of the SOS response pathway which is induced by DNA-damaging agents. Postexposure β-galactosidase activity reflects the genotoxicity of a given sample. Alkaline phosphatase activity provides an indirect measure of bacteriostatic and bacteriocidal effects.

Toxicity tests specific for soils are now available using soil dwelling organisms like earthworms, insects and nematodes.

3.5.5.4. *Other Measurements*

Environmental monitoring programs can draw on measurements from numerous studies, databaes and tabulated phenomena. The tabulation of natural resources and the measurement of the occurrence and distribution of natural and manmade substances are all important parameters. The temporal and spatial characteristics of the collected information constitutes a database that can be invaluable toward the validation of biomarkers (i.e., relevant, sensitive and specific).

Another equally important use for these types of data is for modeling studies. Models generate predictions based on existing data. Biomarker data could be especially useful in helping reduce the degree of uncertainty that exists in modeling studies and vise versa. The reader is referred to Appendix 3.7.4 of this chapter ("Assessment of Atmospheric Pollution on Spruce Canopy" by Dr. Jana Kubiznakova) which details a facile approach that estimates the impact of acid air pollutants to spruce trees in Northern Bohemia and the use of monitoring data to estimate future effects on the forests in this area.

3.6. Recommendations

"Whatever you do, or dream you can, begin it. Boldness has genius, power and magic in it. Begin it now". Goethe

Improvement to the environment has followed changes that were initiated in 1989 by the Czech government to restructure the economy and industry [4]. Nevertheless, the Czech Republic is still faced with the daunting task of addressing the legacy of its environmental problems as well as curtailing new ones. Information compiled and discussed in the previous sections of this chapter has identified air pollution as the most common and prevalent environmental problem in Northern Bohemia. Air pollution affects all types of landscapes and habitats and numerous human activities contribute to the problem. The most obvious activities are those associated with energy. The production of energy by power plants, the consumption of energy by heavy industry and the public, and the supply of energy from coal mining are the major contributors. Risk assessment and the concomitant incorporation of a Biomarker Program must be an integral part of any remedial steps taken to correct the situation and the following recommendations are based on this premise.

1. A large interdisciplinary project (Teplice Program) currently exists in Northern Bohemia designed to improve environmental monitoring and research on the health effects of air pollution. Biomarker research employing sophisticated techniques of genetics and molecular biology to evaluate human exposure to suspected carcinogens and mutagens (respirable particles containing PAHs) has been undertaken [10,11]. Since, environmental health is inextricably linked with human health, the Teplice Program should have a strong biomarker program in this arena as well. The primary use of biomarkers in environmental studies is to assess the health of the species present in order to detect and identify potential problems so that unacceptable and irreversible effects at high levels of biological organization such as disease in humans, mass mortality, and loss of ecologically important

species, can be avoid. Arbitrary separation of human and environmental health issues is neither efficient, intellectually sound, or socially productive [50].

2. Because biomarker studies will be confounded by the various types of pollutants that are widely dispersed in the environment in Northern Bohemia (section 3.5.3. of this chapter), it is important that the criteria for biomarker selection be sensitivity, specificity, technical feasibility and biological relevancy. Preference, where possible, should be given to effects biomarkers (section 3.5.4.2. of this chapter) that a) measure early biochemical or cellular responses, b) are closely related to the toxicity of the pollutants, and c) have been validated (i.e., tested against the criteria cited above). The following biomarkers should be considered:
a) The P450 1A1 enzyme system [38] for organic pollutants, particularly were sites are contaminated with hydrocarbons such as PAHs and PCBs.
b) The metal-binding protein metallothionein (phytochelatin in plants) for heavy metal contamination. This protein is induced in a diverse array of eukaryotic species exposed to toxic metals such as Cu, Zn, Cd, Hg, and Pb [36].
c) The stress proteins for numerous pollutants including heavy metals and organic chemicals. These proteins are highly conserved in nature and their response becomes an integrated signal for environmental stress [36].
d) Structural modifications to DNA such as adducts and strand breakage. The detoxification of many organic compounds and heavy metals can cause structural perturbations to cellular macromolecules such as DNA that if not properly repaired, may result in pathological conditions (i.e., disease, tumors, etc. [39,40].

3. The process of interpreting biomaker data can be difficult; however to facilitate matters, all relevant information for a particular study needs to be identified and incorporated into the process. To ensure that pertinent information is obtained, the implementation of the "integrated approach" (section 3.5.2.2. of this chpater), that includes those supplemental methods and approaches (section 3.5.5. of this chapter) appropriate to the study, should be considered. Other sources of ancillary information that should be consulted, as they are often invaluable as well, are past investigations about the study area that contain evidence of historical use, databases from environmental monitoring activities, and previously identified toxic effects observed in bioindicator species.

4. In an attempt to adequately address various environment problems associated with pollution, in particular the potential adverse effect to living systems, the application of biomarkers has come more and more to rely on a multidisciplinary approach. This is because biomarker-based investigations are being conducted at different levels of biological organization (i.e., molecular, cellular, individual, population, community, ecosystem), in diverse environments (aquatic, terrestrial, polar, and so forth) and where toxic agents, both natural and manmade are present. Furthermore, the situation is complicated by the fact that pollution-induced toxicities are modulated by both biological and physical processes. To address these concerns, individual investigators have incorporated into the design of their

biomarkers, the methods and techniques unique to disciplines outside their field (i.e., molecular biology, analytical chemistry, immunology, etc.). Unfortunately, this has created an undesirable situation. Although adequate descriptions of the methodology for most biomarkers are generally available in the scientific literature [35], use of this information by other investigators is not always easily undertaken. There are many reasons for this, the major one being a reluctance to implement a new and unfamiliar methodology. To overcome this deficiency, it is highly recommended that a transfer of biomarker technologies to potential users be initiated. This can be readily accommodated via courses designed and conducted by the experts in the field where personal knowledge and experiences are communicated directly to the recipient.

5. The importance of archiving biological and environmental samples for retrospective analysis cannot be overemphasized. A pilot study has been proposed for the Czech Republic [8] and is enthusiastically endorsed by the participants of this report.

References

1. Ministry of the Environment of the Czech Republic. (1996) Report on the Environment in the Czech Republic in 1995, Academic Publishing House, Prague.

2. Tomasek, M. (1995) Atlas of Soils of Czech Republic, Czech Geological Institute, Prague (in Czech).

3. Bouska, V. (1980) Geochemistry of Coal, Academia Prague, Prague.

4. Schnoor, J.L., Galloway, J.N., and Moldan, B. (1997) East Central Europe: an environment in transition, *Environ. Sci. Technol.* **31**, 412-416.

5. Ministry of the Environment of the Czech Republic. (1993)Environmental Laws of the Czech Republic, vol. 1, General Environmental Laws, Act On The Environment, Act No 17/1992 S.B., Prague.

6. The Czech National Council. (1992) Czech National Council Act On Environmental Impact Assessment, Act No. 244/1992 S.B., Prague.

7. State Veterinary Administration of the Czech Republic. (1992). *Information Bulletin*, vol. 1, no. 1, Series A, Liberec.

8. Kucera, J. Obrusnik, I., Fuksa, J.K., Vesely, J., Stastny, K., Hajslova, J. Mader, P., Miholova, D., and Sysalova, J. (1997) Environmental specimen banking in the Czech Republic: a pilot study, *Chemosphere* **34**, 1975-1987.

9. Czech Ministry of the Environment (Teplice Program).

10. Binkova, B., Lewtas, J. Miskova, I. Lenicek, J., and Sram, R. (1995) DNA adducts and personal air monitoring of carcinogenic polycyclic aromatic hydrocarbons in an environmentally exposed population, *Carcinogenesis* **16**, 1037-1046.

11. Binkova, B., Lewtas, J. Miskova, I., Rossner, P., Cerna, M. Mrackova, G., Peterkova, K., Mumford, J. Meyer, S. and Sram, R. (1996) Biomarker studies in Northern Bohemia, *Environ. Health Perspect.* **104**, 591-597.

12. Sandhu, S.S., Lower, W.R., de Seres, F.J., Suk, W.A., and Tice, R.R. (1990) *In Situ* Evaluation of Biological Hazårds of environmental Pollutants, Plenum Press, New York.

13. Travis, C.C. (1993) Use of Biomarkers in Assessing Health and Environmental Impacts of Chemical Pollutants, NATO ASI Series A: Life Sciences, vol. 250, Plenum Press, New York.

14. Decaprio, A.P. (1997) Biomarkers: coming of age for environmental health and risk assessment, *Environ. Sci. Technol.* **31**, 1837-1848.

15. Peakall, D.B., and Shugart, L.R. (1993) Biomarkers: Research and Application in the Assessment of Environmental Health, NATO ASI Series H: Cell Biology, vol. 68, Springer-Verlag, Heidelberg.

16. Gabriel, J., Capelare, M., Rychlovsky, P., Krenzelok, M., and Zadrazil, F. (1996) Influence of cadmium on growth of Agrocybe perfecta and two spp and translocation from polluted substrate and soil to fruitbodies, *Toxicol. Environ. Chem.*. **56**, 141-146.

17. Gabriel, J., Kofronova, O., Rychlovsky, P., and Krenzelok, M. (1996) Accumulation and effect of Cd in the wood-rotting basidiomycete *Daedalea quercina., Bull.Environ. Contamin. Toxicol.*. **57**, 383-390.

18. Mejstrik, V., and Lepsova, A. (1993) Applicability of fungi to the monitoring of environmental pollution by heavy metals, in Markert B. (ed.), Plants as Biomonitors., VCH, Weinheim, pp. 365-378.

19. Andel, P.(1985) Using bioindicators of air quality after mining uranium activities in Northern Region. Thesis. Charles University, Prague.

20. Holoubek, E., Caslavsky, J., Vancura, R., Kocan, A., Chovancova, J., Petrik, J., Drobna, B., Triska, J., and Cuddling, P. (1994) The occurrence of persistent

organic pollutants (POPs) in mountain forest ecosystem, *J. Toxicol. Environ. Chem.* **45**,189.

21. Pysek, P. (1994) Pattern of species dominance and factors affecting community composition in areas deforested due to air pollution, *Vegetatio* **112**, 45-56.

22. Mejstrik, V., and Svacha, J. (1988) Concentration of Co, Cd, Cr, Ni, and Zn in crop plants cultivated in the vicinity of coal power plants, *Sci. Total Environ.* **72**, 43-55.

23. Bencko, V., Vasilieva, E.V., and Symon, K. (1980) Immunological aspects of exposure to emissions from burning coal of high beryllium content, *Environ. Res.* **22**, 439-449.

24. Kubiznakova, J., Pospisil, J., Buricova, V., and Laznicka, P. (1989) Impact of Al, Be, Cd, Cu, Pb and Zn into the vicinity of one Czech brown coal power plant, in: J.P.Vernet (ed.), International Conference Heavy Metals in the Environment, vol 2, Geneva, pp. 504-507.

25. Valach, R., and Sedlacek, F. (1989) Die Bioindikation von Fluorals Massstab der fur den Menschen biologisch effektiven Fluoride, in Mengen- und Spurenelemente. Arbeitstagung der Agrarwissenschaftlichen und Chemischen Gesellschaft der DDR, S. pp. 1027-1036.

26. Molokov, D., Sedlacek, F., Bejcek, V., Stastny, K., and Jirous, J., (1985) Changes in the leucograms of the common vole (*Microtus arvalis*) in an industrially polluted and in a background landscape, *Ekologia (CSSR)* **4**, 337-345.

27. Kierdorf, H., Kierdorf, U., Sedlacek, F., and Erdelen, M. (1996) Mandibular bone fluoride levels and occurrence of fluoride induced dental lesions in populations of wild red deer (*Cervus elaphus*) from Central Europe, *Environ. Pollut.* **93**, 75-81.

28. Mankovska, B. (1978) The contents of Pb, Cd, and Cl in needles of Picea excelsa L. caused by air pollution of motor vehicles in *High Tatras, Biologia (Bratislava)* **10**, 775-780.

29. Stary, P., and Kubiznakova, J. (1987) Content and transfer of heavy metal air pollutants in populations of Formica spp. wood ants (*Hymenoptera formicidae*), *Zeit. Angew. Entom.* **104**, 1-10.

30. Stary, P., Kubiznakova, J., and Kindlman, P. (1989) Heavy metal traffic pollutants in small eggar, *Eriogaster lanestris* (L), (Lepidoptera, Lasiocampidae), *Ekologia (CSSR)* **8**, 211-218.

31. Bohac, J. (1995) The effect of air pollution and forest decline on epigeic staphylinid communities in the Giant Mountains, *Acta Zool. Fennica* **196**, 311-313.

32. National Research Council. (1987) Biomarkers in reproductive and developmental toxicology, *Environ. Health Perspect.* **74**, 1-199.

33. Timbrell, J.A., Draper, R., and Waterfield, C.J. (1996) Biomarkers in toxicology: new uses for some old molecules, *Biomarkers* **1**, 1-11.

34. Shugart, L.R., McCarthy, J.F., and Halbrook, R.S. (1992) Biological markers of environmental and ecological contamination: an overview, *J. Risk Analysis* **12**, 352-360.

35. Hugget, R.J., Kimerle, R.A., Mehrle, P.M., and Bergman, H.L. (1992) Biomarkers. Biochemical, Physiological, and Histological Markers of Anthropogenic Stress, Lewis Publishers, Chelsea, MI, USA.

36. Shugart, L.R. (1996) Molecular markers to toxic agents, in M.C. Newman and C.H. Jagoe (eds.), Ecotoxicology. A Hierarchical Treatment, Lewis Publishers, Boca Raton, FL, USA, pp. 133-161.

37. Walker, C.H. (1992) Biochemical responses as indicators of toxic effects of chemicals in ecosystems, *Toxicol. Lett.* **64/65**, 527-533.

38. Guengerich, F.P. (1993) Cytochrome P450 enzymes, *Am. Sci.* **81**, 440-447.

39. Shugart, L.R., Bickham, J. Jackim, G. McMahon, G., Ridley, W., Stein, J., and Steinert, S. (1992) DNA alterations, in R.J. Hugget, R.A. Kimerle, P.M. Mehrle and H.L. Bergman, (eds.), Lewis Publishers, Chelsea, MI, USA, pp. 125-153.

40. Shugart, L.R. (1998) Structural damage to DNA in response to toxicant exposure, in , V.E. Forbes (ed.), Genetic and Ecotoxicology, Taylor and Francis, Washington DC, pp. in press.

41. Qu, S-X, Bai, C-L, and Stacey, N.H. (1997) Determination of bulky DNA adducts in biomonitoring of carcinogenic chemical exposures: features and comparison of current techniques, *Biomarkers* **2**, 3-6.

42. Randerath, K., Reddy, M.V., and Gupta, R.C. (1981) 32P-labeling test for DNA damage, *Proc. Natl. Acad. Sci. USA* **78**, 6126-6129.

43. Pfau, W. (1997) DNA adducts in marine and freshwater fish as biomarkers of environmental contamination, *Biomarkers* **2**, 145-151.

44. Rand, G.M. (1995) Fundamentals of Aquatic Toxicology: Effects, Environmental Fate, and Risk Assessment, 2nd Edition, Taylor and Francis, Washington DC.

45. Tillitt, D.E., and Giesy, J.P. (1991) Characterization of the H4IIE rat hepatoma cell bioassay as a tool for assessing toxic potency of planar halogenated hydrocarbons in environmental samples, *Environ. Sci. Technol.* **25**, 87-92.

46. Kennedy, S.W., Lorenzen, A., and Norstrom, R.J. (1996) Chicken embryo hepatocyte bioassay for measuring cytochrome P450 1A1-based 2,3,7,8-tetrachlorodibenzo-p-dioxin equivalent concentrations in environmental samples, *Environ. Sci. Technol.* **30**, 706-715.

47. Bitton, G., Rhodes, K., and Koopman, B. (1995) Short-term toxicity assay based on daphnia feeding behavior, *Water Environ. Res.* **67**, 290-293.

48. Microbics Corporation. (1992) MicrotoxTM Manual. A Toxicity Testing Handbook. Carlsbad, CA, USA.

49. Quillardet, P., Huisman, O., D'Ari, R., and Hofnung, M. (1982) SOS Chromotest, a direct assay of induction of an SOS function in Escherichia coli K-12 to measure genotoxicity, *Proc. Natl. Acad. Sci. USA* **79**, 5971-5975.

50. DiGiulio R.T., and Monosson, E. (1996) Interconnections Between Human and Ecosystem Health, Chapman and Hall, London.

51. Kubiznakova, J., (1986) Trace metals from coal-fired power station: derivation of an average data base from assessment study of the situation in the Bohemia, in J. Spaleny (ed.) Annual report 1986, ILE CAS, *Ceske Budejovice*, pp. 100-113.

52. Kubiznakova, J., (1991) Atmospheric pollution by Cd, Hg and Pb, in J. Cibulka (ed.) Tansport of Cd, Hg and Pb into Biosphere, Academia Praha, pp. 23-27 (in Czech).

3.7. Appendices

3.7.1. SOIL POLLUTION IN NORTHERN BOHEMIA

LUBOS BORUVKA.
Department of Soil Science and Geology, Czech University of Agriculture, Prague, Czech Republic

Situation in legislation concerning soil pollution in the Czech Republic.

Czech limit values for hazardous elements in soils [1] are set for total content determined in aqua regia digestion, and for content extractable with cold nitric acid ($2mol.L^{-1}$) which is a standard method in the Czech Republic. Till now, criteria of soil pollution with hazardous elements in the Czech Republic are based on soil division into just two groups: sandy soils and other soils. This division is very rough, because behavior of both organic and inorganic pollutants differs according to soil units that have different properties. There are attempts to form more groups of soils with similar properties and to set separate criteria for each group [2]. Moreover, a multi-grade system of soil loading limits has been investigated, including limit of anthropogenic contamination, limit of ecotoxicologically detected loss of soil multifunctionality, limit of threatening of the food chain by transfer of pollutants into crops, and limit of direct impact on human health and remediation needs [3]. However, these approaches have not been officially accepted yet.

The use of total content or content extractable with 2M HNO_3 is questionable as it does not describe the mobility or availability of these elements. It would be perhaps more appropriate to know the forms in which hazardous elements are present in soils. Their speciation is influenced by soil pH, humus content and quality, mineral composition, sorption capacity, microbial activity and other soil properties. One scientific approach considers the content of hazardous elements in soil solution as decisive. However, this value depends on soil moisture content. For this reason, it could be better to choose a mild extractant for characterization of easily or potentially available element content in soil solid phase (e.g. CaC_{12}, $Ca(NO_3)_2$, $NaNO_3$, EDTA and other reagents are used for this purpose in other countries).

Limit values are set also for other polluting materials in agricultural soils [1], for example (in mg.kg-1): B - 40, Br - 20, F - 500, sulphatic S - 2. For limits of organic pollutants, Dutch "A" values [4] were imported by the Ministry of Environment (selected numbers):

- total monoaromatic hydrocarbons and their derivatives - 0.3 $mg.kg^{-1}$
- total PAH - 1.0 $mg.kg^{-1}$
- total chlorinated aliphatic hydrocarbons - 0.1 $mg.kg^{-1}$
- PCB - 0.01 $mg.kg^{-1}$
- total organic chlorinated pesticides - 0.1 $mg.kg^{-1}$.

Recently, limit values specific for the Czech Republic for organic pollutants derived as upper limits of background variability have been investigated [5].

Soil conditions in Northern Bohemia.

The most frequent soil type in Northern Bohemia is Cambisol, relatively large areas are covered also with Chernozems, Luvisols, Planosols, Rendzinas, Podzols and Fluvisols. Regosols, Gleysols, Phaeozems and Histosols are present rather locally on small areas. Brief characteristic of main soil types in relation to pollution and acidification follows.

Cambisols are formed on non-calcareous rocks of differing character with the consequence that soil properties vary very much, ranging from very acidic Dystric Cambisol to relatively fertile Eutric Cambisol. Due to natural acidity and intensive weathering, Dystric Cambisols belong to most problematic soils from the point of view of further acidification, labile aluminum toxicity and heavy metal mobility and bioavailability. This is even more the case with the Podzols, soils formed usually on acidic rocks under spruce and pine forests at relatively higher altitudes.

Chernozems are formed on calcareous rocks, most often on loess. They are characterized by deep humus horizon with humic substances of high molecular weight and strong bindings between humic substances and mineral particles. Microbial activity of Chernozems is fairly high. These soils are neutral or moderately alkaline with high sorption capacity and with good buffering capacity. Most pollutants are strongly bound in Chernozems and acidification usually does not present a big problem.

Luvisols (Orthic Luvisol, Albic Luvisol, Stagno-gleyic Luvisol) are formed in more humid conditions (usually also in higher altitudes) than Chernozems. The principal pedogenic process is leaching of colloids into a deeper horizon. This can lead to a decrease in sorption sites and buffering capacity in topsoil, especially on Albic Luvisols. However, Luvisols are not a problematic soil type in relation to acidification and metal release.

Rendzinas are formed on calcareous rocks. Their properties are inferior to those of Chernozems. Humic substances are less developed, however, quite high pH values can prevent soil acidification and heavy metal mobilization.

Stagnosols and Stagno-gleyic Planosols, developed in conditions of periodical wetting of higher soil layers due to an impermeable (clay) layer in the depth, are rather poor soils with low quality organic matter and the possibility of reducing conditions. They are sensitive to acidification and metal release and migration.

Extension of Fluvisols is limited to alluvia of streams and rivers. They are considered as the most polluted soils, but more due to sedimentation from water than to atmospheric deposition.

A special case is represented by anthropogenic soils on reclaimed areas. They were formed artificially by laying down layers of different materials (some rocks, clays, topsoils removed from other places, mixtures of rock materials with some organic matter etc.). These soils have been widely studied at the Department of Soil Science and Geology of the Czech University of Agriculture in Prague [6,7]. It was found that these soils can exhibit similar properties to naturally formed soils, e.g., if calcareous material is used we can suppose that they will evolve in a similar way to Rendzinas (more probably than Chernozems); if a clay layer is covered by a loamy material, the soil can resemble Orthic Luvisol. However, all these soils are still artificial and their full evolution will need a long time. In some cases, these artificial soils are less polluted than natural soils since they were not exposed to atmospheric emissions and depositions for such a long time as natural soils were.

Situation of soil pollution in Northern Bohemia.

Soil pollution with hazardous elements in the whole Czech Republic has been widely studied by Central Institute for Supervising and Testing in Agriculture (UKZUZ) [8,9]. It was found that percentage of agricultural soils where content of hazardous elements exceeds limit values is generally very low, even in the North-Bohemian region. Limit value of Cd was surpassed at 1.2 % of samples from Northern Bohemia (in the whole Czech Republic it was 1.3 %), for other elements were obtained following percentages: Hg 0.2, Pb 0.4, Be 0.3, Cu 1.1, Ni 0.2, V 0.2, Zn 0.4.

Special attention to soil pollution with both organic and inorganic pollutants in Northern Bohemia has been paid by the Research Institute of Land Reclamation and Soil Protection in Prague [8,9]. By means of retrospective monitoring, no significant increase in risk element content in soils during last 30 years was found [10]. Contents of hazardous elements higher than limit values were found only locally for As, Be, less for Cd, Zn and Pb [11]. It was concluded, that Cd, Be and As are mainly of anthropogenic origin, while higher amounts of Co, Cr, Cu, Mn, Ni and V are mainly of geogenic origin (Podlesakova in Anonymous, 1995). Areas of prevailing anthropogenic and geogenic loads, respectively, were separated [12] Content of hazardous elements in feed crops and cereal grains reflected concentration of these elements in soils, however, they did not reach respective maximum values even when grown on polluted soils [13].

Presence of more than twenty persistent organic xenobiotic substances in soils of Northern Bohemia has been also studied [8,14,15]. Spatial distribution of PAHs, mainly fluoranthene, in soils with maxima in districts of Most and Teplice was similar to that of As and Be, which confirmed their origin in combustion of fossil fuels. Contamination by mononuclear aromatic hydrocarbons is concentrated in the Most district due to the petrochemical industry. PCBs are present especially in soils of highly industrialized districts of Decin and Usti nad Labem. However, the "B" values of Dutch system of limits were not exceeded for any of the studied substances.

Soil acidification.

Soil acidification due to atmospheric emissions especially of SO_2 increases mobility and bioavailability of most hazardous elements. Acidification causes also changes in microbial activity and composition of microbial species, decreasing amounts of bacteria and increasing population of undesirable fungi. However, one of the most serious effects of soil acidification is an increase of labile aluminum content in soil solution [16]. Aluminum is not a pollutant since it originates mainly from natural soil minerals, especially clays and oxides. In normal conditions, i.e. at pH higher than about 4.5, aluminum is present in insoluble forms. As soil pH decreases, aluminum is released to the soil solution in ionic form Al^{3+} and in the form of soluble both organic and inorganic complexes of high toxicity. This effect is more important in forest soils, especially under conifers where conditions are already naturally more acidic.

Influence of soil acidification on toxic forms of aluminum in soil solution in soils of Northern Bohemia has been studied at the Department of Soil Science and Geology of the University of Agriculture in Prague, first on agricultural land [17], when a proposal of limit value of labile Al^{3+} determined according to James et al. [18] after extraction with 1M KCl was given for agricultural soils (0.3 mg.kg-1). Now a project on forest soils and acidification is being worked on. Limit value for Al^{3+} in forest soils has been estimated, however, the most commonly used value is derived from the ratio $Ca+Mg+K/Al^{3+}$ [19], which would be higher than 1.5 for spruce forest. In the Krusene Hory Mountains it was determined to be lower than 1 [20].

References

1. Anonymous (1994) Vyhlaska MZP c. 13/1994Sb., Act by which some details of agricultural land preservation are set. Ministry of Environment of the Czech Republic, Prague.

2. Podlesakova, E., Nemecek, J., and Halova. G. (1996) Proposal of soil contamination limits for potentially hazardous trace elements in the Czech Republic, *Rostl. Vyr.* **42**, 119-125.

3. Podlesakova, E., and Nemecek, J. (1996) Soil contamination and pollution criteria, *Rostl. Vyr.* **42**, 357-364.

4. Anonymous (1988) Technisch Inhoudelijk, Deel II. Leidraad Bodemsanering. San intgeverij, s-Gravenhoge: p. 27.

5. Nemecek, J., Podlesakova, E., and Pastuzskova, M. (1996) Proposal of soil contamination limits for persistent organic xenobiotic substances in the Czech Republic *Rostl. Vyr.* **42**, 49-53.

6. Cibulka, J., Kozak, J., Zeleny, V., Bartak, M. (1992) A preliminary report on the results of pedological, botanical and zoological evaluation of the Bilina Mines area. UNICO AGRIC, Czech University of Agriculture, Prague

7. Kozak, J., Valla, M., Vacek, O., Matula, S., and Donatova, H. (1994) Pedological study of selected dumpsites of the Bilina Mines. Czech University of Agriculture, Prague.

8. Anonymous (1995) 1994. Yearbook of the Ministry of Agriculture of the Czech Republic, Ministry of Agriculture, Prague.

9. Anonymous (1996) 1995. Yearbook of the Ministry of Agriculture of the Czech Republic, Ministry of Agriculture, Prague.

10. Nemecek, J., and Podlesakova, E. (1992) Retrospective experimental monitoring of hazardous elements in soils of the Czech Republic, *Rostl. Vyr.* **38**, 433-436.

11. Podlesakova, E., Nemecek, J., and Vacha, R. (1994) Contamination of soils in the North-Bohemian region by hazardous elements. *Rostl. Vyr.* **40**, 123-130.

12. Nemecek, J., Podlesakova, E., and Vacha, R. (1996b) Geogenic and anthropogenic soil loads. *Rostl. Vyr.* **42**, 535-541.

13. Nemecek, J., Podlesakova, E., and Pastuzskova, M. (1994) Contamination of feed crops and cereal grains in the North-Bohemian emissions impact region, *Rostl. Vyr.* **40**, 555-565.

14. Nemecek, J., Podlesakova, E., and Firyt, P. (1994a) Contamination of soils in the North-Bohemian region by organic xenobiotic compounds, *Rostl. Vyr.* **40**, 113-121.

15. Podlesakova, E., Nemecek, J., and Vacha, R. (1997) Contamination of soils with persistent organic xenobiotic substances in the Czech Republic. *Rostl. Vyr.* **43**, 357-364.

16. Boudot, J.-P., Becquer, T., Merlet, D., and Rouiller, J. (1994) Aluminum toxicity in declining forests: a general overview with a seasonal assessment in a silver fir forest in the Vosges mountains (France), *Ann. Sci. For.* **51**, 27-51.

17. Kozak, J., Boruvka, L., Kristoufkova, S., and Mach, J. (1994b) Toxic forms of Al in soil solution and possibilities of their elimination. Final report of the project Z-783-01-03, Czech University of Agriculture, Prague.

18. James, B.R., Clark, C.J., and Riha, S.J. (1983) An 8-hydroxyquinoline method for labile and total aluminum in soil extracts, *Soil Sci. Soc. Am. J.* **47**, 893 - 897.

19. Sverdrup, H. and Wrafvinge, P. (1993) The effect of soil acidification on the growth of trees, grass and herbs as expressed by the Ca+Mg+K/Al ratio, Report 2:1993, Lund University, Department of Chemical Engineer II, p. 177.

20. Kubiznakova, J. (1997) Application of model FORCYTE for prediction of abatement strategy of SO_2 emission, in A.P.Simon H., A. Grossinhi, J. Brechler,

J. Kubiznákovà, K. Juda-Resler, and J. Zvozdiak (eds.), Emission Abatement Strategy and Environment (EASE).The interdisciplinary study of environmental damage due to sulphur emissions in BT, Final Report for CEC, Project CIPA-CT 93-0071 DG-12HSMU, March 1997, ICSTM, CET, London, UK.

3.7.2. VERTEBRATES AS BIOINDICATORS

The Use of Vertebrates for Bioindication and Ecological Monitoring in Europe's "Black Triangle"

KAREL STASTNY and VLADIMIR BEJCEK
Faculty of Forestry, Czech University of Agriculture, Prague 6 - Suchdol, Czech Republic

Introduction.

The term "black triangle" of Europe refers to its most heavily polluted part between the Czech Republic, Poland and Germany. The condition of the environment in the Czech Republic is indicated by several selected data: with the value of SO_2 emissions attaining 22.4 $t.km^{-2}$.years-1 it occupies the first place in Europe: at the same time, this exceeds 188 $t.km^{-2}.yr^{-1}$ in the most heavily polluted territory of the North Bohemian Region. Although the highest admissible daily concentrations of SO_2 are 150 ng/m^3, they commonly exceed 1000 ng/m^3 in the North Bohemian Region. Besides SO_2 emissions, thousands of tonnes of other noxious substances are emitted by coal burning and chemical production. The emissions exert their heaviest impact on woodlands in the mountainous part of the Krusné Hory Mountains. Here, coniferous forests were completely destroyed in the course of the past 20-25 years. The rate at which these forests are being destroyed in the Czech Republic is continuously increasing: their destruction took 15-20 years in the Jizerské Hory and Krkonose Mountains, but only 12-15 years in the Orlické Hory Mountains, and it was almost instantaneous in a part of the Beskydy mountains..

Further landscape devastation is brought about by mining. Immense surface coal mines and extensive spoil banks, connected with the destruction of thousands of hectares of fields, meadows, pastures and even human settlements, occupy, at present, 250 km^2, i.e. 1/4 of the area of the North Bohemian brown coal basin. They have virtually turned into a "moon landscape". NW Bohemia is generally considered an ecological catastrophe. However, for ecologists this area is really the laboratory in Nature, albeit in a negative sense.

These and other changes have necessarily affected the gene pool in the whole region, including that of vertebrates. The biochemical and somatic changes, development and population changes have been studied there.

Biochemical and Somatic Changes.

The common vole (*Microtus arvalis*) was chosen as an object of the investigations. It has a number of advantages for bioindication: owing to its autecology, it is exposed to the impact of external conditions for a sufficiently long time; its metabolic rate is high in comparison to that observed in farm animals or in man; since its generation time is short, it is possible to trace genetic changes; it lives within a small territory and is bound by its food to this area (herbivorous species); it is possible to obtain a sufficient quantity of material - whenever and wherever required; it is closely related to the laboratory mouse or rat and therefore the results obtained on them may be utilized in both human and veterinary medicine as well [1].

Relation of the spleen weight to the total body weight.

The high frequency of hypertrophy of the spleen was found in polluted areas. In the region of Trebon the proportion of spleen weight ranged from 0.1-1 % of the body weight. In the region of Most which represents heavily polluted area, the proportion of spleen weight was higher than 1 % in 15 % of individuals (in three of them even higher than 3.5 %). The hypertrophy of spleen is always a pathological phenomenon accompanying a number of pathological processes in the organism. Thus, the increased frequency of occurrence of hypertrophic spleens in the population of the common vole indicates an increased frequency of certain diseases (e.g. the hypertrophy of about mentioned organs in laboratory mice is known in the course of experimentally induced leukemia). The results indicate the potential use of the common vole as an auxiliary bioindicator of the industrial pollution [1].

Leukocyte counts.

The best indication is the number of leukocytes. The normal number is of about 4300/mm^3, however in polluted areas it is more than 10 000. Fluctuation of leukocyte counts often represents an accompanying feature of diseases. High numbers of leukocytes - so called leukocytosis - is most frequently provoked by acute infection or intoxication. Leukocytosis is characteristic of lead, mercury and carbon monoxide poisoning. Some components of industrial emissions, e.g. SO_2, F, CO, if affecting the organism, cause similar reaction - an increase in the number of leukocytes in the blood circulation [2].

White blood cell profile.

From the white blood cell profile the total count of leukocytes and the absolute count of lymphocytes appear to be the most suitable parameters for the purposes of the bioindication of industrial pollution. At all times of collection (November - September), the difference between the region of Most and Trebon was highly statistically significant. And the explanation: the result of the common function of leukocytes is the coordinated response of the organism to various microörganisms and harmful substances. In the region of Most, the influence of industrial emissions was evidently the main reason for (either directly or through mediation) the change of the white blood cell profile [2]. Later, similar results were observed by doctors of medicine for the human population - leukocytosis, leukemias. (While we could write about our results because they were obtained on animals, their results were kept secret and it was strictly forbidden to publish them.)

Under the category "biochemical and somatic changes" other phenomena were studied: changes in haematological values in the common vole [3] and common hare (*Lepus europaeus*), changes in the chemistry of the blood serum, the chemistry of the eye corneas and the chemistry of urine in the common hare. Also studies were changes in the level of elements, namely heavy metals in the hair, feathers and droppings of mammals and birds, and changes in the quality of roe deer antlers (*Capreolus capreolus*) [4].

Development and Population Changes.
Devastation of the landscape, deforestation of mountains and other changes have inevitably affected bird and mammal populations. The monitoring of bird populations in the Czech Republic is a separate project that has been under way for 15 years. Besides that, however, it has been found very desirable to monitor in greater detail even the individual large territories in which, due to specific situations, the population trends are often diametrically different.

Bird communities of spruce forests affected by industrial emissions in the Krusné Hory mountains.
The succession of bird communities was studied in spruce forests heavily affected by industrial emissions. The qualitative, quantitative and structural characteristics (number of species, density, species diversity, evenness) are similar in mature (40-80 years old) stands of Norway Spruce (*Picea abies*) and in young (10-15 years old) stands of the same species. They fall sharply with progression from the relatively healthy spruce forest of various ages through the heavily-damaged ones to the totally dead stands without branches. On the other hand they increase from the clearings to freshly afforested areas with young stands of Norway and Blue Spruce (*Picea pungens*), the tree species used for the economic restoration of forests in the Krusné Hory mountains. The reduction in numbers of some bird species, especially those which feed on needles (goldcrests, tits) can indicate the decrease of needles caused by industrial emissions; a rise in numbers of other species feeding on bark-beetles (woodpeckers) can indicate increase of bark-beetles on injured trees [5].

Bird communities on spoil banks after surface brown coal mining.
The primary succession (the succession which we can meet on islands after the eruption of a volcano) was studied in six successional stages of spoil banks after surface brown coal mining: the initial stage, 2-3 years after heaping, the stages 6 and 25 years afterwards, the stages 6 and 25 years after reforestation and the stage of a full-grown oak forest at which the successional series on spoil banks finally should arrive. All qualitative, quantitative and structural characteristics increase from the initial stage to a full-grown oak forest except the decrease in the stage 6 years after forest reclamation. The forest reclamation at the beginning inhibits somewhat the succession of bird communities (decrease both of the number of species and of the total density) but later it appears to be a highly positive factor accelerating the succession towards the final community [6].

Communities of small terrestrial mammals on the spoil banks.
The primary succession of small mammals on the spoil banks after surface brown coal mining was studied on seven successional stages: 2-4 years, 7-9 years, 13-15 years, 19-21 years old, non-reclaimed spoil banks, and 10-12 years, 18-20 years after silvicultural reclamation, and in an acidophilic oak forest about 150 years old outside the spoil banks. The first small mammal species invading the spoil banks is the long-tailed field mouse (*Apodemus sylvaticus*). It quickly migrates onto the freshly dumped spoil banks and represents the only small mammal species in the first years after dumping. In the third year there occurs the common vole (*Microtus arvalis*), and in the fourth year the common shrew (*Sorex araneus*). Very sensitive indicators of succession trends are the changes of dominance, density, biomass, production of individual species. The values of density, production and consumption of plant mass by the small mammal community increased until the stage 19-21 years old. After the silvicultural reclamation they decreased rapidly until the climax. The determining species of these changes was *Microtus arvalis*. The silvicultural reclamation is a very effective method of biological control of *M. arvalis*: at the same time it does not enable the population expansion of any other small mammal species. It is important from the epidemiological viewpoint (*M. arvalis* is a reservoir of zoonoses) as well as from the viewpoint of protection of the surrounding crop cultures (spoil banks - the refuges of *M. arvalis*). The small mammal communities indicate the capacity of both non-reclaimed and by silvicultural methods reclaimed spoil banks to develop towards the function formations in the landscape system of the North Bohemian Brown Coal Basin [7].

In the part "development and population changes" other populations and communities were studied: dependence of nesting density of the house martin (*Delichon urbica*) on environmental pollution, dependence of population indicators on pollution and on the anthropic disturbance of the environment, changes in physical development of the common hare [4].

Acknowledgment.
Financial support was given by the Grants Agency of the Czech Republic, grants numbers 206/97/0850 and 206/97/0771.

References

1. Bejcek, V. Jirous, J., and Stastny, K. (1982) The difference in the spleen weight in the common vole (*Microtus arvalis*) between industrially exposed landscape and background landscape. Proceeding of the conference of young scientists from branch of environment. Brno, June 8-9, 1986, p. 66-72.

2. Bejcek, V. , Hejlková, D., Jirous, J., and Stastny, K. (1982) The difference of the leucocyte counts in the common vole, Microtus arvalis from the landscape heavily burdened by industry and from a background landscape. In: Proc. of the IVth International conference: Bioindicators deteriorisationis regions. Liblice u Prahy, June 28-July 7, 1982.

3. Sedlácek, F., Stastny, K., and Bejcek, V. (1988) Hematocrit values of common vole (Microtus arvalis) from industrially polluted and control areas, Sborn.Okr.muz. v Moste **85**, rada prírodovedná, 55-59.

4. Nováková, E. (1987) Use of wild birds and mammals for bioindicators, biodiagnostics and ecological monitoring, VSZ Praha.

5. Stastny, K. and Bejcek, V. (1985) Bird communities of spruce forests affected by industrial emissions in the Krusné Hory (Ore Mountains), in K. Taylor, R.J. Fuller and P.C. Lack (eds.), Bird Census and Atlas Studies, Proceedings of the VIII International Conference on Bird Census and Atlas Work. BTO Tring., pp. 243-253.

6. Bejcek, V., and Stastny, K. (1984) The succession of bird communities on the spoil banks after surface brown coal mining, *Ekologia Polska* **32**(2), 245-259.

7. Bejcek, V. (1988) Communities of small terrestrial mammals on the non-reclaimed and silviculturally reclaimed spoil banks in the Most Basin, VSZ Praha.

3.7.3. INTEGRATED FIELD METHODS IN SOIL ECOLOGY

Integrated Field Methods in Soil Ecology as Strategies to Determine the Impact of Chemicals on Soil Processes

WERNER KRATZ
Fachbereich Biologie FU Berlin FB Biologie WE5, Grunewaldstrasse 34, D-12165 Berlin, Germany

Introduction.
Integrated field methods in soil ecology are often used to determine the impact of environmental chemicals on soil organisms and soil processes. Soil biological methods in the field are increasingly aiming to determine not only structural parameters in a specific biocenosis but also functional processes like litter decomposition, feeding activity of soil animals and nitrification. The strategy of integrating functional investigations into soil biology at the ecosystem level has the advantage of being able to explain processes of material change and questions of ecosystem management and soil protection. In terrestrial ecotoxicology where effect determinations are required these methods are very helpful. Through the measurement of selected soil processes, such as litter degradation, cellulose decomposition etc., the effects of environment chemicals can be demonstrated [1,2]. Presently, the important integrating methods in soil biology are litter-bag methods [3,4], or litter-container methods [5,6], bait-lamina test [7,8,9] and the mini-container method [13]. Under the "integrating methods" those well

established soil processes that involve mainly the participating soil microorganisms and soil invertebrates, are investigated. These methods do not allow specific interpretation of the influence of the participating group of organisms and their activities. The "integrating methods" used so far in soil biology examine primarily the biological activity in the litter layer and the upper soil layers, which are strongly influenced by litter decomposition. Depending on the thickness of the litter layer, the upper mineral soil layer can also be assesses by the bait-lamina test. The investigation of root degradation in litter-bags is best done in the mineral soil layer. Investigations using litter-bags, litter-containers and mini-containers are very laborious.

The Törne method, called the mini-bait test (Bait-lamina test), has the advantage of providing a screening method which is quick and inexpensive to evaluate biological decomposition processes [10]. Currently the bait-lamina test system, modified by terra protecta GmbH, Berlin, Germany, has wide application using a variant of PVC-stripes and a standard substrate mixture containing cellulose powder, bran flakes (70:30, m/m) and traces of active coal. Through the standardization of the bait-lamina test system it is possible to compare the soil biological decomposition process and the feeding activity of soil organisms between different ecosystem situations (e.g. soil pollution, management measures) [9]. Using the bait-lamina test it is also possible to explain the performance of soil invertebrates in conjunction with soil microorganisms on the nutrient cycle of terrestrial ecosystems. The bait-lamina test system can also supply considerable biometric data. Together with the decomposition activity and feeding performance of soil organisms in undisturbed soil systems, the bait-lamina test also can be employed to determine the influence of chemicals [11] and ecosystem management measures [12] on the decomposition process in soils.

Since 1990, 46 publications on the bait-lamina test system (most of them in German) have appeared. In 1994 a first workshop on bait-lamina test was organized at the Technical University of Braunschweig (AG Prof. O. Larink), Germany with the objective of promoting information exchange amongst users of the test system. A second workshop on the same topic was organized in September 1996 at the University of Mainz (AG Prof. G. Eisenbeis), Germany with the aim of discussing the selective applications of the bait-lamina test.

General Description of Method.

In the bait-lamina test, small bait portions are fixed in small holes bored in PVC strips which are then exposed to biogenic decomposition processes in the soil. Soil invertebrates and soil microorganisms progressively degrade the bait placed in the soil substrate. It is assumed that disappearance of material is brought about by microbial processes and microbiogenic metabolism which is mainly influenced by the feeding activity of soil invertebrates. Exposure time depends on location and is related strongly to the soil moisture content [11], but is usually between 10 (in soils with good moisture conditions and a normal soil biological activity) and 20 days (dryer soils). Evaluations of the exposed baits are made after washing the strips carefully under flowing tap water and examining on a lighted bench, where differentiation is made only between "bait eaten" (light falls through the bait) and "bait not eaten". Comparison of results on different sites can be made using Mann-Whitney U-test.

Advantages and Disadvantages of the Bait-Lamina Test.

With the bait-lamina strips:

a) The procedures are very simple and cheap.

b) Taxonomic knowledge is not required to conduct the test, although a basic knowledge of ecological field work methodologies is required (e.g. determination of soil moisture, micro-climatic parameters).

c) A large amount of data is available in a short time for biometric evaluation with the potential for standardization according to GLP criteria.

d) The simple experimental design of the test does not require disturbance of vegetation and soil.

The bait-lamina test has further advantages in comparison to other known decomposition assessment methods:

a) Experimental error can be determined quickly with little work by placing "blind value bait-lamina" into the soil and immediately removing and assessing the loss of bait material.

b) There are possibilities of selection and modification of bait substances and replicating the experiment several times.

c) Each test gives two sets of results to be evaluated - one due to the local plot area and layer distribution differences occurring between two stochastical independent random samples - and the other comparing different systems in a landscape.

d) The seasonal decomposing activity changes can be determined by placing bait-lamina units successively for long periods of time.

The main disadvantages of the bait-lamina test system are that only powdered substances are suitable as bait material, there is no graded response, and the possibilities of modification are limited.

Application Areas of Bait-Lamina Test System.

The bait-lamina test at present has been utilized in different situations. The important applications are :

a) In ecotoxicological field and laboratory experiments, the bait-lamina test has been used to document the overall effects of chemicals on soil organisms. The pesticides registration authority in Germany (Biologische Bundesanstalt Braunschweig) is presently examining whether a bait-lamina test could be included in registration procedure.

b) In forest ecosystems the bait-lamina tests are employed to monitor deposited pollution gradients (along highways, forests), and effects of pesticides and liming of forest. The differences in feeding activity in small areas (litter, grass, soil chemistry etc.) as well as large cultivation areas can be evaluated by the bait-lamina test and investigated further.

c) In agroecosystems the influence of different cropping pattern on the feeding activity of soil arthropods can be investigated. Also the influence of soil improvement measures (fertilizers, manure etc.), soil workup measures, soil compression and fungicide application on soil biological activity can be determined.

d) In urban ecosystems where the questions of effects of composts, pollution deposition gradients (roads etc.) and polluted soils (cable fire, sewage fields) can be of importance, the bait-lamina test can be employed.

e) The bait-lamina test can also be utilized in peat ecosystem reclamation (denaturalization), in extreme space research, research on raw material growth (Miscanthus), nature conservation (water-meadow landscape) and in research on global climate change.

Outlook.

To determine the potential influence of pollutants on soil ecosystems, the side effects of chemicals should be determined and evaluated along with the biocenosis level at the soil dynamic process. The changes in the processes which take place caused by chemicals give data which, in relation to the influence on a certain environmental compartment, is interpretable. As a result processes, such as degradation of the organic matrix, soil respiration, microbial growth and microbial turnover, enzyme activity of soil, the rate of nutrient change should be selected as well as the feeding activities of soil arthropods. The experience and the necessary standardization of bait-lamina test system gives soil biologists a method which can be employed in areas of applied and theoretical field ecology.

References

1. Bolger, T., and Heneghan, L. (1996) Effects of acid rain components on soil microarthropods: a field manipulation, *Pedobiologia* **40**, 413-438.

2. Edwards, C.A., Subler, S., Chen, S.K., and Bogomolov, D.M. (1996) Essential criteria for selecting bioindicator species, processes, or systems to assess the environmental impact of chemicals on soil ecosystems, in: Straalen, N.M., Kriovolutsky, D.A. (eds.), Bioindicator Systems for Soil Pollution, Kluwer Academic Publishers, Amsterdam, pp. 67-84.

3. Berg, B., Berg, M., Bottner, P., Box, E., Breymeyer, A., Calvo de Anta, R., Couteaux, M., Gallardo, A., Escudero, A., Kratz, W., Madeira, M., Meentemeyer, V., Munoz, F., Piussi, P., Remacle, J., and Virzo de Santo, A. (1991) Litter mass loss in pine forests of Europe: relationships with climate and litter quality, in: A. Breymeyer (ed): Proceedings of Scope Seminar. Geography of Carbon Budget Processes in Terrestrial Ecosystems. Szymbark, Aug. 17-23, pp. 81-109.

4. Berg, B., Berg, M., Box, E., Bottner, P., Breymeyer, A., Calvo de Anta, R., Couteaux, M., Gallardo, A., Escudero, A., Kratz, W., Madeira, M., Mälkönen,

E., McClaugherty, C., Meentemeyer, V., Munoz, F., Piussi, P., Remacle, J., and Virzo de Santo, A. (1993) Litter mass loss rates in pine forests of Europe and Eastern United States: some relationships with climate and litter quality, *Biogeochemistry* **20**, 127-159.

5. Kratz, W. (1991) Cycling of nutrients and pollutants during litter decomposition in pine forests in the Grunewald, Berlin, in: Nakagoshi, N., F.B. Golley (eds.) Coniferous forest ecology from an international perspective, SPB Acad. Publ., The Hague, pp.151-160.

6. Kratz, W. (1992) Ecosystem study in a central European pine forest - studies on the cycling of elements during litter decomposition in pine forests in the Grunewald, Berlin, in Teller, A., Mathy, P., and Jeffers J. (eds.), Responses of forest ecosystems to environmental changes, Elsevier Applied Science, London, pp. 976-977.

7. Törne, E.(1990) Assessing feeding activities of soil-living animals. I. Bait-lamina-tests, *Pedobiologia* **34**, 89-101.

8. Törne, E. (1990) Schätzungen der Freßaktivitäten bodenlebender Tiere. II. Mini-Köder-Test, *Pedobiologia* **34**, 269-279.

9. Larink, O., Kratz, W. (1994) Köderstreifen-Workshop in Braunschweig - ein *Resümee, Braunschw. Naturkdl. Schr.* **4**, 647-651.

10. Hoffmann, H., Kratz, W., and Neinaß, J. (1991) Der Ködermembrantest - ein Screening-Test zur Ermittlung tierischer Freßaktivität in *Böden. Mitteilgn, Dtsch. Bodenkdl. Gesellsch.*, **66**, I, 507 - 510.

11. Larink, O. (1993) Bait lamina as a tool for testing feeding activity of animals in contaminated soils, In: Donker, M.H., Eijsackers, H., and Heimbach, F.(eds.), Ecotoxicology of soil organisms, Lewis Publishers, Baton Rouge, pp. 339-345.

12. Kratz, W. (1994) Der Köderstreifen-Test - ein *Schnelltest zur Bestimmung der bodenbiologischen Aktivität nach Waldkalkungen, Braunschw. naturkdl. Schr.* **4**, 659-663.

13. Eisenbeis, G., Lenz, R., Dogan, H., and Schüler, G. (1996) Zur biologischen Aktivität von Nadelwaldböden: Messung der tierischen Freßaktivität mit dem Köderstreifen-Test sowie Bestimmung von Streuabbauraten mit dem Minicontainer-Test, *Verh. Ges. Ökol.* **26**, 305-311. (summary in English)

3.7.4. ASSESSMENT OF ATMOSPHERIC POLLUTION ON SPRUCE CANOPY.

"Inner Cycling of Spruce Canopy" as Biomarker of Atmospheric Pollution

Jana Kubiznakova, Institute of Landscape Ecology, Czech Academy of Sciences, 37005 Ceske Budejovice, Czech Republic

Forest Decline.

In Czech Forest Law No.289/1995 the multifunctional characteristics of the forest are acknowledged. National policy now subscribes to the fundamental principles that forests are a sustainable renewable resource for the benefit of generations to come, that more ecological forest management must be applied, and that forests are complex holistic interpretation of nature and its function. Some details about sustainable management of Czech forest are in Poleno [1]. Every five years on about 500 established permanent research plots (PRP), inventories of the state of health of Czech forests are conducted. Conditions such as deformity of crown, defoliation, discoloration, ring analysis, foliar and basic soil nutrient analysis are recorded in a database. The estimation of their health state is provided in accordance with international criteria. This visual determination of the level of injury in trees' canopy is given in relation to a presumption of surrounding emission situation. Increasing spruce damage by air pollution has been confirmed in relationship to altitude [2,3].

The causes of forest decline in the central part of Europe has been studied since the 1960's, however, even today it is not understood satisfactorily. The most widely observed symptoms of decline of Norway spruce (*Picea abies*) are loss of foliage, crown thinning, shedding needles, reduced resistance to drought and insects, insufficient water, nutrient deficiency, tree weakening and finally dieback of tree. Symptoms of injury have also appeared as chlorosis of the older needles, particularly on the upper needle surface. Spruce decline may be attributed to five main hypotheses: direct effect of acid pollutants, acidification (aluminum toxicity) and eutrophication (excess-nitrogen) of soils, Mg-deficiency, oxidants effects (ozone, hydrogen peroxide, UV), and climatic changes that would be included in comprehensive multistress theory.

Acid deposition and subsequent acidification of the soil is only one part of the multistress hypothesis. Also the role of photo-oxidants, especially ozone, in plant damage is now widely accepted as a contributing stressor. Hydrogen peroxide is considered the most important oxidant of sulphur dioxide (SO_2) in atmospheric water droplets, converting SO_2 to sulfate (SO_4^{-2}) at pH values <5.0. In the Czech Republic, there are very acidic throughfalls (mean pH=3.75) and bulk depositions (mean pH=4.14). Yearly mean potential total acid deposition (sum dry and wet) was estimated at about 4.15 Keq H+/ha in 1991 and 3.2 Keq H+/ha in 1994.

The risk assessment of spruce decline to acid deposition depends on many variable factors such as microclimatic, ecological and orographical condition and changes in ground level of atmosphere (air pollutants, meteorological condition and running atmospheric chemistry processes). Also important are the physiology of tree, age of tree, morphology, needle density and shape of crowns. Some details about impact of abiotic stress factors to spruce forest and its response are given in [4].

Spruce Canopy as Receptors.
The principal influence on the inner nutrient cycle in canopy (e.g., nutrient uptake or leaching of nutrients) is precipitation (wet deposition). Liquid water can readily mobilize canopy solutes. Aerial precipitation of cloud droplets (mist and fog) has the potential to remove canopy material. A fraction of the deposited matter is stored on leaf surfaces. The captured solution in the canopy undergoes chemical reaction with constituents on the leaf surface (neutralization by foliar exudates) and there may be direct nutrient uptake by the leaf when the supply is exhausted. The deposition of gaseous, aerosol and particulate matter is also enhanced by the canopy in proportion to the leaf area index. Many oxidants play an important role in the acidification of rain-, cloud-, and fog–water. Ozone (O_3) and hydrogen peroxide (H_2O_2) present in acidic fog as precursor of pollution can affect sensitive compartment of spruce canopy (needles). Long-term exposure of Norway spruce by ozone during the growing season may increase transpiration and cause stomatal sluggishness many months later. Low pH value in deposition enhances nutrient leaching from soil and subsequently to cause depletion of foliar nutrient content.

Both wet and dry deposition are important vectors of sulphur input in forest. Dry deposition exceeds wet deposition in polluted regions. Two major forms of sulphur deposited into forest are sulphuric acid (H_2SO_4) and SO_2 which is oxidized to H_2SO_4 after deposition. Uptake and reduction of SO_4^{-2} by plants or microbes consumes H+. H_2SO_4 results in bigger increases of leaching and loss of cation than does nitric acid. Internal processes in the canopy are hardly affected by SO_4^{-2} (foliar leaching and uptake) . The enrichment factor of this ion in throughfall can be used as a suitable "indicator" for dry and wet horizontal deposition. This anion plays the main role in indirect effects (soil acidification). Subsequent fate of the sulfate inputs must include the accompanying cations.

Nitrogen may enter into the nutrient cycles of the forest ecosystem via atmospheric input in a variety of forms: nitrate as a salt or acid, ammonium as salt, nitrous oxides deposited on needles, organic-N, nitrogen via biological N_2 fixation. The H+ effect of nitrate input changes with the fate of nitrate within the forest ecosystem. Nitrogen in the form of ammonium can volatilize or be converted to nitrate by bacteria.

In different regions, precipitation contains variable quantities of base cations. Annual input rates of 2-10 kg/ha are common for K, and Ca inputs are often 5-20kg/ha. The actual effect of cationic deposition depends on the anion composition of precipitation.

Inner Cycle of Canopy to Acidic Impact.

Recently in the Czech Republic a few permanent PRPs have been established, especially in areas with heavily polluted and very damaged forests, to perform long-term, systematic chemical analyses of throughfall (TF) chemistry. The three main reasons for following the chemical composition of TF, bulk (CBULK) and stemflow (SF) depositions are:

To estimate and quantify the total deposition (wet and dry). Sulphur was chosen as an element that is not leached or taken up by the needles.

To estimate wet and dry deposition. Both inputs into forest plots have some relationship to local air-pollution and distance of emission situation.

To find changes in internal nutrient cycling. Increased acidic deposition into forest causes changes of calcium and magnesium in leaching from the crown. Abnormal internal circulation may serve as an indicator of nutrient imbalance or deficiency.

The determination of total deposition in forest stand was based on open CBULK and throughfall measurements. The capture and filtration effect over the canopy was estimated for each forest stands as the sulphur enrichment canopy factor (FS). This factor is based on ratio of sulphur as sulfates in total deposition onto forest (TF) divided by total deposition in open field (INC):

$$FS = TF\,(SO_4^{-2}) \,/\, INC\,(SO_4^{-2})$$

The total deposition into forest ecosystem:

$$(CTOTAL)ion = (FS) \times (CBULK)ion$$

then canopy exchange is equal to inner cycle:

$$CCE = CTF - CTOTAL$$

The canopy may either increase or decrease the acidity of incident precipitation. The flux of soluble matter in TF and SF is a normal part of the forest nutrient cycle. Thus solutes of nutrients removed from plant canopy become available to be taken up by the plants from soil and litter and subsequently release again.

The processes that alter solution chemistry in the canopy can be summarized in the following mass balance equations (where NTS = net deposition; L = leaching; U = uptake; L - U = net leaching from canopy):

For open area:

INC (bulk) = WD + DD (wet and dry deposition)

Total deposition for cover area:

TD = WD + NTS = DD + (L - U)

Then:

NTS = (TF + SF) – INC and also NTS = (L – U) + DD

During inner canopy exchange, three possible processes exist:

a) (L - U) > 0, then L > U and leaching prevails. This is a "natural process" in healthy spruce tree and dominates in most cases.

b) (L-U) = 0, then L = U. This situation prevails when the uptake of nutrient by fine roots of trees is not available (an imbalance of nutrient may occur).
c) (L-U) < 0, then L < U and uptake is preferred. This is an unusual process and will be observed when the ecological stand condition is exposed to stress factors (physical, chemical and biological)

The total deposition of most of the important cations and anions associated with the acidic impact on forests at the permanent PRPs was estimated for the year 1996 by incorporating the data from the chemical composition of TF into the inner canopy cycling approach described above. It should be noted that the data was derived from areas heavily polluted with sulphur and containing mature spruce trees and that uptake or leaching of sulphur from needles was neglected. Therefore, the TF of $SO_4^{-2}$was equal to total deposition. This approach neglects some canopy-derived materials (particulate litter) that can be wash off and some particulates of dry deposited material that can be retain in the canopy and never appear in throughfall (e.g. fixed in wax layer).

Furthermore, data obtained from the inner canopy cycling approach can be used to estimate both the health condition of forests (e.g., physiological state or adaptability to changing stand situations) and the supply of nutrients which can reflect the degree of acidification of the soil due to increased weathering rate or changes in ecological and microclimate conditions.

Use of Models to Predict Success of Bioremediation.

Over the past three decades there has been a dramatic growing interest in the development and use of various types of forest growth models. Several tens of models have been developed to examine theories of succession and to assess the effects of growth-modifying factors on the species composition of forest communities from 37 permanent research plots (PRPs). Hybrid simulation models are those in which a simulation of the actions of certain growth-determining processes are added to historical bioassay models of forest stand growth. This type of model is one which uses the historical patterns of biomass accumulation and ecological processes to establish the expected patterns for an unchanging future, and adds process-based simulation to permit the yield estimates to reflect expected changes in growth conditions.

The impact of air pollution on mountain forests was simulated by using the hybrid stand forest ecosystem simulation model FORCYTE-11 [5]. It is a hybrid empirical/process stand model and also a hybrid between a stand model and an individual tree model. For our application, the following parameters were used: a) nutrient limitation of tree growth, b) air pollution as "insufficient simulated nutrient", and c) biomass production as an indicator of forest condition. Data for spruce trees from 37 PRPs was used to predict conditions in the year 2010. In Table 1, are shown the past and expected changes in productivity of biomass at the 37 PRPs.

TABLE 1. Historical, and Predicted Forest Production

Category	1978-1986	1986-1994 reduction	25% sulphur reduction	50% sulphur
Good	12	5	12	25
Fair	19	15	17	9
Bad	6	17	8	3

Spruce trees (between ages of 80 and 120 years) are divided into three categories according to productivity of biomass. The model was calibrated for forest conditions and age of trees from all PRPs to predict the effect in the year 2010 in the case of 25% and 50% reduction of sulphur emission. At a 25% reduction of sulphur emission, the forest conditions will be essentially restored to the those of 1978-1986, while a 50% reduction of sulphur emission would result in a significant improvement [6]. Simulation for PRPs in the Krusne Hory predict that a 25% reduction of sulphur emission will not prevent the forests from dying, while a 50% reduction short of year 2000 would rescue the forests. Consequently, the similar model simulations for the Krkonose Mountains indicate that a 25% reduction of sulphur would result in a slight increase in productivity and a 50% reduction would result in good productivity by the year 2010.

Conclusions.
The aim of this study was to develop simple and inexpensive monitoring methods to estimate the impact of acid air pollutants to spruce trees in the Czech Republic. It was found that open and throughfall deposition determinations could be obtained rapidly using calculations of internal circulation of nutrients in the canopy of spruce trees. This approach proved to be more exact than visual inventory of spruce damage. Simulation modeling was used to predict future forest damage. Although modeling is more convenient than long-term biological monitoring, the application of the method would be enhánced if more basic information and data about all compartments of landscape were available.

References

1. Poleno Z. (1996) Sustainable management of forests in the Czech Republic. Ministry of Agriculture of the Czech Republic. Prague.

2. Palat, M., Vasicek, F., Henzlik, V., and Kasperidus H.D. (1994) Condition of damage to Norway spruce stands in the Czech Republic, *Lesnictvi - Forestry*, **40**, 217-237.

3. Kubiznakova J., and Kubiznak J. (1995) Input of atmospheric deposition to mountain forest ecosystems, in Matejka K. (ed.), Investigation of the Forest

Ecosystems and of Forest Damage. Processes in Forest Ecosystems and their External Functions, Proceedings of the Proc. Seminar, April 1995, Opocno, Czech Republic, pp. 80-85.

4. Lindberg, S.E., Page, A.L., and Northon, S.A. (1990) Acid Precipitation. Vol.3, Sources, Deposition and Canopy Interactions, Springer Verlag, New York.

5. Kimmins, J.P., Scoullar, K.A., and Apps, M.J. (1990) "FORCYTE - 11 user's manual for the benchmark version", For. Can., Northwest Reg., North. For. Cent. Edmonton, Alberta.

6. Kubiznakova, J. (1997) Application of model FORCYTE for prediction of abatement strategy of SO2 emission, in A.P.Simon H., A. Grossinhi, J. Brechler, J. Kubiznákovå, K. Juda-Resler, and J. Zvozdiak (eds.), Emission Abatement Strategy and Environment (EASE).The interdisciplinary study of environmental damage due to sulphur emissions in BT, Final Report for CEC, Project CIPA-CT 93-0071 DG-12HSMU, March 1997, ICSTM, CET, London, UK.

4. REPORT OF THE WORKING GROUP ON THE STRATEGY FOR THE IMPLEMENTATION OF A BIOMARKER BIOMONITORING PROGRAMME IN THE KOLA PENINSULA, RUSSIA.

J. M. WEEKS (Rapporteur), G. EVDOKIMOVA (Chair), W.H.O. ERNST, J. SCOTT-FORDSMAND, J.P. SOUSA, V. E. SERGEEV and M. NAKONIECZNY.

4.1. Summary

This chapter considers the utility of biomarkers for monitoring environmental health in an extremely contaminated and polluted locality, the Kola Peninsula, in Northern Russia. Due to the extreme nature of the pollution, biomarkers have little place as predictors or to forewarn of likely damaging effects of contaminants in their respective ecosystems; the damage has been and continues to be done. Biomarkers can, however, be used to monitor and map the worst affected areas, and look for signs of recovery in areas that have been remediated or where pollution prevention controls have been put in place. There are various caveats in using biomarkers in such a fashion and these are discussed. The following text serves to give a simple overview of the implementation and reasoning for using biomarkers, suggest the type of material that one may wish to consider to monitor, and outlines a simple procedure or strategy for implementing such a survey.

Keywords: Kola Peninsula, Biomarkers

4.2. Introduction

Science-based tools for determining environmental and ecological risk following exposure to contaminants still remain in their infancy. Biological tools can be used to characterise and assess pollutants. However, the success of these techniques depends largely upon the choice of biological assays which should be both sensitive and reflect the relevant ecological effects of toxicity at the site.

Recent advances in toxicity and biomarker tests allow one to estimate with greater confidence the ecological risk to certain components of the ecosystem. For example, the increasing use of relevant toxicity endpoints (e.g. reproduction) in earthworms and the use of exposure/toxicity biomarkers in earthworms and field exposure approaches using mesocosms have all recently emerged as more

D. B. Peakall et al. (eds.),
Biomarkers: A Pragmatic Basis for Remediation of Severe Pollution in Eastern Europe, 279–298.

appropriate technologies for the ecological assessment of harm from contaminated soils than traditional chemical analyses alone. Many of these techniques have yet to be fully developed for pollutant assessment purposes, or for monitoring the change in risk during and following the remediation of such localities. However, such information would offer greater precision on the ecological risk of the environmental pollutants and the health of restored sites, not only to soil-dwelling organisms such as earthworms, but also to the health of the environment *per se*, and subsequent estimates of risk to human health.

The potential socio-economic benefits of such research in the Kola Peninsula will be in the implementation of quantitative biomarkers that do not rely on the established lethal dose protocols. This will facilitate the grading of habitats for remediation or other treatments so that a more informed judgement as to the appropriate use of severely contaminated localities may be made by the regulatory authorities and site owners/managers.

In recent years there has been an increasing interest in the use of biomarkers for the assessment of the potential adverse effects of chemicals to the environment. Biomarkers have been used extensively to document and quantify both exposure to, and effects of environmental pollutants. As monitors for exposure, biomarkers have the advantage of quantifying only biologically available pollutants. Some aspects of biomarkers are covered in the plenary lectures of this workshop (chapters 9 to 12). Additionally there is a vast literature on biomarkers [1,2,3].

The overall aim of this chapter is to review the types of systems that may be utilised with effect in a chronically and seriously contaminated landscape, the Kola Peninsula. In the past it was sufficient to simply detect and determine the presence and amount of accumulated pollutant residues. Even with the ever increasing sensitivity and complexity of modern analytical techniques, few earlier studies established the need to interpret the biological significance of this data. Today, however, it is considered to be of fundamental importance to relate such information to disturbances at high levels of organisation such as the individual, population and community [4,5 and Chapter 11 of this book]. Most ecotoxicological studies involve single species testing under controlled conditions in the laboratory, this leads to the development of biomarker assays which can be used in predictive biomonitoring, or ecological risk assessment. The use of biomarkers avoids the pitfalls involved in the interpretation of chemical measurements due to differences in bioavailability.

Ecotoxicological biomarkers have been developed, predominantly for vertebrate species, but with increasing emphasis on invertebrates (Chapter 11, this volume) and in particular aquatic invertebrates. The biochemical parameters that have been most widely used are inductions of detoxification enzymes. These include the mixed function monoxygenases and the glutathione S-transferases which are responsible for metabolising many classes of organic pollutants. However, conflicting reports of enzymatic activity in invertebrate species have complicated the interpretation of these enzymes in the biotransformation of pollutants. Also studied in vertebrates (and to a lesser extent in invertebrates) are the activities of acetylcholinesterase (Chapter 12, this volume) and non-specific cholinesterase in response to carbamate and organophosphate pesticides. Despite the number of investigations carried out on these biomarkers few direct relationships between activity and pesticide exposure have been elucidated. Numerous examples of other biomarkers can be found in the literature. Clearly, however, this chapter does not serve to investigate established biomarker mechanisms, but to develop a strategy for their use in the Kola Peninsula.

The remainder of this chapter will describe the locality of concern – the Kola Peninsula and suggest different approaches from molecular through to analysis of functional groups and communities to develop an overall strategy for the monitoring of the environmental status through the use of biomarkers.

4.3. The Local Environment

The Kola subarctic region is undeniably faced with a severe degree of chronic pollution, resulting from many metal-smelting and related mining activities. More than 8 % of the sulphur dioxide and more than 30% of the metals emitted into the air in the Northern part of Russia are deposited within the Kola Peninsula. The largest smelter "Severonikel" is located in Monchegorsk and has been in production since 1935.

It is quite likely that, in addition to the levels of SO_2 and heavy metals emitted by this smelter, PAHs are also produced by the coal-fired heating source. It is anticipated that the gradient of PAHs (if any) would mimic that of the metal deposition gradients [6]. In addition, much mining activity takes place in the eastern part of the peninsula, where ores typically contain 24% nickel. In recent years, much richer ores have been transported from further east by sea, because of their richer mineral content (*circa* 30%). However, these ores are also much richer in pyrite. In addition to this large Ni/Cu smelter, there is a similar but smaller smelter located in the town of Nikel. In the south there is a nuclear power station and a small aluminium smelter. Smaller deep-mining activities occur in the eastern part of the peninsula. These industrial activities and the rugged landscape of the tundra have resulted in serious contamination in the vicinities of these large industrial complexes (see Figure 1).

There are examples of effects of elevated soil metal concentrations in the immediate vicinity of the Severonikel smelter on microbial communities. At a distance (>50 km) there are around 18 groups or species complexes consisting of bacteria, actinomycetes, algae and fungi and these decrease to around 8 species or groups, dominated by Fungi e.g. *Pencillium* and *Trichoderma* as one moves towards the smelter [7]. Fungi are generally more tolerant of the toxic actions of heavy metals. In the tundra, there is an extensive development of sexually sterile mycelia indicating a reduction in the life-cycle of this group, potentially a means of overcoming stress by economising energy use. In addition, within a 5 km radius form the centre of the smelter, in a North/South ellipsoid, soil copper concentrations range between 600-1000 mg/kg and nickel from between 1700-2600 mg/kg in the upper soil horizon. At 15 Km from the epicentre, soil metal

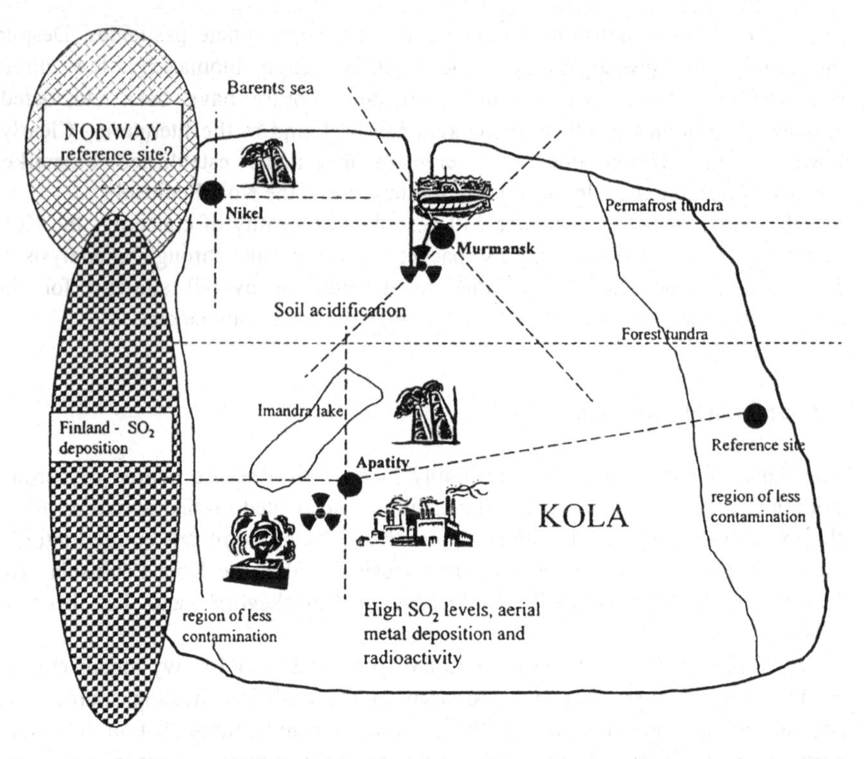

Figure 1. Pictorial view of Kola Peninsula showing the major environmental impacts and a simplified monitoring strategy along gradients from central industrial foci. Not to scale, based on a map provided by MSU, Dept. Geography, Moscow, 1996. - - - - - indicates possible sampling transects.

concentrations have come down to between 200-400 for copper and between 500-1000 mg/kg for nickel. At 50 km from the smelter, soil metal concentrations approach background, with an average of 40 mg/kg for copper and nickel, an average of 80 mg/kg. Concomitant with the change in soil metal concentrations, there is also a change in surface vegetation. Close to the smelter large areas of woody trees are absent and only depressed regrowth of silver birch (*Betula pendula*) may be seen. Ground vegetation covers only about 20% of the surface at this location, and is composed mainly of *Empetrum hermaphroditum* and *Deschampsia flexuosa*. Mosses and lichens are absent. At the intermediate zone, vegetation is of the pine-forest shrub type with the addition of birch and predominance of *Vaccinium vitis-idae*, *Vaccinium myrtillus* and *Arctostaphylos uva-ursi*, as ground cover. Lichens (*Cladonia rangiferina, Cladina mitis*) are present. Further from the smelter normal pine and shrub-lichen types with the addition of birch become predominant. *Vaccinium vitis-idae* and *Arctostaphylos uva-ursi* dominate among the shrubs, representatives of *Cladonia, Cladina* and *Cetria* dominate among the lichens and *Hylocomium* and *Pleurozium* prevail among the mosses (Evdokimova, Chapter 1, this volume).

Since 1995, the smelter has cut SO_2 emissions by 50 % from 200,000 tonnes sulphur dioxide per annum. This was brought about by a change in the volume throughput through the smelter rather than by any guidelines or regulatory control. Indeed, the smelter has very strict emission guidelines and values, although the operators of the plant are unable technologically to meet them.

4.4. Marine Environment

The Kola Peninsula in Northwest Russia is the location of the most concentrated collection of nuclear waste in Europe. Off the coast of the Barents Sea lie some 80 scrapped nuclear submarines, half of which still contain fuel rods. Thousands of tonnes of liquid and solid radioactive waste lies stored in boats and onshore tanks at naval bases and shipyards (see Figure 1). The Norwegians launched an action plan in 1995 to help to reduce the threat of pollution or a serious nuclear accident (New Scientist, No. 2102, 4 October, 1997, p.22 *Arctic stand-off*). In a further report in the New Scientist (May, 1998, No 2133, p.11) it was suggested that this radioactive waste has begun leaking into the sea. The area, home to Russia's Northern fleet, has massive potential waste problems. Researchers from the Russian Academy of Science's Marine Biology Institute in Murmansk collected over 100 samples of sediment from 100 km of coastline. They found that levels of cobalt-60 outside the naval base at Poljarny on Pala Bay rose from 10 to 80 becquerels per Kg between 1995 and 1997. According to the authors of the report this was because a nearby store for liquid waste from submarine reactors is corroding. Levels of caesium-137 have also increased indicating recent leaks.

4.5. Human Health

At least half a million people on the peninsula drink water that is contaminated with raw sewage and heavy metals. Norwegian environmental engineers also warn that residents of Murmansk may suffer major epidemics of diarrhoea because urine from chicken farms is leaking into the River Kola, a major source of the city's drinking water.

A two year study of drinking water quality in the region has revealed that sewage works are old and broken, water filtration plants inadequate and chemical treatment ineffectual. The study concludes that an investment of £50 million in new treatment plants, pollution controls and sources of clean water is needed urgently. The investigation was commissioned by the Barents Euro-Arctic region, an organisation representing nine regional governments from the north of Russia, Norway, Finland and Sweden. The stomach problems caused by dirty water may contribute to the high death rate among children in the area and could fatally weaken the elderly. Chicken urine leaks into the River Kola upstream from Murmansk from riverside farms housing 4.5 million birds. Untreated sewage is dumped in the river Kola at Kildinstroy, just 10 km upstream from the main water intake for Murmansk. In Kirovsk, 150 km south of Murmansk, aluminium from factories has made the water strongly alkaline, and a lake that supplies 70,000 people in nearby Apatity (see Figure 1) with water contains high levels of iron, nickel, copper and cobalt. A report argues that money to make the water safe must come from Russia's neighbours, but the task could take decades (New Scientist, 1 November, 1997, p. 14).

4.6. Strategies for using biomarkers

The following text outlines various strategies for selecting which biomarkers might be measured within each habitat type. The following sections are not exhaustive as this would entail a lot of repetition of previous works, and serves only as a guide to the types of measurements that may prove useful.

4.6.1. REMEDIATION

It is worthwhile considering that the Kola Peninsula is rather a special case. It is unlikely that emissions from the smelter and other diffuse and point-sources will be reduced, and thus the use of biomarkers to measure recovery or the success of remediative technologies or actions is compromised. The definition given of a biomarker in the chapter by Peakall (Chapter 9, this volume) refers to deviation away from normality. However, this concept cannot be easily applied in regions where habitat destruction is chronic and unrelenting, the damage has been done, and will for the foreseeable future continue. Therefore, the principal utility of biomarkers lies in their successful use to measure the ongoing current "health" of selected indicator species, and to note improvements upon this lower initial status (see Figure 2). Invariably, preventative action will be required to halt the demise of species nearing the limits of any compensatory homeostatic mechanisms.

In addition, the question of resistance in some species, may hide the true picture of the well-being of an animal or plant if biomarkers alone are used. If, for example, we take the grass *Deschampsia flexuosa,* it is known to become resistant to heavy metals, rather quickly and survive in quite seriously contaminated soil [8,9]. However, with time, this resistance is finite, so although we may be measuring a biomarker of heavy metal stress in such plants due to changes in their genetic composition signs of "improvement" would not be noted. However, there is a cost to developing resistance and ultimately this trade-off can show through with a loss of reproductive output (See Figure 3). Thus, it is recommended that simple biomarker measurements are made in tandem with physiological/reproductive indicators, particularly in longer-term studies. If we look to deviations away from normality for our biomarker responses we may become misguided as we cannot expect such resistant organisms to show "normal" values for the species.

4.6.2. PLANT BIOMARKERS

Plant species under consideration are typical for the Northern Taiga and tundra. The silver birch is an important tree species with moderate metal resistance that can be usefully employed in a biomonitoring biomarker scheme in the Kola Peninsula. Bleeding sap in this species is a good biomarker, hence the level of metal accumulated in the bleeding sap just before leaf flushing is highly correlated with the degree of metal resistance of tree. By contrast measurement of phytochelatins is less suitable. *Deschampsia flexuosa* is one of the dominant grass species. It may have evolved a high Cu and Ni resistance and cell division cycle evaluation [10], and may be tested by a modified rooting test as described by Schat et al. [11]. It is possible that such resistant grasses may be used for phytostabilization ("greening") of bare soil. There may be a negative impact of SO_2 on survival of metal resistant grasses. With regard to human health it will be necessary to look to metal speciation in lettuce (*Lactuca lativa; polychelatins for Cu, histidine for Ni, carbonic anhydrase for Zn*) and mushrooms: the latter from accumulation of heavy metals and radionuclides and it is advisable they are not collected for human consumption [12].

A list of plant biomarkers for the chemicals of concern in the Kola Peninsula are given in Table 1. A broader and more detailed considerations of plant biomarkers is given in chapter 10.

With regard to reindeer food-chains, lichens (*Cladonia* sp. *Cladina* sp., *Cetraia* sp.) have to be analysed for metal concentrations and speciation (binding of copper and nickel to specific lichen acids). The development of new biochemical biomarkers for lichens has to be undertaken.

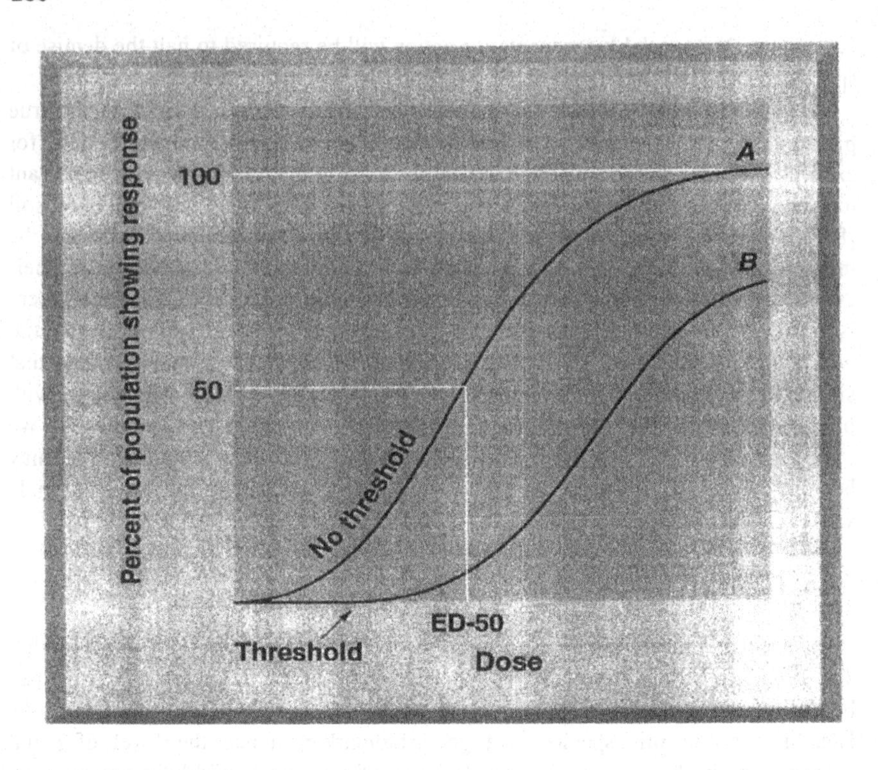

Figure 2. In this hypothetical toxic dose-response curve, toxin A has no threshold; even the smallest amount has some measurable effect on the population. The ED-50 for toxin A is the dose required to produce a response in 50% of the population. Population B, has some inherent tolerance and thus small concentrations of toxin have no effect, the concentration causing 50% mortality in this population is much higher than for population A.

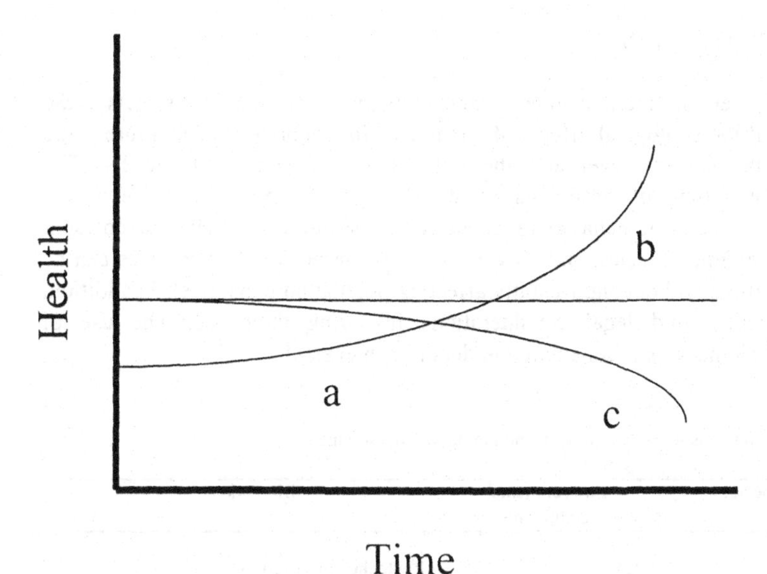

Figure 3. Three hypothetical scenarios outlining the problem of tolerant or resistant species used for biomarker biomonitoring. In scenario a, the population is seen to recover with time following remediative action or cessation of pollutant impact and such improved health would presumably be represented by concomitant biomarker measurements. In scenario b, however, this species has become tolerant or resistant to the pollutant and can survive in a contaminated environment. Thus, there would be no changing biomarker measurement, nor necessarily a change in apparent population health measured by other parameters. In scenario c, the clean-up of a site may now in fact be deleterious to a previously resistant population and it may now begin to decline.

4.6.3. INVERTEBRATES

Due to their essential functions in ecosystems, invertebrates are useful targets in the assessment of the ecological effects of chemicals. Invertebrate tests measure acute toxicity at the species level and the population and community levels. The preliminary endpoints of these tests are survival, growth (measured as biomass), reproductive success and behavioural changes. Furthermore, invertebrates represent >95% of all animals species, and they are often numerous (so that samples can be taken for analysis without significantly affecting population dynamics). In addition, there are ethical and legal considerations favouring their use. The use of invertebrate biomarkers is considered in detail in chapter 11.

Table 1. Selected plants for biomonitoring and biomarker measurements

Plant species	Environmental compound	Biomarker
Deschampsia flexuosa and *Betula pendula*	Cu	Phytochelatins (PC) Rooting test for resistance Cell division cycle
	Ni	Histidine Rooting test for resistance cell division Cell division
Betula pendula	Cu, Ni	Bleeding sap leaching
	SO_2	SO_2 accumulation Glutathione(GSH)
	F	Necrotic spot quantification Fluorocitrate
Lichens	Cu, Ni	PC Metal complexing lichen acids
	Cs^{137}, Cs^{139}, Sr^{90}	Residue
Mushroom	Cu, Ni	Metallothioneins (MT)
	Cs^{137}, Cs^{139}, Sr^{90}	Residue
Lactuca sativa	Cu	PC, residue
	Ni	Residue, organic acids
	F	Fluorocitrate
	SO_2	SO_2 accumulation

Invertebrates such as earthworms are useful targets in the assessment of the ecological effects of chemicals. Earthworm tests measure acute toxicity at the species level and the population and community levels. The preliminary endpoints of these tests are survival, growth, reproductive success and behavioural changes. In earthworms it is possible to measure the effects of both inorganic and organic contaminants by determining changes in the permeability of the lysosomal membrane of earthworm coelomocyte cells using the neutral red retention assay as described by Weeks & Svendsen [13]. The neutral red assay is an established technique for the evaluation of cell toxicity. It has been used to test the cytotoxicity of chemical compounds, pharmaceutical agents, surfactants, food additives, pesticides, solvents, complex mixtures and a variety of miscellaneous chemical agents, using cells from mammals, humans and fish as targets. The assay is based on the ability of viable cells to accumulate a dye, neutral red. The ability to retain this dye is dependent upon the lysosomal viability or function (how damaged the cell is). The relative release of neutral red is therefore often used as a specific indicator of lysosomal function or activity (the time taken for the dye to leak from the cell).

This assay makes use of the fact that healthy lysosomes take up and retain the dye (neutral red) indefinitely, whereas the dye in lysosomes of coelomocyte cells taken from "stressed" earthworms, gradually leaks into the surrounding cytoplasm. The time taken for 50 % of the cells to show "leaking lysosomes" is termed the neutral red retention time. This response has been quantified in both the laboratory and field using various model organic and inorganic contaminants [14,15,16]. This marker is non-specific, responding equally sensitively to organic or inorganic contamination, however, if used in combination with an earthworm immuno-competance marker [17], such as total immuno-activity of the coelomocytes, then it is possible (with caution) to be more specific as to the likely nature of contamination. This technique also has applications in freshwater and marine invertebrates.

4.6.4. VERTEBRATES

4.6.4.1. Birds

The presence of mature plant species following phytostabilisation (planting of grass species and birch trees) makes it possible to use nesting insectivorous passerine birds to monitor the exposure to the different chemical compounds present in the area. Some species of this group are successfully and widely used in monitoring programmes and some species occurring in the area are quite suitable for these purposes.

Monitoring hole nesting birds and ground nesting birds will give us important information on the different resources being used by these birds, thus possibly on different types of exposure pathways. The nesting success of hole nesting birds can be enhanced by the placement of nest boxes; this procedure will allow the researcher to have better control of the situation and an easy access to potential sample material.

Besides measuring nesting success (number of occupied nests) it is necessary to measure breeding success; not only should normal breeding parameters be measured (e.g. number of eggs, number of eggs hatched, development of young etc.), but also several biochemical/physiological parameters should be measured:

1. To measure the exposure to heavy metals, residue analysis can be undertaken non-destructively on several biological materials (feathers, blood and eggs).
2. To check the exposure to SO_2 deposition, the measurement of egg shell thickness is advisable. SO_2 causes the elution of Ca from the environment, which may result in a deficiency of this element in birds. The same measurement can be used to check exposure to fluoride, as the F/Ca exchange process is similar in its mechanism [18]. However, before attributing eggshell thinning to the above causes it is necessary to check the level of DDE as this is a potent agent for causing eggshell thinning in some species [19 and Chapters 9 and 12, this volume].
3. Exposure to PAH's can be monitored by measuring several biomarkers (MFO's P450, CYP1A and DNA disruptions) (see Chapter 12, this volume). This can be achieved by non-destructive methods (e.g., taking blood samples) or by collecting eggs. In this case, particular attention should be given to the sampling procedure in the field, in order to reduce the variability in the results. When retrieving one egg per nest, one should take care to always remove (if possible) the same "order" egg (always the first or the second egg). It is also important, especially in the case of CYP1A induction, to have samples in the same developmental stage. This can be achieved by incubating the egg in the laboratory and taking measurements when they are "about to hatch" [20,21].
4. In the case of exposure to radionuclides, measurements of histopathological and DNA alterations are advisable.
5. The sampling design should be accomplished giving consideration to the dimensions of space and time. Besides following the established gradients, samples must be taken, at least, during the breeding season (before and after egg deposition). Depending on the evolution of the emissions (maintaining or decreasing with time) monitoring should be done on an annual or biannual basis.
6. Although insectivorous birds are the most accessible species for this type of biological monitoring, some other species may deserve attention. Species either located at the top of the food chain or those well established in human settlements can also be used for monitoring purposes. In such cases, non-destructive methods should be used to monitor exposure.

4.6.4.2 Mammals

Mammalian exposure to the several sources of chemical stress can be monitored using a set of representative species in the area.
1. Generally small mammals are good indicators of mammalian exposure to both inorganic and organic compounds. A monitoring plan for small mammals can be put in place unobtrusively in the area. After selection of the most representative sites a sampling scheme using removal methods (e.g. line traps) can be established. This type of sampling scheme can be used not only to monitor population or community changes (numbers, age class structure, reproduction), but one can measure on captured animals a suite of biomarkers (MT's MFO's, P450, CYP1A and DNA alterations), and also determine heavy metal residues. To relate these measurements to soil residue content would enhance their relevance. Other measurements like fat content, particularly in females, serve as good indicators of

the health of the animals and also as good indicators of food availability/quality in the immediate area.

2. Larger mammals can be used as indicator organisms in the Kola Peninsula. Emphasis should be placed with the top predators (e.g. arctic fox) and herbivores (reindeer). Using non-destructive methods in these animals (blood or hair samples, and also the antlers - in the case of the reindeer) can also be useful for any monitoring programme. In this latter case, these animals can be considered as realistic monitors of exposure to chemical pollution in this area, mainly due to their diet of lichens (see previous discussion).

4.6.4.3. Biomarkers for the aquatic habitat

The Kola Peninsula covers a large area of the northern part of Scandinavia. A large proportion of this area is covered by lakes and running water. A range of aquatic species are found in this area, for example freshwater invertebrates and many fish species. In addition aquatic plants and algae are also present.

Samples to be collected from contaminated streams, rivers and lakes should include water, sediment and biota. Residue measurements should be extended for a full and comprehensive range of contaminants together with measurements of water and sediment biotic and abiotic parameters such as pH, BOD, dissolved oxygen etc.

A biomonitoring scheme should be set up for the assessment of the community/population parameters for the selected organisms at certain points in time. These population parameters should include density of animals, age structure and possible estimates of biomass.

The summary of suggested biomarkers for this type of habitat (see Table 2) should cover the range of anticipated contaminants likely to be present. Effort should be made to link residue determinations to the presence or absence of particular species assemblage of benthic invertebrates (in addition measurements such as total glycogen will give some information on the availability of energy for reproduction and subsequent scope for growth type measurements).

4.7. Sampling strategies

First and foremost it is essential that a residue survey is undertaken to determine environmental pollutant concentrations in the major components of the ecosystem - soil, water, snow, plant and animal. It is important that the effect of season is noted during this survey, particularly for vegetation samples, i.e., time to leaf abscission. It is necessary to undertake such measurements in the assumed reference site(s) also.

4.7.1. RESIDUE MEASUREMENTS

The determination of chemical residues in selected biotic compartments will be a fundamental exercise in the overall survey procedure. Data from such measurements will inevitably underpin the requirement to make further biomarker selection and pollutant residue measurements. What should one do if there is no detectable residue? If there is no elevated inorganic residue then perhaps there is no requirement for subsequent biomarker analyses? If, however, there is no organic residue but the compound is very bioactive there may still be a requirement to undertake biomarker analysis as there may well be secondary transformation and metabolism of the parent compound. If residue analysis is to be undertaken suitable representatives would include, soils, plants, animals, snow and water. If samples are to be taken for residue analysis there is a requirement for a system of data archiving and storage, with associated Good Laboratory Practice and Quality Assurance throughout all analytical procedures. Samples must be able to be stored in good condition (either by e.g. freezing/drying), and transported to suitable local laboratories.

Table 2. Biomarkers to be employed on freshwater organisms in the Kola Peninsula region of Russia.

Organism	Biomarker
Gammarus	Histopathology of gills (all)
	Lysosomal changes in blood/midgut gland (Heavy Metals /PAHs)
	P450/GST in midgut gland (PAHs/PCBs/DIOXINS)
	DNA-alterations (Ni/ PAHs/PCBs/DIOXINS/radionuclides)
	Glycogen in midgut gland (all)
Fish	as for *Gammarus*, except using liver and blood samples (possibility of non-destructive methods)
Microalgae	Respiration (all)
	Cell division (all)
Macroalgae	Respiration (all)
	Cell division (all)
	Histopathology (all)
Higher Plants	Histopathology
Fish-eating birds	DNA-alterations (Ni/ PAH/PCB/DIOXINS/RADIONUCLIDES) - possibility of non-destructive methods
	Histopathology (all)
	P450/GST (PAHs/PCBs/DIOXINS) - possibility of non-destructive methods
	Glycogen of liver (all)

4.7.2. CHOICE OF TAXA

This choice will be largely dependent on the availability of the species present in sufficient abundance in the area where the study or monitoring scheme is to be carried out. In practice, however, it appears that the order of preference for ease of use and precision of estimates appears, in the case of plants, to be moss followed by lichens and then pine bark or needles. For animals, invertebrates and then birds and small mammals are favoured. The alternative is the use of introduced indicator species, or species transplanted from clean environments to monitor change particularly in areas where suitable organisms do not occur naturally or have been destroyed through the severe pollution present at their locality. It should be remembered that pollutant levels in mosses and lichens and lower order animals (the invertebrates) are likely to respond more quickly than higher plants or higher animals as their physiological systems will buffer change to some limited extent. Consideration should be given to the avoidance of vertebrates when invasive biomarker are needed. Where possible non-invasive samples should be taken, or invertebrate alternatives sought (ref. Chapter 12, this volume).

Furthermore, plant and animals community assemblages should be analysed, noting the effects of seasonal changes, and inclusive of reference sites. This initial survey will show obvious disparities in species composition for predicted versus observed faunal and floral analyses thus guiding the final selection of appropriate biomarker/bioindicator species, and also foretelling of changes to ecosystem function.

It is important, in the first instance to determine aspects of soil health such as soil processing and functionality (including turnover of nitrogen, carbon and decomposition rates). When assessing the effects of chemicals in terrestrial systems, not only should measurements at structural level be taken (e.g. changes in population and community composition etc.), but also information at the level of soil processes is required. Soil processes, such as organic matter decomposition, are determined by the activity of the resident fauna and soil microorganisms mediating them and are key measurements in any survey, since they truly reflect the "status" of the soil. We recommend that the following measurements be undertaken:

Litter decomposition, using the litter-bag or litter-container techniques, of the most relevant plant species in the area; it is advisable to compare exposed and non-exposed vegetable material of two different types of plant species widespread in the area. Since decomposition of organic matter is the main gateway of essential elements to the soil, information given by these measurements are important to understand the processes of nutrient cycles in this area. Connected to the measurements of mass loss over time, measurements of chemical residues, nutrients and also some biological parameters (e.g. fungal biomass and litter respiration etc.) are to be recommended, allowing a better understanding of the processes occurring.

Soil respiration is a key measurement in these cases, giving us information about total soil activity. This can be done by measuring the substrat- induced respiration or basal respiration, and comparing polluted and reference soils which are representative of different soil types. Soil microbial biomass can also be estimated by this method. However, special care should be taken when measuring soil respiration by this technique because of the temperature extremes that the Kola soils are subjected to. Therefore, some previous assays to assess the natural fluctuations in these parameters should be done. Continuous measurements, taking into account the temperature regime, should be made whenever possible.

Microbial and faunal activity in soil. These parameters can be measured using the well known "cotton-strip" and "bait lamina" methods. The first is a good indicator of microbial activity, and the latter has produced promising results in measuring feeding activity of soil fauna, especially in forested areas. Measurements must be made over time, particularly when the soil fauna is more active (seasonality).

4.7.3. RECONNAISSANCE

Before a full scale monitoring programme is started, some form of reconnaissance or exploratory study should be carried out. This would provide information on the maximum distance at which the pollutant(s) can be detected above ambient levels, so that sampling sites can be positioned to cover the extent of pollution deposition. It is worthwhile considering the area immediately surrounding this established zone so that any increase in the area affected by the pollutant can be established. This preliminary survey should also attempt to collect information as to the distribution of the chosen biomarker monitoring organisms(s) and hence the distribution of sampling sites. It would also provide information on the practicalities that are essential for planning a full-scale monitoring programme, such as the number of sub-samples required to give the desired level of precision.

4.7.4. STATISTICAL CONSIDERATIONS

As mentioned above, the sampling sites should be positioned to represent the whole area over which the pollutant in question can be said to have originated from the source being monitored. In addition, the sampling of areas immediately outside the area where the pollutant can be detected would be useful for giving warning of cumulative impacts or progression of continued pollutant loading.

The number of sampling points deployed will depend on the resources available. However, the more points sampled, the greater is the likelihood of detecting change, and the greater the reliability (statistical accuracy) of data to enable one to do so. Subsampling at each station will increase the precision of pollutant estimation.

At the simplest level, a transect of ten to twelve points along a downwind gradient of the source (point-source) should be sufficient to show the distance over which the pollutants are likely to travel and be deposited to allow accurate curve fitting (in the case of the Kola Peninsula the overall distance of these transects may be many hundreds of kilometres). Further transects upwind and crosswind (see Figure 1) would increase the understanding of the local deposition patterns, and improve the estimation of the effects of environmental factors such as altitude on deposition. For some point sources in the Kola Peninsula a gradient sampling strategy will result in a set of straight line transects emanating from each epicentre at 90 degrees for at least 200-250 km with sampling stations located according to a logarithmic series. Accessibility may be a consideration at some sites, and an initial satellite survey (remote sensing) image would help to locate potential problems, or a short aerial survey by low flying aircraft. Sites should be matched to Global Positioning Satellite localities. Historical records of past mining activities and other decommissioned industrial plants may be used to locate local sites of potential contamination or interference to the overall transect. Information from monitoring stations may be used where available (only available from the 1970's in the Kola Peninsula) in a similar fashion.

Possibly the optimum sampling scenario would be to monitor sites at varying and increasing distance from the source in all directions (Figure 1). These could be stratified to include different community types. The effects of seasonality of sampling must be known, particularly during the harsh weather extremes of the Kola Peninsula, and the representativity of samples collected should be understood. Any survey activity must be long-term to measure the effects of any reductions in pollutant emissions; a period of 5-10 years is recommended, with repetitive sampling.

The most commonly employed methods should be adopted for the species or groups of plant or animals selected during the initial survey and reconnaissance. All reasonable precautions should be taken to prevent further contamination of the collected samples. Samples should be taken from similar situations (habitat). Samples should be taken away from local sources of potential contamination that may give unrepresentative readings. A representative subset of samples should be taken at each site to give a measure of precision. It would be more informative to measure both a biochemical marker in association with a physiological/ reproductive biomarker for the species selected.

For simple transects it may be possible to compare best-fit lines for different sets of data. For more comprehensive surveys it may be enough to compare best-fit lines for a number of transects separately. Problems will arise if different transects show different trends. If sampling is carried out over a whole region, or at many different sites scattered around a point source, then the use of some type of geostatistic would be useful.

Such techniques are designed to allow interpolation between scattered data points. Though it is not possible to directly compare interpolations derived from different data sets, it is possible to derive estimates of total deposition, for example, in addition to determining error distributions. The most commonly used method for doing this is a modelling technique based on interpolation between mapping ordinates known as kriging. Kriging has been used for mapping soil pollution by heavy metals and for risk assessment in planning procedures.

Where measures of community composition have been made, then the use of multivariate measures, especially canonical correspondence analysis, could be used to identify the influence of different pollutants and environmental factors.

4.8. Conclusions

This chapter has attempted to briefly show the utility of using biomarkers to enable scientists to define the effects of contamination on environmental and human health in the Kola Peninsula and to address the problem of establishing ecological effects of contaminants. These tools include chemical analyses, toxicity tests, field trials, field surveys and special studies such as biomarkers and simulated mesocosm studies. Each tool provides particular information which, if used alone, are certainly useful.

However, the greatest value would be obtained by using tools of all types together, each providing a vital piece of the overall environmental puzzle. Once the tools have been deployed and the results assessed, the final step is to demonstrate which contaminants (if any) have caused the damage. In most cases the extent and nature of the toxic effects do not unequivocally demonstrate which agent was responsible for the injury. Correlations between a substance and a form of damage may be obvious, but causation is most difficult to prove, and often can be established only indirectly. The critical role of biomarkers for establishing the toxic effect of particular typoes of pollutant should be emphasised. Manipulative studies remain the best way of establishing a causal relationship between contaminants and responses at higher levels of organisation. Furthermore, due to the inconsistencies among various approaches, it is also important (particularly at moderately polluted sites) to develop integrated approaches to address contaminant impacts. The very nature of the contamination present in the Kola Peninsula makes monitoring an essential prerequisite for maintaining the current health or monitoring the improvement in the health of this harsh environment. It is hoped that the strategy outlined will serve to enable the development of a fully integrated approach to reduction of pollutants in this area.

References

1. McCarthy, J.F. and Shugart, L.R., 1990. *Biomarkers of Environmental Contamination.* Lewis Publishers, Boca Raton, FL.457pp.

2. Huggett, R.J., Kimerle R.A., Mehrle P.M. Jr. and Bergman H.L., (1992). *Biomarkers. Biochemical, Physiological, and Histological Markers of Anthropogenic Stress.* Lewis Publishers, Boca Raton, FL, USA 357pp.

3. Peakall, D. (1992). Animal biomarkers as pollution indicators. Chapman & Hall, London 291 pp.

4. Weeks, J.M. 1995. The value of biomarkers for ecological risk assessment: Academic toys or legislative tools. *Applied Soil Ecology* **2**, 215-216

5. Weeks, J.M., (1998). Effects of pollutants on soil invertebrates: links between levels. In: *Ecotoxicology, Eds Shuurmann G. and Markert, B. Ecological Fundamentals, Chemical Exposure and Biological Effects.* John Wiley and Sons, New York, pp. 645-662.

6. Van Brummelen, T.C., Verweij, R.A., Wedzinga, S.A., Van Gestel, C.A.M. (1996). Enrichment of polycyclic aromatic hydrocarbons in forest soils near a blast furnace plant. *Chemosphere* **32**, 293-314.

7. Evdokimova, G.A. (1995). Effects of pollution on microorganism biodiversity in arctic ecosystems. In: *Global Change and Arctic terrestrial ecosystems., Section 6, Ecosystem Research Report*.pp. 315-320.

8. Bush, E.J., Barrett, S.C.H. (1993). Genetics of mine invasions by *Deschampsia cespitosa* (Poaceae). *Can. J. Botany* **71**, 1336-1348.

9. Cox, R.M., Hutchinson, T.C. 1980. Multiple metal tolerances in the grass *Deschampsia cespitosa. New Phytologist* **84**, 631-647.

10. Davies, M.S., Francis, D., Thomas, J.D. (1991). Rapidity of cellular changes induced by zinc in a zinc tolerant and non-tolerant cultivar of *Festuca rubra* L. *New Phytologist* **117**, 103-108.

11. Schat, H., Vooijs, R., Kuiper, E. 1996. Identical major gene loci for heavy metal tolerances that have independently evolved in different local populations and subspecies of *Silene vulgaris*. *Evolution* **50**, 1888-1895.

12. Ernst, W.H.O., Leloup, S. (1987). Perennial herbs as monitor for moderate levels of metal fall-out. *Chemosphere* **16**, 233-238.

13. Weeks, J.M. and Svendsen, C. (1996). Neutral-red retention by lysosomes from earthworm (*Lumbricus rubellus*) coelomocytes: A simple biomarker of exposure to soil copper. *Environ. Toxicol. Chem.* **15**, 1801-1805.

14. Svendsen, C., A.A. Meharg, P. Freestone and J.M. Weeks (1996). Use of an earthworm lysosomal biomarker for the ecological assessment of pollution from an industrial plastics fire. *Applied Soil Ecology* **3**, 99-107.

15. Svendsen, C. and Weeks, J.M. (1997). Relevance and applicability of a simple earthworm biomarker of copper exposure: I. Links to ecological effects in a laboratory study with *Eisenia andrei*. *Ecotox. Environ. Safety*. **36**, 72-79.

16. Svendsen, C. and Weeks, J.M. (1997). Relevance and applicability of a simple earthworm biomarker of copper exposure: II Validation and applicability under field conditions in mesocosm experiment with *Lumbricus rubellus*. *Ecotox. Environ. Safety* **36**, 80-88.

17. Svendsen, C. Spurgeon, DJ, Zvezdelin BM., and Weeks, JM. (1998) Lysosomal membrane permeability and earthworm immune-system activity; field-testing on contaminated land. In: *Advances in Earthworm Ecotoxicology*, SETAC Press, Ed. Sheppard, S. pp.225-232.

18. Graveland, J. 1995. The quest of calcium. Calcium limitation in the reproduction of forest passerines in relation to snail abundance and soil

acidification. Doctorate Thesis, University of Groningen, NL. Published in *J. Animal. Ecol.* **66**, 279-288. 1997; *Physiol. Zool.* **70**, 74-84. 1997; *Can. J.Zool.* **74**, 1035-1044. 1996.

19. Peakall, D.B. (1993). DDE-induced eggshell thinning: an environmental detective story. *Environ. Rev.* **1**, 13-20.

20. Eeva,T. Lehikoinen,E. (1995). Egg shell quality, clutch size and hatching success of the great tit (*Parus major*) and the pied flycatcher (*Ficedula hypoleuca*:) in air pollution gradient. *Oecologia* **102**, 312-323.

21. Eeva, T., Lehikoinen, E. (1996). Growth and mortality of nestling great tits (*Parus major*) and pied flycatchers (*Ficedula hypoleuca*) in a heavy metal pollution gradient. *Oecologia* **108**, 631-639.

ABSTRACTS OF POSTERS

1. CONCENTRATION OF HEAVY METALS IN SELECTED GRASSES IN RELATION TO THEIR SOIL CONTENT IN A HEAVILY POLLUTED AREA

LUBOS BORUVKA
Department of Soil Science and Geology
Czech University of Agriculture in Prague
Czech Republic

Besides the region of Northern Bohemia where the pollution is widespread, there are some places of point contamination in the Czech Republic. One of them is the alluvium of the Litavka river in the district of Príbram, an area polluted with heavy metals due to metallurgical production. Samples of soils and grasses of genera *Festuca* and *Poa* were collected in order to determine content of Cd, Pb and Zn in plant roots and shoots and study to which extent is metal concentration in plants dependent on content of different metal forms in soils and on soil properties. Plant samples were mineralised in a dry mode mineraliser APION, the resulting ash was diluted using 1.5% HNO_3 and content of Cd, Pb and Zn in the solution was measured by means of atomic absorption spectrometry (AAS). In soil samples, a kind of sequential extraction with 0.1M $Ca(NO_3)_2$ for exchangeable metals and with 0.05M $Na_4P_2O_7$ (pH 12) for organically bound metals was used. Also, content of metals extractable with cold 2M HNO_3 in soils was determined.

According to main source of contamination and to metal content in soils and plants, two areas with different levels of contamination were distinguished. The first one was an area slightly polluted by atmospheric deposition from lead-smelting plants, the second one was an area heavily polluted with floods of water contaminated by wastes from metallurgical plant settling pits. In the former, contents of Cd, Pb and Zn in plant shoots were 0.312, 4.42, and 43.4 $mg{\cdot}kg^{-1}$, respectively, in the latter, these contents were 1.865, 19.4, and 326.5 $mg{\cdot}kg^{-1}$, respectively. Root concentrations were much higher. Content of all the three metals in soil of the latter area reached extreme values, too. The highest mobility in the system soil - plant roots - plant shoots was found for zinc, while mobility of lead was very low. In addition, lead was taken up through foliar uptake so that correlation of lead concentration in plant shoots with its topsoil content was rather poor. Cadmium and especially zinc concentration in plant shoots was well correlated with content of metals extractable with nitric acid and with exchangeable metals in topsoil. By means of multiple regression it was shown that among soil properties soil pH and organic matter content were the most important for metal

D. B. Peakall et al. (eds.),
Biomarkers: A Pragmatic Basis for Remediation of Severe Pollution in Eastern Europe, 299–300.

uptake by plants. It can be concluded that for zinc and to a lesser extent also for cadmium studied grasses can serve as a rough indicator of metal pollution of soil, while for lead this is not so clear.

2. USING BIOMARKERS TO MONITOR REMEDIAL ACTIONS AT A SUBTIDAL ESTUARINE SITE SEVERELY CONTAMINATED WITH POLYNUCLEAR AROMATIC HYDROCARBONS

TRACY K. COLLIER and MARK S. MYERS
NOAA, NMFS, Northwest Fisheries Science Center
Environmental Conservation Division
Seattle, WA 98112. USA

Extensive scientific evidence indicates that polycyclic aromatic hydrocarbons (PAHs) and their derivatives are responsible for a variety of serious biological effects in marine fish exposed to such compounds. While much of this information is derived from laboratory studies, there is an increasing body of evidence from field studies supporting these relationships between PAH exposure and biological damage. Several of the most definitive field studies in this regard have been done in Puget Sound, Washington, where one of the most heavily PAH-contaminated sites is Eagle Harbor, on the east side of Bainbridge Island. Studies done by NOAA scientists over the past decade have shown that flatfish residing in Eagle Harbor are at increased risk of developing serious toxicopathic liver diseases, including cancer, and of exhibiting impaired reproductive function and altered physiological responses. The USEPA completed a Remedial Investigation/Feasibility Study, and in December 1991 proposed a phased cleanup of the harbor. In June 1993, under Superfund Removal authority, the USEPA recommended placement of uncontaminated material over the most contaminated area of Eagle Harbor; in late September of 1993 the USEPA began placement of this sediment cap as a means of 1) controlling transport of contaminants; 2) isolating contaminants from marine biota; and 3) providing clean habitat for benthic organisms, thus reducing overall risk of exposure to the contaminants contained in the sediments. This capping process was completed in March 1994, resulting in placement of an approximately three foot thick cap of clean sandy material and associated wood debris over an original area of ca. 54 acres of the most highly contaminated subtidal sediments within Eagle Harbor. The purpose of this work was to assess whether the sediment cap has reduced PAH exposure in flatfish residing in Eagle Harbor. To accomplish this objective, several indirect measurements of PAH exposure were combined, and measured in three epibenthic flatfish species: English sole (*Pleuronectes vetulus*), rock sole *(Lepidopsetta bilineata*) and starry flounder (*Platichthys stellatus*). These measures, also referred to as biomarkers of PAH exposure, are described below. Because of the high rate of metabolism and depuration of PAHs by many species of fish, direct measurement of parent compounds in tissues is not always the best

D. B. Peakall et al. (eds.),
Biomarkers: A Pragmatic Basis for Remediation of Severe Pollution in Eastern Europe, 301–304.

indicator of exposure of fish to PAHs. In fact, the focus on tissue analyses has made it difficult to document exposure of fish to PAHs in contaminated habitats. In recent years, however, methods have been developed for determining such exposure, based on our increased understanding of the mechanisms and pathways of PAH metabolism in fish. In this investigation three methods were used.

The first is based on the measurement of fluorescent aromatic compounds (FACs) in bile by high performance liquid chromatography (HPLC) fluorescence. Many polar and fluorescent metabolites of PAHs are primarily excreted via the hepatobiliary system in fish. This method provides a semi-quantitative means for rapidly assessing recent PAH exposure in fish and has the added advantage of being able to be 'fine-tuned' for different classes of PAHs, such as 2-3 ring, 3-4, ring, or 4-5 ring structures.

The second method relies on the finding that hepatic cytochrome P4501A (CYP1A), together with the associated monooxygenase activity of aryl hydrocarbon hydroxylase (AHH), can be rapidly induced in many fish species after exposure to a wide variety of organic contaminants, including PAHs. Thus, the induction of CYP1A represents an early biological effect which is also useful in estimating PAH exposure of fish; however, CYP1A induction is not specific for PAH exposure. Moreover, the induction of CYP1A is initiated by the interaction of compounds with the Ah receptor, a step believed to be integral to the development of toxic effects, including reproductive dysfunction, teratogenicity, mutagenicity, and carcinogenicity.

The third method we employed is based on findings that electrophilic intermediates of high molecular weight PAHs can form covalent adducts with nucleophilic centres within biological macromolecules, including hepatic DNA. This recent method is highly sensitive for detecting bulky hydrophobic xenobiotic-DNA adducts utilising 32P-postlabeling of deoxynucleotides and chromatographic separation of adducted nucleotides from normal nucleotides. These methods have been employed by the staff of the ECD in determining PAH exposure of fish. The time- and dose- responsiveness of each of these measures has been determined after exposure to either single PAHs or to complex mixtures of PAHs and other organic contaminants and each of these measures has been extensively field tested by our group, both in Puget Sound and elsewhere in the USA. The results of these studies have shown that the combined use of several indicators is the optimal approach, because each method focuses on a different aspect of PAH toxicokinetics, and provides useful information concerning possible toxic effects of PAH exposure. The main objective of this project was to evaluate whether PAH exposure of flatfish in Eagle Harbor was reduced following the completion of the sediment cap. However, as data were collected and analysed, other specific hypotheses were also formulated and tested, such as whether reductions in exposure differ among the three species, or in classes of PAHs. There were also some limitations inherent in this investigation. Because fish are able to move freely over the capped and uncapped areas of Eagle Harbor, and because we could not trawl within the capped area, the project could not be designed to delineate reduced exposure specifically in the capped area. However, our results were able to assess whether there was an overall reduction of

PAH exposure of fish in Eagle Harbor, subsequent to the capping of a substantial portion of the most contaminated sediments in the harbor.

This investigation suggested that the placement of the sediment cap over a portion of eastern Eagle Harbor may have contributed towards an overall reduction in PAH exposure of bottomfish. However, factors other than exposure associated with the capped area may have been equally or more important in determining total PAH exposure and associated biological effects in resident flatfish species. Generally, indicators of short-term exposure to PAHs (e.g. CYP1A induction and concentrations of FACs in bile) were considerably reduced in flatfish from Eagle Harbor even before cap placement began, in comparison to historical values. This indicates that previous source control measures enacted between 1988 and September 1993, such as control of upland and groundwater PAH contamination, may have contributed to reduced PAH exposure in epibenthic fish in Eagle Harbor. Nonetheless, the placement of the cap appears to have enhanced habitat for flatfish species, by converting an area which had been too grossly contaminated to allow longer term residence into an area in which flatfish could inhabit. This hypothesis is supported by two observations. First, in laboratory studies conducted in 1987, in which English sole were placed into aquaria that contained a layer of undiluted Eagle Harbor sediment collected from the capped area, there was an immediate (i.e., within 10 minutes) mortality of all fish. Thus, it is likely that the most contaminated portion of Eagle Harbor did not provide usable flatfish habitat in its uncapped state. Second, we observed in the field that there was a noticeable increase in the catch per unit effort of rock sole after placement of the sediment cap, especially in relation to English sole and starry flounder. Rock sole prefer a coarser substrate (e.g., sand, gravel, and cobble) relative to other flatfish species common on the Pacific Coast, including Puget Sound . The cap material used was substantially coarser than the silt/mud sediments that generally characterise Eagle Harbor; thus it is possible that the increased proportional catch of rock sole noted after sediment capping is due to an expanded habitat for this species provided by a more suitable substrate, represented by the exogenous capping material. This hypothesis is further supported by the finding that, following cap placement, rock sole exhibited the most consistent temporal decreases among the biomarkers of PAH exposure employed in this study (hepatic CYP1A, hepatic DNA adducts, and biliary FACs), with both hepatic CYP1A and hepatic DNA adducts clearly showing a decreasing trend. The exceptions to this pattern in rock sole were found in levels of FACs. The reason for this exception is unknown, but of the biomarkers used, biliary FACs respond most rapidly to PAH exposure, and may be more subject to variability due to the occurrence of episodic exposure. The trend toward decreased exposure of rock sole relative to other species was especially evident for the longer-term dosimeter of PAH exposure and response, hepatic DNA adducts, where the decline seen in the post-capping sampling points was most pronounced in rock sole. Preferential occupation of the capped area by rock sole and reduced exposure to PAHs relative to English sole is further supported by the finding of substantially and consistently lower concentrations of high molecular weight PAHs in stomach contents of rock sole compared to English sole sampled at several timepoints subsequent to completion of sediment capping operations (unpublished data). The data on PAHs in stomach contents together with data on temporal changes since capping for the three major biomarkers used here, strongly suggest that English sole and starry flounder may have occupied PAH- contaminated areas peripheral to the cap. It can be speculated that if the cap material had contained a higher proportion of silt/clay, English sole and starry flounder may have occupied the capped area to a greater extent and thus exhibited clearer reductions in PAH exposure. However, because we did not conduct trawling operations over the sediment cap proper, and consequently lack adequate data to assess relative densities of flatfish species in the area capped by exogenous sediments, the above hypothesis must be regarded as conjectural. The current investigation did not begin until after onshore remedial activities had been conducted, and perhaps more questions regarding reduction of PAH exposure in the nearshore marine environment could have been answered had appropriate measurements been made prior to these onshore activities. Nonetheless, the data presented in this report will provide a useful reference for evaluating the effectiveness of future efforts to restore essential habitat for living marine resources. Clearly, there are many areas of Eagle Harbor peripheral to the capped area which are still contaminated with PAHs. We recommend that, as remedial actions are planned for these other units, they be accompanied by biological surveys of marine biota as well as biochemical assessment of fish and invertebrate populations prior to, during, and subsequent to these actions. As agencies take actions to redress past insults to marine resources resulting from chemical contamination, the timely use of a combination of assessment tools will allow for an intelligent and objective evaluation of the efficacy of such actions.

3. HEAVY METAL POLLUTION OF THE UPPER SILESIA REGION: BIOMONITORING STUDIES OF MAGPIE (*PICA PICA*) TISSUES

KRZYSZTOF DMOWSKI
Department of Ecology
Warsaw University, 00-927 Warsaw
Poland

Magpies fulfil most criteria as biomonitoring species. They have widespread occurrence, resident behaviour and possibility of accumulation of trace elements within a relatively small individual territory. During the period 1989-1993 the magpies from 21 Polish regions were captured in living-traps. The regions were different in type of industrialisation and urbanisation. There were also included the surroundings of the 3 largest Polish zinc smelters. Two of them are located in the Upper Silesia region (Szopienice and Miasteczko Slaskie). The Bukowno area is placed at the south-east edge of the region and can seriously influence the Upper Silesia by air-borne dust emissions. Tail feather samples were taken from all of captured 165 birds and about 60 samples of internal organs from the sacrified birds.

Twenty elements were determined (Pb, Ba, Cd, Zn, Mn, Fe, Hg, Tl, Sb, Sn, Mo, Rb, Se, As, Ge, Cu, Ni, Co, Cr, V) with inductively coupled plasma mass spectrometry (ICP-MS).

In the feather samples from the vicinities of all the studied zinc smelter areas greatly enhanced concentrations of sȯme elements, especially Pb, Cd, Zn, Sn, Ge, were found. These were at least a dozen and sometimes scores of times higher than those in the samples from other industrialised areas, so far considered as seriously polluted with heavy metals. There were also found very high levels of: copper and selenium in Szopienice, barium in Miasteczko and thallium in Bukowno. If the areas surrounding the zinc smelters can be recognised as the most polluted with heavy metals in Poland, the primacy of pollution degree among them belongs to the Szopienice area. The concentrations of many elements were the highest there, especially in 1989.

The analyses of magpie kidneys and livers have revealed much more thallium in Szopienice than in Bukowno, and its permanent presence in 1989 and 1993. Although the emissions have decreased in Szopienice and Miasteczko between 1989 and 1993 changes in tail feather contents indicated that the heavy metal amounts stored in soils and transferred through food chains were at the same level. There were unchanged (or even rising) contents of Pb and Cd in magpie livers and kidneys.

D. B. Peakall et al. (eds.),
Biomarkers: A Pragmatic Basis for Remediation of Severe Pollution in Eastern Europe, 305.

4. ENERGY METABOLISM IN THE BUTTERFLY *Parnassius apollo* FROM PIENINY MOUNTAINS (SOUTHERN POLAND) AND ITS POSSIBLE USE FOR BIOTOPE ASSESSMENT

A. KÊDZIORSKI, M. NAKONIECZNY, J. BEMBENEK,
K. PYRAK, G.ROSIÑSKI*
University of Silesia, Dept. Human & Animal Physiology;
Bankowa 9 PL 40-007 Katowice, Poland;
**The A. Mickiewicz University, Dept Animal Physiology;*
Fredry 10, PL 61-701 Poznañ, Poland

In Europe the apollo butterfly - considered to be the symbol of insect protection - is a relic of inerglacial fauna. Presently occurence of this rare species is limited to areas with rather severe climatic conditions. In Poland it inhabits screes covering mountain-sides of Pieniny and Tatra Mts. These biotopes occupy rather small areas, nevertheless numerous plant and animal species – particularly insects – can be found there. Nowadays, due to human activities influencing the whole landscape, these formerly stable habitats are submitted to destructive factors of both natural and anthropogenic origin. The most severe natural threat is overgrowing of the screes by bushes and trees, due to lack of large herbivores, and planned aforestation with conifers (also new ecotypes of spruce) and deciduous species. For this reason artificial deforestation of these biotopes is necessary for their further existence. In the last few decades another threatening factor seems to be of growing importance. Numerous chimneys, exhaust-pipes of cars, pesticide use and so on, release to the environment increasing amounts of pollutants. Moreover, the lay-out of Pieniny massifs creates a natural barrier for air-borne pollutants coming from the south and west. The screes, covering mostly southern or south-western slopes, are particularly exposed to their influence.

Vulnerability of the screes to all these factors raises a question how to assess the actual condition of these biotopes. Many ecological and ecotoxicological criteria can be applied although, in fact, we do not know the 'blank' situations, i.e. what should be the values of various ecological or toxicological parameters in "healthy" habitats. One of the useful indicators seems to be so called "umbrella species". This term means a species typical for the biotope and responding to its slight changes. Hence, the stable population of the species indicates healthy condition of its habitat. Moreover, if the population suffers from various impacts, actions aimed at its conservation protect simultanously other biota. *Parnassius apollo* is considered to be such an indicator species for the screes. In the last few decades it has suffered rapid decline in almost all Europe, although in Poland the process became particularly severe. It coincided with the above-mentioned shrinkage of its biotopes.

D. B. Peakall et al. (eds.),
Biomarkers: A Pragmatic Basis for Remediation of Severe Pollution in Eastern Europe, 306–307.

As a result of this, in the begining of the 1990s there were only two small apollo populations within Polish borders - one in Pieniny and the other in Tatra Mts. To stop further extinction a reintroduction programme aimed at preserving the subspecies (*P. apollo ssp. frankenbergeri*) in Pieniny Mts was launched in 1991. It has involved rearing and breeding apollo in semi-natural colony (being a constant supply of individuals for the wild population) accompanied with population studies in the field. Hence it became desireable to study also physiological condition of the species. Analysis of energy budget during apollo development seems to provide vital information on the individuals requirements and condition. So far, in the first step of these investigations, we could only use individuals taken from the colony. We were also interested whether, on this basis, it was possible to select the most prospective reproducers (particularly females) among the reared individuals for the needs of reintroduction programme.

Results suggest, that the energy budget of the apollo butterfly do not differ markedly from that of other lepidopteran species examined so far. The most energy was consumed and retained in the last larval instar. The larvae that hatched later (march-april) had lower values of energy budget (particularly consumption and assimilation) in comparison with early-hatching individuals. Whatever the reason (less favourable abiotic conditions, lower quality of food, higher toxicity of food-plant due to accumulation of pollutants and secondary metabolites) it resulted in higher mortality of the "later" individuals, and lower fecundity of the females. Further, more detailed studies (in the progress) should shed more light on the links between external factors and energy budget in this species.

Despite significant differences between energy metabolism of males and females in the last larval instar (but not in younger larvae and pupae) no convenient indicator has been found so far, that could allow selection of the best reproducers.

5. SMALL MAMMALS AS BIOINDICATORS FOR TERRESTRIAL ECOSYSTEMS IN BOHEMIA

E. KOROLEVA, D. MIHOLOVA, J. CIBULKA, P. MADER, A. SLAMEVA
Moscow State University, Laboratory of Bioindication
Department of Biogeography
119889 Moscow, Russia

The concentration of heavy metals in the environment, their deposition and long range transport are becoming one of the more prominent issues for some regions of Central and Eastern Europe. Heavy metals pollution can be monitored by using several basic methods. One is to measure the accumulation and concentration of heavy metals in tissues of indicator species. Biological indicators (or bioindicators) are organisms which - with reference to chemical, physiological, ethological or ecological factors - provide information on the state of ecosystems. They must be sensitive enough to indicate environmental changes as soon as possible (Novakova, 1987). Small mammals represent one of the best groups as ecological indicators in terrestrial ecosystems. Mammalian body, especially kidney, bone and liver responses are realistic monitors of mammalian exposure to heavy metals. The problems associated with conducting open cast coal mining operations in an environmentally friendly and acceptable manner are important issues for some regions in Central and Eastern Europe. An indivisible part of industrial activity is subsequent rehabilitation of all areas of abandoned and closed mines. Such revitalisation of landscape (restoration of soil fertility, creation of basic green cover, reforestation and re-joining it with neighbouring biotops) requires both biological and chemical monitoring. Small mammals are often used as monitors of environmental contamination (Talmage and Walto, 1991). Study of the input of risk elements from the environment into the animal and human organisms requires as one of the most important factors the knowledge of natural levels (as background or so-called reference values) of the elements in organs of animals living in uncontaminated areas. Three regions of the Czech Republic with expected minimal and maximal heavy metal contamination were selected for the research. The first region was the area in South Bohemia in the vicinity of Temelin; the second is the protected area of the water reservoir Klicava in the Central Bohemia, and the third region (Bilina) was the polluted industrial area in North Bohemia. As bioindicators of environmental monitoring, small mammals were used. Pb, Cd, Cu and Zn contents in liver, kidney, spleen, stomach and bones of small mammals living in different biotypes were determined by atomic absorption spectrometry. The results

D. B. Peakall et al. (eds.),
***Biomarkers: A Pragmatic Basis for Remediation of Severe Pollution in Eastern Europe*, 308–309.**

show the significant differences between metal concentrations in tissues of animals from contaminated and uncontaminated areas as well difference between animals with mixed and plant diets. The data produced from Klicava region can be used as reference values for small mammals as bioindicators in environmental monitoring in the Czech Republic.

6. ACCUMULATION OF MUTAGENIC XENOBIOTICS IN WATER ECOSYSTEMS WITH SPECIAL REFERENCE TO LAKE BAIKAL

SERGEY V. KOTELEVTSEV, LUDMILA I. STEPANOVA, VADIM M. GLASER
Biological Department, Moscow State University
119899, Moscow, Russia

Mutagenic substances are amongst the most dangerous compounds of the antropogenic pollution. Their action may be delayed and later manifested in the subsequent generations of animals (including fish) and people. Recent sharp increases in the number of tumours in fish in most cases can be attributed to the pollution of aquatic ecosystems with mutagenic and carcinogenic xenobiotics (MCX), with a smaller number of malformations caused by viral diseases. Chemical methods of detection are not sensitive enough to monitor MCX which can cause biological effects at extremely low concentrations in tissues of animals. The advantages of biological testing are high sensitivity and possibility to identify compounds which acquire mutagenic and carcinogenic properties after metabolic activation, for instance with cytochrome P-450. We assayed tissue extracts of aquatic organisms (aquatic plants, decapoda, molluscs, fish, Baikal seals, fish-eating birds and their eggs) from Lake Baikal and the Selenga River estuary as well as fresh and waste water samples using *Salmonella*/microsome Ames test. Presence of MCX was followed in the main food chain in Lake Baikal which consists of a primary zooplankton producer *Epishura baikalensis* (90%) and amphipods, endemic sculpins (the main species of Baikal oilfish) and Baikal seal. We did not find mutagenic activity in the fish tissues. However some extracts from seal fat and muscles demonstrated genotoxic activity. Frame shift mutations were caused by mutagenic samples from seal fat (*Salmonella typhimurum*, strain TA 98). Generally seal fat extract has a toxic effect on the TA 100 strain even without metabolic activation. Weak mutagenic activity was found in the eggs of birds. Mutagens were present in both fish eating (*Larus ridibundus*) and mollusc eating (*Anas plathyrhynchos*) birds. Most of the birds examined come to the Selenga estuary only at breeding time and can accumulate mutagens not only from Lake Baikal ecosystem. The activities of cytochrome P-450 and the enzymes of phase II detoxification were studied in the liver of fish and birds. The relationship of bio-accumulation to the levels of enzymes possessing both detoxification and activation activities in fish and birds will be discussed. In general, the level of the mutagen accumulation depends on the activity and the level of induction of detoxification and metabolic activation enzymes in the liver of birds and fish.

The study was supported by INTAS (Ref. No. 94-0531) and NATO Environment.

D. B. Peakall et al. (eds.),
Biomarkers: A Pragmatic Basis for Remediation of Severe Pollution in Eastern Europe, **310.**

7. THE SITUATION REGARDING AIR POLLUTION IN THE CZECH PART OF BLACK TRIANGLE (CZ-BT) AND ITS IMPACT ON SPRUCE CANOPY USING INNER CYCLING

JANA KUBIZNAKOVA
Institute of Landscape Ecology Czech Academy of Sciences
37005 Ceské Budejovice, Czech Republic

CZ-BT region is heavily damaged by large anthropogenic activities, including, emission sources, and extended emission situation. In the whole Czech Republic, there is about 78 thousand km^2 area with a population of 10.3 mil. inhabitants with very heterogeneous environmental conditions, land cover, geological bedrocks, soil types, orography, and the biggest emission sources in a relatively small area. CZ-BT constitutes one fifth of the Czech Republic, but permissible levels of air pollutants for urban, rural environment for the health of man are exceeded. This can be demonstrated by health statistics, changes in biodiversity species, decline of forest ecosystems, decreased yield of agricultural crops, corrosion of materials, affecting of weathering. The mobility of toxic elements in water (see e. g. Al, As, Be, Cd, F and Pb) enhances the likelihood of inputs to the food chain. CZ-BT has a highest yearly deposition of: >4g $S{\cdot}m^{-2}$; >1.5 g $N{\cdot}m^{-2}$; >0.4 mg $Cd{\cdot}m^{-2}$; >50 mg $F{\cdot}m^{-2}$; >5 mg $Pb{\cdot}m^{-2}$. Since 1960's in the Ore Mts., first symptoms have shown new types of damage to Norway spruce - yellowing, loss of needles, loss of fine roots, decreased growth, and premature death. This phenomenon is widespread in Europe and has complex causes. Spruce decline has been intensively studied and has been attributed to five main causes: aluminium toxicity (indirect acidification), ozone effect, Mg-deficiency, excess-nitrogen (eutrophication) and comprehensive stress theory. The stress hypothesis will not be discussed, but impact assessment of air pollution on forest will be emphasised in the context of other stress factors.

Some ecotoxicological criteria and criteria for assessment of acidification and eutrophication in spruce forest reflecting the direct and indirect effects of air pollution were chosen. Coniferous canopies are used as relatively efficient and remedial collectors removing some particles and undesirable gases from polluted atmosphere. The change in inner circulation of spruce tree will be applied as an indicator of stress in complex interactions of environmental factors (cold injury, drought stress, soil acidity, aluminium toxicity, insect attacks, high concentration of sulphur dioxide, nitrogen oxides, ozone, acid rain etc.). This response occurs before there is any visible evidence of damage.

Our new approach estimates the impact of acid air pollutants on spruce trees, and their physiological behaviour and consequently reflects the morphological response of trees. Some results from own measurements of atmospheric depositions will be presented in a case-study, which is part of an international join project EASE.

D. B. Peakall et al. (eds.),
Biomarkers: A Pragmatic Basis for Remediation of Severe Pollution in Eastern Europe, 311.

8. CONTAMINATION OF LAKE BAIKAL WITH ORGANIC PRIORITY POLLUTANTS

ALBERT LEBEDEV, OLGA POLYAKOVA, NADEJDA KARAKHANOVA, VALERY PETROSYAN
Organic Chemistry Dept., Moscow State University
119899 Moscow, Russia

Lake Baikal is the largest reservoir of natural fresh water in the world. Nevertheless industrial enterprises on its banks as well as river effluents and transboundary transfer contaminate this unique basin. However an overall estimation of the presence of all classes of organic pollutants has never been done. The present work was launched to identify organic toxicants, especially the priority pollutants of lake Baikal and to find out their routes of appearance. We have studied the presence of organic priority pollutants (US EPA list) in samples including water, soil, snow and biota like molluscs, fish, birds and seals in order to clarify also the accumulation of the pollutants in trophic chains. The chemical studies were accompanied by biotesting. The results obtained in 1996-1997 throw light upon the presence of organic toxicants (PAHs, phenols, organochlorines, phthalates) in birds and seals, i.e. the representatives of the highest trophic levels. The results demonstrate high level of contamination of the studied animals. The food of the most polluted species consists mainly of molluscs (not fish); thus we propose that the major role of transfer of toxicants into the higher trophic levels is via water and aquatic invertebrates. The identification of more than 70 organic compounds in the samples was the first step for the creation of the priority pollutants list for the lake. Besides that the data array obtained allows differentiation between contaminants really characteristic for Baikal and those which are accumulated by birds during their winter stay in China.

D. B. Peakall et al. (eds.),
***Biomarkers: A Pragmatic Basis for Remediation of Severe Pollution in Eastern Europe*, 312.**

9. MONITORING AND BIOLOGICAL INDICATORS IN KUZBASS

VLADIMIR E. SERGEEV
Dept. Zoology & Ecology
Kemerovo State University
Krasnaja 6, 650043 Kemerovo, Russia

The peculiarities of relief and climate give a unique character to the Kuzbass nature, manifesting itself in the richness of flora and fauna, and diversity of biosystems. But intensive technogenic development of the region brought about an ecological crisis due to industrial discharges and run-offs, and also the degradation of soil cover in natural biosystems because of mining. About 80% of the land has been changed due to direct environmental impact. Progressive heavy damage to *Abies sibirica*, the main forest species, has been reported. The taxonomic diversity of the flora and fauna and the structural diversity of biosystems were found to decrease. Nevertheless the assessment of the technogenic impact on the environment is determined mainly by the level of the contamination of air, water and soil. The use of biological indication is less developed. Only a few specialists making individual studies are engaged in it. In spite of the current situation it can be stated that the accumulated empirical data in Kuzbass, e.g. the list of plant, forest and grass communities with a different level of anthropotolerance, can facilitate the development of the environmental monitoring system and biomarkers. Also, the populations and communities of the vertebrates have been studied as biomarkers for 20 years and their effectiveness as indicators has been estimated by the indices of biological diversity and anatomical and morphological structures. The assessment of health quality of population and community structure and environmental quality assessment were determined by means of ecological screening methods used in human and animal physiology. For this purpose the semiautomatic two-channel blood express-analyser developed at the Department of Human and Animal Physiology, Kemerovo State University was used. It enabled us to consider more than 30 structural and functional indices of blood as biomarkers. The studies mentioned above can facilitate the development of practical actions to remedy the situation.

D. B. Peakall et al. (eds.),
Biomarkers: A Pragmatic Basis for Remediation of Severe Pollution in Eastern Europe, 313.

10. ASSESSMENT OF WATER POLLUTION FROM OIL SHALE PROCESSING USING BIOMARKER RESPONSES IN FISH *IN VIVO* AND IN PLHC-1 FISH CELLS *IN VITRO*

ARVO TUVIKENE[1,2], SIRPA HUUSKONEN[3], KARI KOPONEN[3], OSSI RITOLA[3], ÜLLE MAUER[4] & PIRJO LINDSTRÖM-SEPPÄ[3]
[1] *University of Tartu, Institute of Zoology and Hydrobiology EE-2400 Tartu, ESTONIA*
[2] *Limnological Station, Institute of Zoology and Botany EE-2454 Rannu, ESTONIA*
[3] *University of Kuopio, Department of Physiology, P.O.B. 1627, FIN-70211 Kuopio, FINLAND*
[4] *University of Tartu, Institute of Physical Chemistry EE-2400 Tartu, ESTONIA*

The aquatic pollution of River Narva was monitored by measuring activities of biotransformation enzyme in caged rainbow trout and in feral perch and roach, as well as in a fish hepatoma cell line (PLHC-1) exposed *in vitro*. R. Narva passes the oil shale mining and processing area and thus receives elevated amounts of, e.g., polycyclic aromatic hydrocarbons (PAHs), heavy metals and sulphates as pollutants.

In fish studies, xenobiotic metabolism was detected by measuring cytochrome P4501A (CYP1A) dependent monooxygenase (EROD, AHH) activities as well as conjugation (GST, UDP-GT) enzyme activities. CYP1A induction was further studied by detecting the amount and occurrence of CYP1A enzyme protein in fish tissues. PLHC-1 cells were exposed to the sediment extracts and CYP1A induction was detected by measuring EROD activity in the exposed cells. Total protein content of the cells was assayed as a measure of cytotoxicity. Selected PAHs and heavy metals (Cd, Cu, Hg and Pb) from fish muscle and liver tissues, and from sediment samples were analysed to point out the major contaminants.

EROD and AHH activities did not show CYP1A induction in contaminated fish when compared to references. However, total content of PAHs in the same fish was quite high. The lack of CYP1A induction in fish from R. Narva was not expected due to the high content of PAHs and was probably caused by elevated content of some heavy metals, which are known to be inhibitors for monooxygenases. Another possibility is that through adaptation the fish had lost some of their sensitivity to PAHs. GST activities were slightly increased at exposed areas in both caged and feral fish. The sediments collected from contaminated sites caused cytotoxicity and EROD induction in the PLHC-1 cells. The effects seen in the cells represented the potential rather than the actual hazard of the pollutants in the sediments. The PLHC-

D. B. Peakall et al. (eds.),
***Biomarkers: A Pragmatic Basis for Remediation of Severe Pollution in Eastern Europe*, 314–315.**

1 cells could be used to roughly categorise aquatic sediments and to screen waters prior to other field experiments. Either complex pollution, caused by oil shale processing, masked part of the harmful effects measured in this study, or oil shale industry did not have any severe effects on fish in R. Narva. Our study illustrates the difficulties in estimating environmental risk in a case where there are numerous various contaminants affecting the biota.

11. THE EFFECT OF ZINC ON DETOXIFYING ENZYMES IN THE GROUND BEETLE *POECILUS CUPREUS*

G. WILCZEK, A. BABCZYNSKA, P. KRAMARZ*, R. LASKOWSKI*, P. MIGULA
Department of Human and Animal Physiology, University of Silesia, Bankowa 9, 40-007 Katowice, Poland
**Institute of Environmental Biology, Jagiellonian University Ingardena 6, 30-060 Kraków, Poland*

Carnivorous insects are important for the functioning of terrestrial ecosystems. Xenobiotics tend to accumulate especially in higher trophic levels of food chains. The secondary consumers are particularly vulnerable to various pollutants, including heavy metals. The ground beetle *Poecilus cupreus*, as a predator, is potentially exposed to elevated levels of toxicants. However, its body contains relatively low concentrations of heavy metals even in highly polluted areas.

In this study, by studying a simplified fragment of a basic food chain under the laboratory conditions, we evaluated whether excessive amounts of consumed zinc may affect detoxification abilities of this carnivorous insect.

Activities of non-specific carboxylesterases (CarE) and glutathione S-transferases (GST) have been studied in pupae and in adult females and males of ground beetle *Poecilus cupreus* subjected to prolonged intoxication (80 days) with zinc. Carabids were kept under laboratory conditions from the egg stage. Immediately after hatching larvae were fed every two days with house fly pupae kept on an artificial medium enriched with 400, 1600, 6400 and 8000 mg $Zn \cdot kg^{-1}$. The control group was offered pupae reared on uncontaminated food. Activities of CarE were measured against p-nitrophenyl acetate (pNPA). The GST was assayed by measurements of the conjugation rate of 3,4-dichloronitrobenzene (DCNB) and 1-chloro-2,4- dinitrobenzene (CDNB) with glutathione.

The effect of zinc on carboxylesterase and glutathione S-transferase activities in adults of *Poecilus cupreus* depended significantly on the metal concentration in the food. In groups which were moderately intoxicated (1600 mg $Zn \cdot kg^{-1}$ and 6400 mg $Zn \cdot kg^{-1}$) the enzymes activities were strongly stimulated. At the highest zinc concentration (8000 mg $Zn \cdot kg^{-1}$) used for feeding the prey, the activity rates of both enzymes were insignificant in relation to control. This means the involvement of other (probably nonenzymatic) detoxifying mechanisms.

The GST assays showed the highest rates of activity in pupae both against DCNB and CDNB and were always stimulated by zinc present in their diet. The conjugation reactions are probably important in xenobiotic biotransformation at this developmental stage.

D. B. Peakall et al. (eds.),
Biomarkers: A Pragmatic Basis for Remediation of Severe Pollution in Eastern Europe, **316.**

12. BIOTESTING AS A METHOD OF ENVIRONMENTAL ASSESSMENT IN INDUSTRIAL AREAS

S.V. CHERNYSHENKO, A.N. MISYURA, V.Y. GASSO, A.V. ZHUKOV
Dniepropetrovsk State University,
13 Nauchny Lane
320625 Dniepropetrovsk-10
Ukraine

In some parts of the world, as with the steppe Dnieper region, anthropogenic influence on the environment has intensified to a large extent. Although there are other methods for assessment of populations and ecosystems, indices of metabolism are also able to indicate the state of populations.

Research on soil invertebrates, amphibians, reptiles and mouselike rodents were carried out in differently polluted ecosystems of the Dnieper region. Data obtained showed that most of the animals have similar alterations of metabolism indices to pollution factors, which mirror the stage of their adaptations. These alterations were increasing nucleic acids (RNA, DNA), phospholipids and cholesterol and decreasing triglycerides, that were most pronounced in liver, skin, lungs and kidneys of amphibians and reptiles. As this takes place, differences of mixed function oxygenases b-5 and P-450 between earthworms and amphibians under pollution impact were found. One consequence is the increase or decrease of animal resistance to toxicants. As a whole, it is necessary to note that for the biotesting system such biochemical indices as nucleic acids content, some lipids fractions and liver microsomal cytochromes b-5 and P-450 are mostly suitable. These parameters may be used in biomonitoring system not only for polluted but for clean and protected territories.

Existing earthworm toxicity test use *Eisenia andrei* and artificial soil substratum (Van Gestel *et. al.*, 1989). In this method earthworms are exposed to chemical substances for 3 weeks to assess effects on growth and reproduction. But, it is obvious that before toxic effects occur on the growth and reproduction capacity, biochemical changes will take place. Therefore biochemical parameters of earthworm may be considered to be more sensitive then growth and reproduction response. The aim of the investigation is to study biochemical changes in the earthworm related to the degree of the soil contamination by heavy metals. To decide this problem the laboratory experiment was made. The animals used were adult with a well developed clitellum and were of the species *Lumbricus rubellus* and *Octolasion lacteum*. Earthworms were put into vessels with different degree of

D. B. Peakall et al. (eds.),
***Biomarkers: A Pragmatic Basis for Remediation of Severe Pollution in Eastern Europe*, 317–318.**

soil contamination (20, 40, and 80 mg Cd/kg dry soil). Soil was treated by toxicant in the form of chloride (Cd $Cl_2 \cdot 2H_2O$).

Biochemical analyses were performed after 2, 5, 10, and 20 days of experiment. Earthworms in both experimental and field condition were tested to estimate the total protein, nucleic acids (RNA, DNA), lipids and their fraction, cytochrome P-450 and RNA content in microsomal fraction. The bioconcentration factor (BCF) was found to reach a maximal value of more then 1.00 for *L. rubellus* (1.74-3.30) and less or equal 1.00 for *O. lacteum* (0.79-1.08). This difference may be due to ecological properties of these species of earthworms. The *L. rubellus* is known to be a litter dwelling invertebrate and the earthworm of species *O. lacteum* is soil dwelling animal. On the other hand, the metabolism of *L. rubellus* is more intensive than *O. lacteum*. Therefore, the litter dwelling earthworm shows a higher level of cadmium accumulation, as soon as heavy metal consumption and excretion depended on metabolism. The total RNA amount reflecting protein synthesis intensity was decreasing for 2 days of cadmium exposure. After this period the fall of intensity of protein synthesis process was substituted by compensatory gain of RNA content. However, it is noteworthy that maximal RNA concentration (2.81 mg/g) was shown at the 40 mg Cd/kg level of contamination and at the 80 mg Cd/kg this parameter was lower (2.42 mg/g). For 5 days of experiment the DNA content increase was detected at the 20 and 40 mg Cd/kg, but at the 80 mg Cd/kg a sharp decrease of this parameter was observed. After 10 days of experiment the changes of biochemical parameters maintained the stable direction. Toxic cadmium effect led to considerable changes of protein synthesis. Detoxifying processes involving energy costs were revealed by changes of TAG content and other lipid fraction. Experimental data were used to carry out mathematical treatment by means of factor analysis methods. The factors 1, 2, and 3 embrace 71.3% of the total variance (33.9, 22.6, and 14,8 respectively). The factor 1 has a close relation with such parameters as DNA, cytochrome P-450, total lipid, free fatty acids, triacylglycerols. This factor may be considered as being a total stress response of the earthworm. Factor 2 correlated with total and microsomal fraction RNA and steroid content. Factor 3 is determined by phospholipids and unidentified fraction. The total protein content in the body of the earthworm is a quite stable parameter.

INDEX